E. Nevo · A.B. Korol · A. Beiles · T. Fahima

Evolution of Wild Emmer and Wheat Improvement

Springer

Berlin
Heidelberg
New York
Barcelona
Hong Kong
London
Milan
Paris
Tokyo

E. Nevo · A.B. Korol · A. Beiles · T. Fahima

Evolution of Wild Emmer and Wheat Improvement

Population Genetics, Genetic Resources, and Genome Organization of Wheat's Progenitor, *Triticum dicoccoides*

With 59 Figures and 85 Tables

Springer

Professor Eviatar Nevo
Professor Abraham B. Korol
Dr. Avigdor Beiles
Dr. Tzion Fahima

Institute of Evolution
University of Haifa
Mount Carmel
Haifa 31905
Israel

ISBN 3-540-41750-8 Springer-Verlag Berlin Heidelberg New York

Library of Congress Cataloging-in-Publication Data
Nevo, Eviatar.
Evolution of Wild emmer and wheat improvement : population genetics, genetic resources, and genome organization of wheat's projenitor, Triticum dicoccoides / E. Nevo, A.B. Korol, T. Fahima. p. cm.
Includes bibliographical references (p.).
ISBN 3540417508 (hardcover)
1. Emmer wheat – Evolution. 2. Crop improvement. I. Korol', A.B. (Abram Bentsionovich) II. Fahima, T. (Tzion), 1959– III. Title.

Springer-Verlag Berlin Heidelberg New York
a member of BertelsmannSpringer Science+Business Media GmbH
http://www.springer.de

Cover design: D & P, Heidelberg
Cover photo: Wild Emmer wheat, *Triticum dicoccoides*, Almagor, North of the Sea of Galilee, Israel (Photo by Tzion Fahima)
Typesetting: Best-set Typesetter Ltd., Hong Kong
SPIN 10795445 31/3130-5 4 3 2 1 0 – Printed on acid-free paper

Preface

This book is about the contribution to evolutionary theory and agricultural technology of one of humankind's most dramatic imitations of the evolutionary process, namely crop domestication, as exemplified by the progenitor of wheat, *Triticum dicoccoides*. This species is a major model organism and it has been studied at the Institute of Evolution, University of Haifa, since 1979. The domestication by humans of wild plants to cultivated ones during the last ten millennia is one of the best demonstrations of evolution. It is a process that has been condensed in time and advanced by artificial rather than natural selection. Plant and animal domestication revolutionized human cultural evolution and is the major factor underlying human civilization.

A post-Pleistocene global rise in temperature following the ice age, i.e., climatic-environmental factors, may have induced the expansion of economically important thermophilous plants and in turn promoted complex foraging and plant cultivation. The shift from foraging to steady production led to an incipient agriculture varying in time in various part of the world. In the Levant, agriculture developed out of an intensive specialized exploitation of plants and animals. Natufian sedentism, followed by rapid population growth and resource stress, induced by the expanding desert, coupled with available grinding technology, may have triggered plant domestication.

The earliest signs of domestication found so far in the archeological records appear 9,000–10,000 years ago in the Fertile Crescent of the Near East, Central America, and southern China, involving different crops and independent cradles of domestication. Each was founded on cereal staples whose reliability, yield, and suitability for storage may have been crucial for early commitment to an agricultural way of life and the domestication of other plants. It is generally believed that plant domestication first took place in the Jordan Valley and adjacent areas of the southern Levant (in present-day Israel and Jordan). However, botanical, genetic, and archaeological evidence point to a small core area within the Fertile Crescent – near the upper reaches of the Tigris and Euphrates rivers, in present-day southeastern Turkey/northern Syria – as the cradle of agriculture. Further evidence is needed to clarify when and where wheat and barley domestication and agriculture originated driving modern civilization.

The genetic changes required for domestication to occur were relatively straightforward and rapid, most of them arising from selection for higher yield. Considerable progress has been achieved in the field of the wild ancestry of Old World crops, including wild cereals. The wild progenitors of most of these cultivated plants have now been satisfactorily identified by comparative morphology and genetic analysis. The distribution and ecological ranges of the wild relatives have been established; furthermore, comparisons between wild types and their cultivated counterparts have revealed the evolutionary changes brought about by domestication. Research in this field has enabled us to assess the relative importance of the evolutionary forces driving wheat evolution, hybridization, migration, drift, and natural selection, interacting in generating the final wheat genotype. Our studies suggest that besides hybridization, which contributed to the tetraploidy of *T. dicoccoides*, natural selection played the largest role and it has oriented wheat evolution primarily through the mechanisms of diversifying and balancing selection regimes.

The evolution of domestication can be considered in two contexts: (1) the evolution of new species of crop plants by humans through strong artificial selection. This provides fascinating insight into the evolutionary process. (2) The evolution of human civilization and the current consequences of population explosion and increasing world hunger in developing countries of the southern continents.

The accumulated information provides partial answers to the following questions: (1) How much and what kind of genetic changes were involved in domestication? (2) How fast was this change? (3) Can the crops be regarded as new species created by humans? (4) Could they survive without the ecological niche created by humans? (5) Can they be further improved to reinforce their qualitative and quantitative agricultural traits?

In its five parts and nine chapters, this book attempts to answer some of these questions and open new horizons in our attempts to answer both theoretical and applied problems related to the evolution of domestication, the population-genetic structure of the progenitor species and their genetic resources, genome organization and evolution, and the future of wheat improvement. Part I deals with the origin and evolution of wheats, focusing on domestication of cultivated wheats (Chap. 1); the origin, classification, ecology, and domestication of wild emmer wheat, *Triticum dicoccoides*, the progenitor of most wheats (Chap. 2); and the centers of origin and diversity of wild ancestors and crop improvement (Chap. 3). Part II deals with population genetics of wild emmer wheat. This part is subdivided into a discussion of macro-geographic allozyme and DNA polymorphisms (RAPDs, SSRs) in the Near East Fertile Crescent (Chap. 4), and an examination and discussion of four microsites north of the Sea of Galilee (Qazrin, Yehudiyya, Tabigha, and Ammiad) that differ in their climatic, soil, and topographic-lithological divergences at the microscales. Each of the four microsites was analyzed with respect to allozymes and DNA polymorphisms (RAPDs, SSRs; Chap. 5). We intentionally expanded this part including extensive tabular data sets to demonstrate that the patterns of the noncoding genome (primarily RAPDs and SSRs) are very similar to those of the coding genome (allozymes) suggesting that both nuclear genomes are adaptive and subjected to natural selection. This has in-depth theoretical and practical implications. Part III presents a series of genetic and agronomic profiles of the genetic resources of wild emmer for wheat improvement. These include the phenotypic and genotypic diversity of morphological, physiological, and agronomic traits; heat production; responses to stresses, abiotic (temperature, salinity, drought, herbicides), and biotic (viral and fungal pathogens); amino acids, amylases, photosynthetic yield, chlorophyll a/b binding protein (CAB) circadian rhythms, and general discussions of the importance of wild emmer to wheat production (Chap. 6). Part IV deals with genome organization and genetic mapping, firstly, genome structure and evolution (Chap. 7), and secondly, genetic maps of agronomically important traits (Chap. 8). A discussion then follows of molecular evolution and ecological stress in wild emmer wheat at regional and local levels: natural selection in action (Chap. 9). Finally, Part V presents the conclusions of two decades of studies of wild emmer wheat as an evolutionary and agronomically model organism, and the prospects of future studies in the genomic and post-genomic era.

Wild emmer, *Triticum dicoccoides*, is a plant of fundamental importance. Deciphering the messages of its exceedingly large genome (12,000 and 17,000 Mb in tetraploid and hexaploid wheats, respectively), will substantially contribute to understanding genome structure, function, and evolution. This will lead to optimizing the utilization of the rich genetic resources for wheat improvement, increased food production, and coping with spreading starvation in the developing countries of the Southern Hemisphere.

No research program evolves in isolation, and the one described in this book is no exception. Numerous friends, colleagues, collaborators, and students of evolutionary biology and agricultural genomics are represented here, and we do apologize to many who were not included or cited because of space limitations. The spectrum of researchers cited here who have contributed to our inter- and multidisciplinary research program of wild cereals is extensive. It includes evolutionary, molecular and organismal biologists, geneticists, ecologists, agronomists, mycologists, physiologists, mathematical geneticists, experimentalists, and theoreticians, all attempting to explain the patterns and evolution of wild cereals. We acknowledge some, though not all, who contributed to this book in the following Acknowledgements section.

We are indebted to all our numerous friends, colleagues, collaborators, and students who have shared with us the exciting journey towards understanding the evolution and agronomic potentials of wild emmer wheat, and to the University of Haifa which enabled us to conduct these studies.

Acknowledgements

This book is based on a long-term ongoing research program, now celebrating its 25th anniversary, at the Institute of Evolution, University of Haifa, Israel. The program embraces research of wild cereals, wild barley (*Hordeum spontaneum*) and wild emmer wheat (*Triticum dicoccoides*) the latter being the focus of this Volume. The program combines field, laboratory, and theoretical studies on two model organisms that represent the progenitors of cultivated wheat and barley, major crops for feeding humans and animals. The evolution of domestication of these wild cereals, which occurred 10,000 years ago in the Fertile Crescent, extended by humans the evolution of wild cereals which proceeded for millions of years in our geographic area. An intimate acquaintance with the ecology of wild cereals has enabled us to link the genetic and agronomic studies with their evolutionary cradle and dynamic ecological driving forces. Climatic factors, particularly the aridity index, are central players in genome evolution of wild cereals and provide the best insight into understanding their evolutionary history.

This book could not have been created without the help of many colleagues, collaborators, assistants, and generous foundations. We deeply thank Daniel Zohary, who first introduced one of us (E.N.) to the world of wheat and barley.

We also deeply thank Tamar Krugman, whose constant collaboration, assistance, supervision of our field, lab, and wild cereal gene banks have been indispensable for the advancement of our wild cereal research program. The field and gene bank assistance of Suad Chalifa was indispensable. We acknowledge the work of many graduate students and postdoctoral students who have contributed to this book. They include: A.H.D. Brown, E.M. Golenberg, V. Chague', G.L. Sun, Y.C. Li, J.H. Peng, A. Dahan, D. Lewinsohn, J.P. Cheng and Y. Jun.

We thank our colleagues and collaborators: J.G. Moseman, Y. Gutterman, N. Storch, D. Kaplan, the late Z. Gerechter-Amitai, A. Grama, M.A. El-Morshidy, R.B. Flavell, J. Snape, I. Payne, B.F. Carver, R.M. Hunger, J. Gorham, I. Noy-Meir, Y. Anikster, J. Manisterski, I. Raskin, Y. Furuta, K. Nishikawa, T. Kawahara, B. Piechulla, R.A. McIntosh, T.T. Thé, B. Rubin, E. Porceddu, M.A. Pagnotta, L.R. Joppa, J. Dubcovsky, K. Gill, H. Burstrmyer, K. Kato, M.S. Röder, K. Wedenhake, A. Beharav, Y.I. Ronin, V.M. Kirzhner, A. Belyayev and O. Raskina.

We also wish to acknowledge the financial support of the Wolfson Foundation, the United States-Israel Agricultural Research and Development Fund, BARD; the United States-Israel Binational Science Foundation, BSF; Israel Science Foundation; Israeli Ministries of Science and Agriculture and the Israeli Plant Gene Bank; the Israel Discount Bank Chair of Evolutionary Biology; the Ancell-Teicher Research Foundation for Genetics and Molecular Evolution; the Humana Foundation; and the German-Israeli Cooperation Project (DIP project funded by the BMBF and supported by BMBF's International Bureau at the DLR).

Our profound thanks are extended to Irit Fenton, Esther Kabulansky, Michael Margulis, and Murray Rosovsky whose devoted assistance in typing and editing the text, arranging illustrations and tables, and generating the graphics of this book made its birth possible. Words cannot express our deep gratitude for their enormous efforts and hard work contributing to the production of this book. Special thanks to Avi Morgenstern and the late Itzhak Fisher for their help in aiding the research laboratories.

We acknowledge with thanks the continuous financial support of the University of Haifa, which has fostered the Institute of Evolution in the domain of the wild cereal research program.

Eviatar Nevo, Abraham B. Korol, Avigdor Beiles, Tzion Fahima
Institute of Evolution, University of Haifa, Haifa, Israel

Contents

Abbreviations

σ^2	variance
6Pgd	6-phosphogluconate dehydrogenase
A, or NA	number of alleles per locus
Aat	aspartate aminotransferase
Acph	acid phosphatase
Adh	alcohol dehydrogenase
AFLP	amplified fragment length polymorphism
Al	altitude
ANOVA	analysis of variance
ARN	average repeat number
ARO	agricultural research organization (Israel)
AVA	angle to rock top
AVH	average height to rock top
BA or Ba	basalt soil
*Bam*HI	a restriction enzyme
CAB	chlorophyll-a/b binding protein
Cat	catalase
Ci	internal concentration
cM	centi-Morgan
cpDNA	chloroplast DNA
CS	Chinese spring (bread wheat cultivar)
CT	chlorotolurone (herbicide)
CV	coefficient of variation
D	genetic distance
*Dde*I	a restriction enzyme
df	degree of freedom
Dm	absolute degree of gene differentiation among subpopulations
D_{AS}	allele-sharing distance index
Dm	average of interpopulation diversity
Dst	measure of the average degree of gene differentiation
Dw	mean number of dew nights in summer
ELISA	enzyme-linked immunosorbent assay
EST	expressed sequence tag
Est	esterase

Ev	mean annual evaporation
F	F-statistics for variance
Gdh	glutamate dehydrogenase
GISH	genome in situ hybridization
Gluc	glucosidase
GM	genetically modified (plants)
Got	glutamate oxaloacetate transaminase
GPC	grain protein content
Gst	relative degree of gene differentiation among subpopulations
GWH	100-grain weight
GWM	Gatersleben wheat microsatellite markers
HAVDB	distance to rock base
HD	heading date
He	gene diversity
*Hinf*I	a restriction enzyme
Hk	hexokinase
HMW	high molecular weight (proteins)
Hs	mean gene diversity in a population
HT	plant height
Ht	total diversity; genetic diversity of a subdivided population
Hu14	mean humidity at 14:00
Huan	mean annual humidity
I	genetic identity
Ipol	indophenol oxidase (leaf)
Ipor	indophenol oxidases (root)
IT	infection type
KNL	kernel number per spikelet
KNP	kernel number per plant
KNS	kernel number per spike
LD	linkage disequilibrium
Ldh	lactate dehydrogenase
LDN	Langdon durum (tetraploid wheat cultivar)
Ln	longitude
LOD	logarithm of the likelihood-ratio test statistic
Lt	latitude
M	soil moisture content
Ma	marginality
MANOVA	molecular analysis of variance
Mdh	malate dehydrogenase
MGP	multilocus genotype pattern
MR	multiple regression analysis
mtDNA	mitochondrial DNA
MX	metoxuron (herbicide)
NA, or A	number of alleles
Nadhd	lipoamide diaphorases

NFS	north facing slope
NOR or Nor	nucleolus organizer
nuDNA	nuclear DNA
P	proportion of polymorphic loci
p	statistical significance
PCA	principal component analysis
PCR	polymerase chain reaction
Pepc	phosphoenol pyruvate carboxylase
Pept	peptidase
Pgi or PGI	glucosephosphate isomerase
Pgm	phosphoglucomutase
PL	number of polymorphic loci
PSII	photosystem II
QTL	quantitative trait locus/loci
PAGE	polyacrylamide gel electrophoresis
R^2	coefficient of multiple determination
RAPD	random amplified polymorphic DNA
Rd	mean number of rainy days
rDNA	ribosomal DNA
Ren	rendzina soil
RFLP	restriction fragment length polymorphism
RGA	resistance gene analog
Rn	mean annual rainfall
Rr	mean relative interannual variability of rainfall
Rst	inter- vs intrapopulational degree of gene differentiation among subpopulations
RUL	mean values for ^{22}Na uptake in *T. dicoccoides* relative to ^{22}Na uptake in the Langdon control
Rv	mean interannual rainfall variability
r_s	Spearman's rank order correlation
satDNA	satellite DNA
SCAR	sequence-characterized amplified region
SDS	sodium dodecyl sulfate
SFS	south facing slope
Sh	mean number of Sharav days
SLS	spikelet number/spike
SNP	spike number/plant
SNPs	single nucleotide polymorphisms
So	soil type
SS	sums of squares
SSA	smallest space analysis
SSD	single-seed descent
SSR	short sequence repeats (microsatellites)
SSW	single spike weight
SWP	spike weight/plant

Sz	estimate of population size
Ta	mean August temperature
*Taq*I	a restriction enzyme
Td	seasonal temperature difference
Tdd	day-night temperature difference
TF	translocation frequency
TG	transgenic plants
Th	Thornthwaite's moisture index
Tj	mean January temperature
Tm	mean annual temperature
Tr or TR	terra rossa soil
Trd	mean number of tropical days
UPGMA	unweighted pair group method with arithmetic average
Vc	valley center
Vn	narrow valley
WMS	wheat microsatellite (marker)
Ws	spike glaucousness
WSBMV	wheat soilborne mosaic virus
WUE	water-use efficiency
YLD	grain yield/plant
Yr	yellow rust

Part I

Origin and Evolution of Wheat

Chapter 1

Domestication of Wheats

1.1 Introduction

Wheats are the universal cereals of Old World agriculture (Harlan 1992; Zohary and Hopf 1988, 1993) and the world's foremost crop plant (Feldman et al. 1995), followed by rice and corn. It is the most widely cultivated food crop and is the staple food in more than 40 countries and for over 35% of the global population (Williams 1993). The earliest present evidence for the utilization of wheat comes from the Ohalo II site in Israel, where the wild, brittle tetraploid wheat *Triticum dicoccoides*, dated as 19,000 years old, was found, suggesting the initial steps towards sedentism and cereal agriculture (Kislev et al. 1992). Wheat and barley constituted the principal grain stock upon which Old World agriculture was founded. These were among the earliest domesticated crop plants, 10,000 years ago, in the Pre-pottery Neolithic Near East (Harlan and Zohary 1966; Harlan 1976, 1992; Lupton 1987; Evans 1993; Price and Gebauer 1995; Harris 1996; Valdes et al. 1997; Hillman and Colledge 1998; Ladizinsky 1998; Lev-Yadun et al. 2000). Wheat predicated sedentism and was responsible for the increase in the population by enabling humans to produce food in large quantities, thereby contributing to the emergence of human civilization. Today, wheats rank first in the world's grain production and account for more than 20% of the total food calories and protein in human nutrition. Wheats are now extensively grown across the temperate, Mediterranean, and subtropical parts of both hemispheres of the world, from 67 °N in Norway, Finland, and Russia, to 45 °S in Argentina. However, in the subtropics and tropics, the cultivation of wheat is restricted to higher elevations. The world's main wheat-producing regions are in temperate and southern Russia, the central plains of the United States, southern Canada, the Mediterranean basin, north-central China, India, Argentina, and south-western Australia.

Wheat cultivars are superior to most other cereals in their nutritive value (Zohary and Hopf 1993). Besides 60–80% starch, their grains also contain 8–15% protein, which may rise in elite wild genotypes up to 17–28% (Avivi 1978, 1979; Avivi et al. 1983; Grama et al. 1983; Nevo et al. 1986a; Levy 1987). The gluten proteins in the seed endosperm impart to wheat dough stickiness and unique bread-baking qualities. Wheats were and are the staple food in the ancient and modern world for billions of people.

Modern wheat cultivars belong primarily to two species: (1) hexaploid bread wheat, *Triticum aestivum* (2n = 42 chromosomes), and (2) tetraploid, hard or durum-type wheat, *T. turgidum* (2n = 28) used for macaroni and low-rising bread. Other species are relict. Wheats are almost fully self-pollinated; hence genetic diversity is represented in the wild by numerous clones and in cultivation by some 25,000 different cultivars. Cultivated primitive wheat forms have hulled grains, whereas advanced cultivated wheats are free-threshing. Likewise, wild wheats have brittle ears that disarticulate at maturity into individual spikelets. Each spikelet, with the wedge-shaped rachis internode at its base, constitutes an arrow-like device that inserts the seed into the ground (Zohary 1969). By contrast, all cultivated wheats have non-brittle ears that stay intact after maturation, depending on humans for reaping, threshing, and sowing (Figs 1.1, 1.2). The non-brittleness and nakedness of cultivated wheats depend on the Q locus (Luo et al. 2000), located on chromosome 5 of genome A, and it may have arisen from the *q* gene of the hulled varieties by a series of mutations (Feldman et al. 1995).

1.2 Cytogenetic and Taxonomic Background

The tribe Triticeae is economically the most important group of the family Gramineae. It has given rise to cultivated wheats, barleys, ryes, oats, and a number of important range grasses. Hybridization among genera within the tribe has allowed the exchange of genetic material and given rise to polyploidy in the form of amphiploidy. The wheats (genus *Triticum*) comprise a series of diploid, tetraploid, and hexaploid forms, the polyploids having arisen by amphiploidy between *Triticum* species and diploid species of the genus *Aegilops* (Kihara 1954; Riley 1965; Sears 1969; Miller 1987; Kimber and Feldman 1987; Dvorak et al. 1993; Van Slageren 1994; Feldman et al. 1995; Caligari and Brandham 2001; Figs. 1.1 and 1.2).

The wild diploid species, some of which have contributed to polyploid wheats, are presumably monophyletic in origin, though they have diverged considerably from each other. This divergence is particularly evident in the morphologically well-defined seed-dispersal units of the species and their specific ecological requirements and geographic distributions. Cytogenetic data have corroborated the taxonomic classification by showing that each diploid species contains a distinct genome (Kihara 1954). The related chromosomes of the different genomes show little affinity with each other and do not pair regularly in interspecific hybrids, so the result is complete sterility and isolation of the diploid species from each other.

The polyploid species are a classic example of evolution through amphiploidy. They behave like typical genomic amphiploids; that is, their chromosomes pair in a diploid-like fashion and the mode of inheritance is disomic. The allopolyploid nature of the *Triticum* polyploids has been veri-

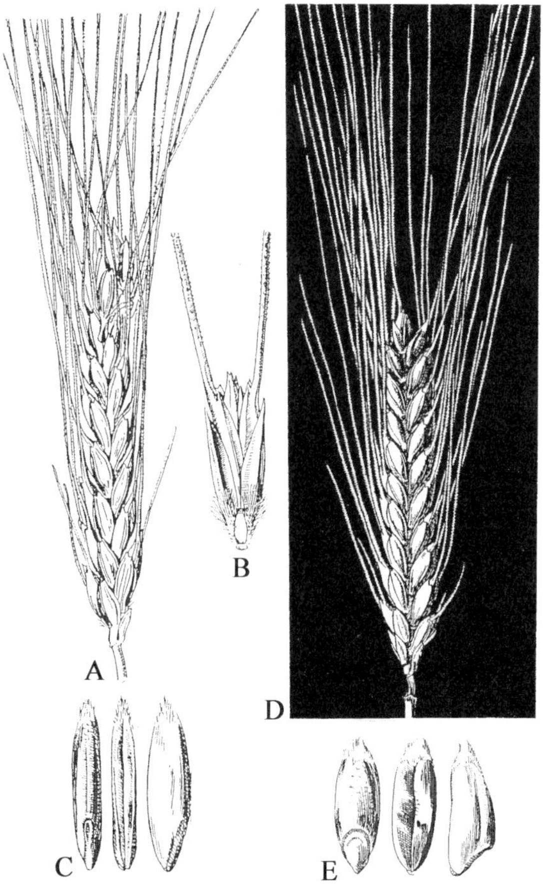

Fig. 1.1A–D. Tetraploid *Triticum turgidum* wheat. **A** Ear (1:1), **B** spikelet (2:1), **C** grain (3:1) of wild emmer wheat, *T. turgidum* subsp. *dicoccoides*. **D** Ear (1:1), **E** grain (3:1) of cultivated emmer wheat, *T. turgidum* subsp. *dicoccum*. (Zohary and Hopf 1993)

fied by cytogenetic analysis of hybrids between species of different ploidy levels. Each polyploid species can be identified as a product of hybridization followed by chromosome doubling.

At the diploid level there are two species of einkorn wheat, *Triticum monococcum* L. and *T. urartu* Thum. The sterility of their hybrids (Johnson and Dhaliwal 1976) indicates that they are valid biological species. *T. monococcum* includes cultivated ssp. *T. monococcum monococcum* (*T. monococcum* L.) and wild ssp. *T. monococcum aegilopoides* (Link) Thell. *T. urartu* presumably exists only in its wild form. At the tetraploid level there are two species, *T. turgidum* L., which includes wild ssp. *T. turgidum dicoccoides* (Korn.) Thell (here designated *T. dicoccoides*) and several cultivated subspecies (Morris and Sears 1967), and *T. timopheevi* (Zhuk., which includes wild ssp. *T. timopheevi*

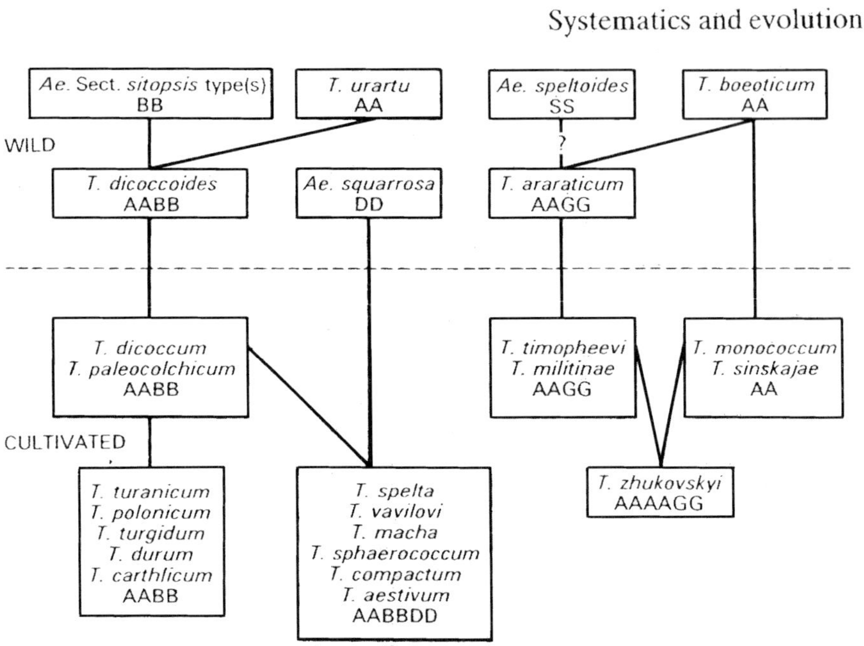

Fig. 1.2. Evolution of the polyploid wheats from current evidence. (Miller 1987)

araraticum (Jakubz.) Mac Key and cultivated ssp. *timopheevi* =*T. turgidum* ssp. *timopheevi* (Zhuk.). At the hexaploid level there are also two species, *T. aestivum* L., which has several subspecies (Morris and Sears 1967), and *T. zhukovskyi* Menab. et Ericz (Fig. 1.2, Table 1.1).

Cytogenetic studies revealed that the polyploid species constitute two evolutionary lineages. *Triticum turgidum* (genomes AABB) and *T. aestivum* (genomes AABBDD) comprise one lineage. *Triticum aestivum* evolved from the hybridization of *T. turgidum* with *T. tauschii* (Coss.) Schmalh (=*Ae. squarrosa*) which contributed the D genome (Kihara 1944; McFadden and Sears 1946). *Triticum timopheevi* (genomes AAGG) and *T. zhukovskyi* (genomes AAAAGG) comprise the other evolutionary lineage. *Triticum zhukovskyi* evolved from the hybridization of *T. timopheevi* with an einkorn wheat, which contributed its second A-genome (Upadhya and Swaminathan 1963; Fig. 1.2). For wild tetraploid wheat phylogeny determined by RFLP analysis of nuclear DNA, see Mori et al. (1995).

1.2.1 The Origin of the A Genome

It is apparent from these relationships that the key to understanding the evolution of the polyploid wheats is in the elucidation of the evolution of the two tetraploid wheat species. Early cytogenetic studies led to the conclusion that

Table 1.1. Classification of cultivated wheat and closely related wild species. (Feldman et al. 1995)

Species	Genomes	Wild	Cultivated	
		Hulled	Hulled	Free-threshing
Diploid (2*n* = 14)				
Aegilops speltoides	S(G)	All		
Ae. bicornis	S^b	All		
Ae. longissima	S^l	All		
Ae. searsii	S^s	All		
Ae. squarrosa	D	All		
T. urartu	A	All		
T. monococcum	A	Var. *boeoticum* (wild einkorn)	Var. *monococcum* (cultivated einkorn)	Var. *sinskajae* (cultivated einkorn)
Tetraploid (2*n* = 28)				
T. timopheevi	AG	Var. *araraticum*	Var. *timopheevi*	Var. *militinae*
T. turgidum	AB	Var. *dicoccoides* (wild emmer)	Var. *dicoccum* (cultivated emmer)	Var. *durum* Var. *turgidum* Var. *polonicum* Var. *carthlicum* Var. *turanicum*
Hexaploid (2*n* = 42)				
T. aestivum	ABD		Var. *spelta* Var. *macha* Var. *vavilovii*	Var. *aestivum* Var. *compactum* Var. *sphaerococcum*

the A genomes of both tetraploid species were contributed by *T. monococcum* (Sax 1922; Kihara 1924; Lilienfeld and Kihara 1934). As it became apparent that what was known as einkorn wheat actually comprised two biological species, *T. monococcum* and *T. urartu*, a reexamination of the sources of the A genomes in the polyploid wheats became necessary. Chapman et al. (1976) determined that the A genome originated from *T. urartu*. Konarev et al. (1979) concluded from studies of the immunological properties of seed-storage proteins that the A genome in the *T. turgidum* lineage was contributed by *T. urartu*, whereas the A genome of *T. timopheevii* was contributed by *T. monococcum*. From the variation in esterases, Nishikawa (1984) suggested that the A genomes in both lineages were contributed by *T. urartu*. Dvorak et al. (1988) showed that variation in repeated nucleotide sequences can be used in the investigation of the phylogeny of polyploid species of wheat and also concluded that the A genomes in both tetraploid wheats are more related to the A genome of *T. urartu* than to the A genome of *T. monococcum*. Dvorak et al. (1993) reported on variation in repeated nucleotide sequence families. Their extensive data elucidate the relationships between the A genomes of diploid and polyploid wheats and identify the sources of the A genomes in *T. zhukovskyi*.

1.2.2 The Origin of the B Genome

The second genomes of the tetraploid wheats, B in *T. turgidum* and G in *T. timopheevi*, are closely related (Feldman 1966; Chapman et al. 1976; Dvorak and Appels 1982) and both appear to have been contributed by *T. speltoides* (Tausch) Gren. (Dvorak and Zhang 1990). Variation in several repeated nucleotide sequences, and particularly the structure of the rRNA gene families (Dvorak et al. 1989), indicated that the G genome of *T. timopheevi* is more related to the genome of *T. speltoides* than to the B genome of *T. turgidum*. A similar situation was found in the relationship of the chloroplast genomes (cpDNA) of the tetraploid wheats to the chloroplast genome of *T. speltoides*. cpDNA of *T. timopheevi* more closely matched the cpDNA of *T. speltoides* than did the cpDNA of *T. turgidum* (Ogihara and Tsunewaki 1988). In the future it would be useful to test representatives of A and B genome donors from across their geographic ranges. This may settle some of the past and present controversies over their origins. [See the discussion in Chapter 7 of the evidence generated at the Institute of Evolution by Belyayev et al. (2000) based on GISH molecular cytogenetics; see also Yen et al. (1996).]

1.3 Emmer and Durum Wheats: *Triticum turgidum*

The varied aggregate of cultivated tetraploid wheats shares the AABB genome and all of these wheats are interfertile (Zohary and Hopf 1993). They can be classified, according to their threshing properties, into two groups:

1. Hulled emmer wheat, *T. turgidum* L. subspecies *dicoccum* (Schrank) Thell. (traditionally designated *T. dicoccum* Schubl), in which the threshing products are individual spikelets (Fig. 1.1). Hulled emmer represents the primitive situation in cultivated *T. turgidum* wheats.
2. Durum wheat, *T. turgidum* conv *durum* Desf. MacKey (synonym *T. durum* Desf), rivet wheat, *T. turgidum* conv. *turgidum* (L), and several others represent the more advanced free-threshing tetraploids (Fig. 4 in Zohary and Hopf 1993). Threshing in these cultivated tetraploids frees the naked grain.

Hulled emmer *T. turgidum* subsp. *dicoccum* was the principal wheat of Old World agriculture in the Neolithic and early Bronze ages and was used for food and beer. It was later replaced by more advanced free-threshing tetraploid and hexaploid types. Emmer is still sporadically grown in the Balkans, Anatolia, western Asia, and India, and is massively grown in Ethiopia.

Durum wheats are the main representatives of the free-threshing AABB tetraploid wheats. They were domesticated in Neolithic times as the major wheat crop in the summer-dry, warm Mediterranean region. In temperate areas with summer rains and in more continental climates, traditional bread wheat production depended heavily on another free-threshing wheat, hexaploid *Triticum aestivum*, the major current bread wheat across the planet, with many thousands of land-races and modern lines.

1.4 Genetic Erosion of Cultivated Wheats and the Need to Restore Diversity

Until the late nineteenth century, all wheats existed as highly heterogeneous "land-races", that is, mixtures of inbred lines and hybrid segregates, the products of a low level of random crossing in earlier generations. Selection was primarily for increased yield, large seed size, better flour quality, and adaptation to a wider range of climatic and farming regimes (Feldman et al. 1995). Many land-races still exist today, displaying intermediate or high levels of genetic diversity. However, over the past century of modern breeding, the land-races have been almost wholly displaced in advanced agricultures by pure-line, genetically uniform cultivars. These were initially selections from land-races, but later they derived from successive cycles of crosses between established pure-lines, sometimes of diverse geographic origin. The result has been a marked narrowing of the genetic base in many, probably all, advanced agricultures, partially concealed by a great diversity of cultivar names (Harlan 1975, 1976, 1992). While wheat yield increased in advanced agricultures (e.g., Avery 1985), genetic homogeneity also increased due to modern pure-breeding practices (Frankel and Bennett 1970; Frankel and Hawkes 1975; Harlan 1975, 1976, 1992; Frankel and Soulé 1981; Nevo et al. 1982; Nevo 1983a, 1986a, 1989a, 1995, 1998a, 2001b; Plucknett et al. 1983, 1987; Lupton 1987; Nevo and Beiles 1989; National Academy of Sciences 1993).

Advanced cultivars have been pure-bred throughout the last hundred years for high performance under intensive high-input cultivation, which tends to even out environmental and genetic differences of site and season. Modern varieties, in contrast to either their progenitors or to land-races, are purebred cultivars. They are deliberately selected for performance over a range of environments, thus reducing the need for specific local adaptation. Hence, they are selected for a high degree of genetic homogeneity. Consequently, the genetic bases of cultivated crops have been narrowed and placed under serious risk, and their future evolution is in jeopardy (Frankel and Soulé 1981; Plucknett et al. 1987).

The loss of genetic diversity of some of the world's crops has accelerated in recent decades, with many crops becoming increasingly susceptible to diseases, pests, and environmental stresses (Plucknett et al. 1983, 1987). A global network of gene banks has been established to provide plant breeders with the genetic resources for developing more resistant and tolerant crops that will maintain stable and high yields (Lupton 1987; Plucknett et al. 1987; Brown et al. 1989, 1990). Dynamic in situ conservation of wild relatives, the best hope to improve cultivated plants (Feldman and Sears 1981), is being actively discussed as an optimal conservation strategy (e.g., the International Workshop held in Israel, on 2–3 April, Ministry of Science and Technology 1990; Hawkes 1991; Heyn and Waldman 1992; Valdes et al. 1997; Maxted et al. 1997; Nevo 1998a). However, not only conservation of the diverse germplasms is needed. To achieve more efficient and comprehensive utilization of the conserved wild gene pool, it is essential to be able to predict, screen, and evaluate promising genetic diversity and resources, in natural populations and genotypes, of wild relatives, including emmer wheat, the progenitor of bread wheat, thereby optimizing breeding (Nevo 1983a, 1989a, 1992, 1995, 2001b; Peng et al. 2000a–c, 2001). A reliable prediction calls for the analysis of genetic diversity across the geographic range of the progenitor, at both the macro- and microscale, covering the full spectrum of its ecological genomics. This would unravel patterns and forces driving its genome evolution and lay open the full potential of its genetic resources for utilization.

Chapter 2

Wild Emmer, *Triticum dicoccoides,* Wheat Progenitor: Origin and Evolution

2.1 Origin

Genetic and morphological evidence clearly indicates that the cultivated tetraploid *turgidum* wheats (both hulled *dicoccum* forms and free threshing *durum* varieties) are closely related to the wild wheat that is native to the Near East and traditionally called *T. dicoccoides* (Korn) Aaronsohn (wild emmer wheat) (for review see Zohary 1969; Chapman et al. 1976; Miller 1987, 1992; Harlan 1992; Zohary and Hopf 1993; Feldman et al. 1995).

Triticum dicoccoides (Figs 2.1, 2.2, 2.3) includes the A and B genomes of *T. aestivum*, with which it makes fertile hybrids (Table 1.1, Fig. 1.2). *T. dicoccoides* is the origin of all cultivated bread wheats, *T. aestivum*, which constitutes the most important wheat, involving several primitive hulled types (spelta wheat) and numerous modern free-threshing forms (see Figs. 2–5 of diploid, tetraploid and hexaploid wheats in Zohary and Hopf (1993) and wheat classification based on both traditional morphological description and modern cytogenetics (see Table 1.1 and Fig. 1.2).

Wild emmer wheat is an annual, predominantly self-pollinated, tetraploid wheat with large and brittle ears and big elongated grains (Figs. 1.1c, 2.1, 2.2, 2.3), similar to cultivated emmer and durum wheat. It is the only wild ancestor in the genus *Triticum* that is cross-compatible and fully interfertile with cultivated *T. turgidum* wheats. Hybrids between wild *T. dicoccoides* and all cultivated members of the *T. turgidum* complex show normal chromosome pairing in meiosis, indicating that all these tetraploid wheats contain homologous chromosomes of the AABB genomic constitution. Hybrids do form occasionally on the plot borders between cultivated tetraploid wheats, so *T. dicoccoides* has been regarded as the wild race of the cultivated tetraploid and durum wheats, and it is sometimes ranked as the wild subspecies of the *T. turgidum* complex. Clearly, *T. dicoccoides* occupies a central place in wheat evolution.

Fig. 2.1. Wild emmer wheat, *Triticum dicoccoides*: individual plants, including several black spike morphs, near Nahf, Upper Galilee. (Photo by E. Nevo)

2.2 Classification

The nomenclature and classification of wheat are varied and confused (Miller 1987, 1992). In this review, we adhere to the traditional view of *T. dicoccoides* as a valid biological species, as Miller (1992) did, for the following reasons. We now know that speciation can occur with very little genomic and mor-

Fig. 2.2. Wild emmer wheat, *Triticum dicoccoides*, ripe natural population from Qazrin, Golan Heights with some black morphs. (Photo by E. Nevo)

Fig. 2.3. Wild emmer wheat, *Triticum dicoccoides*, spike colour polymorphism: black, green, and yellow morphs. (Photo by E. Nevo)

phological changes (Ayala 1975; Avise 1976; Nevo and Cleve 1978; Nevo 1991; Orr 1991; Coyne 1992; Orr and Smith 1998). *T. dicoccoides* has a unique ecological niche in nature. Likewise, a conspicuous diagnostic difference exists between the wild type and the cultivars safeguarded by human, i.e., in the seed dispersal mechanism (Zohary 1969). *T. dicoccoides* has brittle ears which shatter upon maturity into individual spikelets. Each spikelet operates as an arrow-like device disseminating the seeds by inserting them into the ground. The "wild type" rachis disarticulation and the spikelet morphology reflect adaptive-specialized traits in seed dissemination that ensure survival in nature. Under the human-made system of reaping, threshing, and sowing, this adaptation broke down and non-brittle types were selected. The shift from a brittle spike in *T. dicoccoides* to a non-brittle spike in *T. dicoccum* wheats is determined by more than one major gene (Miller 1992).

Wild and cultivated forms also differ in kernel morphology (Van Zeist 1976). In the cultivars, the grain is wider, thicker, and rounder in cross-section than in *T. dicoccoides* (Fig. 1.1). This trait can distinguish the wild type from the cultivars and at archaeological sites helps to determine the start of domestication (Zohary and Hopf 1993). Likewise, important unique chromosomal translocations (Kawahara et al. 1993; Nishikawa et al. 1994; Joppa et al. 1995; Kawahara and Nevo 1996) and genetic polymorphisms (Nevo et al. 1982; Nevo and Beiles 1989; Fahima et al. 1998, 1999, 2001; Nevo 1998a, 2001b; Li et al. 1999, 2000a–d, 2001a–d) characterize *T. dicoccoides*. This combined evidence justifies its traditional ranking as a biological species which was the progenitor, not the derivative, of *T. turgidum*, as is implied in the name *T. turgidum* subsp. *dicoccoides*. What is the unique ecological niche of *T. dicoccoides* in nature?

2.3 Ecology

Triticum dicoccoides (Figs. 2.1, 2.2, 2.3) is more restricted in distribution and ecology than wild einkorn, *T. boeticum*, and wild barley, *Hordeum spontaneum*. It is found in Israel and Syria, which are its centers of distribution based on genetic diversity (Nevo and Beiles 1989; Nevo 1998a), Jordan, Lebanon, south-east Turkey, northern Iraq, and western Iran (Fig. 2.4). It was first discovered in 1906 in eastern Galilee and on the slopes of Mt. Hermon by Aaronsohn, who already then recognized its potential importance for wheat improvement (Aaronsohn and Schweinfurth 1906; Aaronsohn 1910, 1913; Schiemann 1956; Feldman 1977; Nevo 1983a, 1989a, 1995, 2000). The discovered genetic resources of wild emmer, *T. dicoccoides*, for wheat improvement far exceeded Aaronsohn's vision (Nevo 2001b; Peng et al. 2000a–c, 2001). Notably, the Institute of Evolution at the University of Haifa is currently experimenting with *T. dicoccoides* at the very same Atlit Agricultural Station established by Aaronsohn in 1910 for studying *T. dicoccoides*.

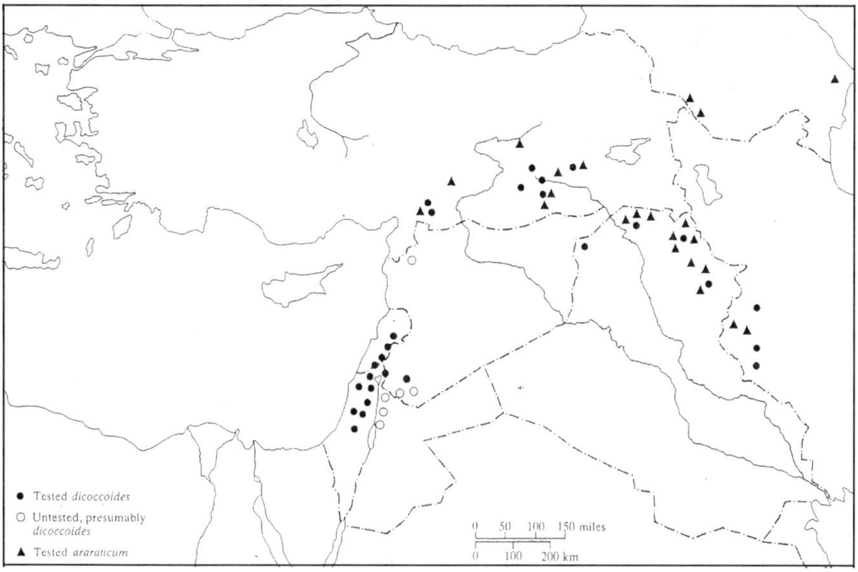

Fig. 2.4. Distribution of the wild tetraploid wheat: (●, ○) wild emmer wheat, *Triticum turgidum* subsp. *dicoccoides* (*T. dicoccoides*); (▲) wild Timopheev's wheat, *T. timopheevi* subsp. *araraticum* (*T. araraticum*). ●, ▲ Collections that were tested cytogenetically. (Zohary and Hopf 1993, and references therein)

T. dicoccoides grows as a common annual component in the herbaceous habitat of the Tabor oak (*Quercus ithaburensis* Decaisne), open park forest belt, together with related steppe-like herbaceous plant formations such as wild barley, *Hordeum spontaneum* (Nevo 1992), and wild oats, *Avena sterilis* L. In peripheral western populations, it may grow in openings in the macquis of the live-oak *Quercus calliprinos* Webb, or in the open park forest of *Ceratonia siliqua* L. Wild emmer is an ecological specialist primarily confined to basaltic and hard limestone bedrocks, weathering into basalt and terra rossa soils, respectively. It grows infrequently on hard chalks and is absent from marls and sandstones. Its central area of distribution is the catchment area of the Upper Jordan Valley (eastern upper Galilee and Golan Heights), where it grows in large continuous stands, becoming sporadic, semi-isolated, and isolated in central Israel as well as in Turkey, Iraq, and Iran (Nevo and Beiles 1989). In the sympatric area with *T. araraticum* in the northeastern distribution of *T. dicoccoides*, the two species are separated by strong sterility barriers (Maan 1973), though morphologically they are similar and practically indistinguishable without cytogenetic analysis (Fig. 1.2).

T. dicoccoides is genetically highly polymorphic in its center of distribution in northern Israel and southern Syria. Morphologically, it is polymorphic for glume hairiness and spike color (Fig. 2.3), and genetically it is polymorphic allozymically (Nevo et al. 1982; Nevo and Beiles 1989; Li et al. 2000c) and for various DNA diversities: rDNA diversity (Flavell et al. 1986), RAPDs (Fahima et al. 1999; Li et al. 1999), microsatellites (Fahima et al. 1998, 2001; Li et al. 2000a,b). Diverse ecotypes relate to climatic and altitudinal variations. *T. dicoccoides* ranges from 100 m below sea level north of the Sea of Galilee up to 1600 m on Mt. Hermon, only 60 km apart, and up to 1600 m on the Zagros mountains of western Iran. The mean annual temperature ranges from 11 to 24 °C; mean rainfall ranges from 360 to 1400 mm, and mean annual number of rainy days from 39 to 75. For the range of other climatic, soil and biotic parameters see Table 1 in Nevo and Beiles (1989). Robust early-maturing ecotypes occupy the winter-warm basin north of the Sea of Galilee. Slender ecotypes occur, either late-blooming forms in the steppic high-altitude habitats of Mt. Hermon or early-blooming forms in low-altitude habitats in the eastern Samaria steppes.

2.4 Domestication

Triticum dicoccoides, like *T. boeticum* and *H. spontaneum*, was probably collected from the wild, long before its domestication (Kislev et al. 1992; Zohary and Hopf 1993; Lev-Yadun et al. 2000; Nesbitt in Caligari and Brandham 2001). Brittle *T. dicoccoides*-like remains with relatively narrow grains appear in several early Neolithic and Natufian Near Eastern settlements. However, nine to ten millennia ago, they coexisted with non-brittle seeds in Jarmo, Iraq, Cayonu, Turkey, Tel Aswad and Tel Abu Hureira (Hillman and Colledge 1998), northern Syria (see archaeological evidence and citations in Zohary and Hopf 1993; Nesbitt 1998; Nesbitt and Samuel 1998; Lev-Yadun et al. 2000) and in Tell Mureybet I and II (9,000–8,000 B.C.). *T. dicoccoides* has also been discovered in Neolithic sites at Jerf el-Ahmar, Mureybet III, Djade, and Cayonu (8,000–7,500 B.C.). Wild forms have been unearthed at sites near the core area, such as pre-Neolithic Hallan Cemi Tepesi and Neolithic Qermez Dere and M'lefaat (Nesbitt 1998) (see the figure in Lev-Yadun et al. 2000). Domesticated forms of einkorn wheat, emmer wheat, and barley appeared in core-area Neolithic sites at Tell Abu Hureira 2A and Cafer Huyuk about 7,500 B.C., and soon thereafter at Cayonu and Nevali Cori (Kislev et al. 1992; Nesbitt and Samuel 1998). From the early beginnings of agriculture in the Near East, 10,000 years ago, and through the Chalcolithic and Bronze times, emmer was the principal wheat of the newly established farming settlements, subsequently spreading to Egypt, the Indian subcontinent, and Europe seven thousand years ago from its Near Eastern area of origin.

The evolution of hexaploid bread wheat, *T. aestivum*, originated outside the Fertile Crescent area when emmer reached the range of *Aegilops squarrosa* (=*Ae. tauschii*) (Van Zeist 1976; Van Zeist and Bakker-Heeres 1985). This probably happened in the Caspian Sea region 8,000–7,000 years ago. Notably, *T. durum* wheats are adapted to Mediterranean environments, whereas hexaploid *T. aestivum* grows in cooler and more continental parts of Europe and western Asia. Thus, *T. dicoccoides*, the progenitor of all wheats, deserves a particular in-depth study, as rightly suggested by Aaronsohn (1910, 1913), as a potential genetic resource for improving all wheats, both tetraploid and hexaploid. In this book, we summarize theoretical and applied studies on *T. dicoccoides* that are important for wheat improvement, including genetic structure across its range (Part II) and its genetic resources (Part III; Nevo 1983a, 1989a, 1995, 2001b). These topics are critically important to overcome the dangerous process of homogeneity of cultivated wheat. Likewise, we discuss genome organization, evolution, and genetic mapping (Part IV), as well as the theory explaining the evolutionary genetics of *T. dicoccoides* (Part V). These studies encompass the entire research program on wild emmer wheat that has been carried out at the Institute of Evolution, University of Haifa, Israel, for two decades (1980–2000; Fig. 3.1).

Chapter 3

Centers of Origin and Diversity of Wild Ancestors and Crop Improvement

The best hope for future crop improvement seems to lie in exploiting the rich gene pools of wild relatives from their centers of origin and diversity (Vavilov 1926, 1951; Zohary 1970a, Feldman and Sears 1981; Nevo 1983a, 1986a, 1987, 1989a, 1995, 1998a, 2001b; Brown et al. 1989, 1990; Hawkes 1991; National Academy of Sciences 1993; Gustafson et al. 1993).

The center of origin and diversity of many wild progenitors of cultivated plants is the Near East Fertile Crescent, particularly Israel (Zohary 1983; Nevo 1986a; Lev-Yadun et al. 2000). Here, wild relatives of many cultivated plants have long evolved across a wide ecological range. This generated in these wild ancestors a wide range of adaptive diversity to multiple diseases, pests, and ecological stresses. Israel, despite its small size and because of its geographic location and geological history, appears to be unique in the Near East Fertile Crescent in its physical and biotic heterogeneity, which generated high adaptive genetic diversity in the ancestors of many cultivated plants (Zohary 1983; Nevo 1986a). These wild genetic resources are as yet mostly untapped and are crucial for crop improvement by traditional plant breeding or modern biotechnology (e.g., Austin et al. 1986; Plucknett et al. 1987; Lupton 1987; Schell 1987; Gasser and Fraley 1989; Murray 1991; Grieson 1991; Shewry 1992; Kung and Wu 1993; Prakash and Pierik 1993; Evans 1993; Henry and Rolands 1994; Nevo 2001b). Unravelling the genome structure and evolution of the cultivars' progenitors can shed light on evolutionary theory, the evolution of domestication, and crop improvement.

3.1 Israel as a Natural Laboratory of Wild Genetic Resources

Israel is central and unique, due to its physical and biotic diversity, as a natural laboratory for wild genetic resources for improvement of cereals and other crops (Zohary 1983; Nevo 1986a, 1998a,b, 2001b). The evidence relevant to wheat improvement derives from studies of population genetics and ecology, climatology, phytopathology and immunology, biochemistry, physiology, agronomic traits, and breeding practices (Fig. 3.1). The following questions have been explored in our analysis of wild emmer in Israel, paralleling our analysis of wild barley (Nevo 1992):

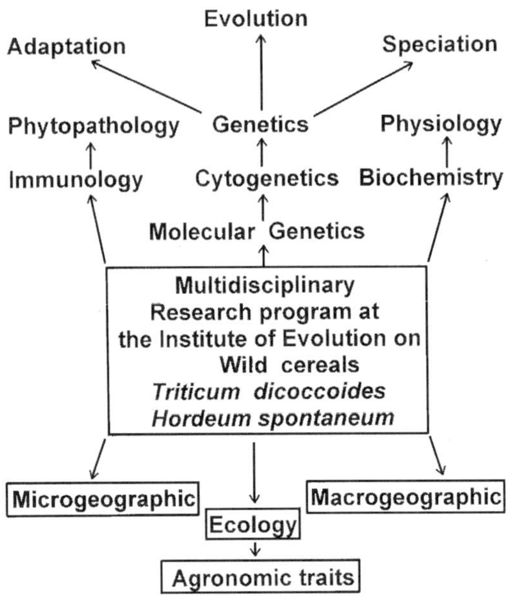

Fig. 3.1. Multidisciplinary long-term research program of wild cereals at the Institute of Evolution, University of Haifa, Israel

1. What is the genetic structure and geographic differentiation of populations at macro- and microscales?
2. What are the ecological-genetic correlates and mechanisms generating population single and multilocus genetic structures of proteins and DNA at the macro- and microgeographic levels?
3. To what extent are the levels and patterns of genetic diversity and differentiation determined by physical (climatic and edaphic) and biotic (pathogens, parasites, competition) natural selection as opposed to migration and random factors?
4. What is the importance of multilocus and linkage disequilibria organization in the dynamics of population genetics?
5. What are the phenotypic and genotypic diversities of economically significant quantitative agronomic traits within and between populations, i.e., at micro- and macroscales?
6. What are the geographic patterns of disease resistance to pathogens, and can they be predicted, either singly or in combination, for better screening of populations and genotypes for breeding?
7. What are the patterns and ecogeographic correlates of seed-storage proteins? Of amino acid composition? Of photosynthetic yield? Of ecological abiotic stresses and resistances (e.g., to temperature extremes, drought, salt, pH)?

8. What is the pattern, origin, and evolution of herbicide resistance?
9. What is the chromosomal location of specific genes? What is the pattern of chromosomal translocations?
10. How does the *T. dicoccoides* genome-wide map differ from that of *T. durum*? Can the genetic difference between the two illuminate the evolution of domestication?
11. In nature, genetic variation often underlies a continuous phenotypic range rather than discrete classes. The genetic variation in nature underlying quantitative traits results from the segregation of numerous interacting quantitative trait loci (QTLs) whose expression is modified by the environment. What is the genetic basis of the differences between QTLs in agronomically important traits? Can QTL mapping identify candidate genes for fitness and agronomic traits to improve wheat, and track its course of domestication?

3.2 Multidisciplinary Research Program of Wild Emmer Wheat, *Triticum dicoccoides*, for Wheat Improvement, at the Institute of Evolution, University of Haifa, Israel

The present review leans heavily on a multidisciplinary research program of wild emmer wheat, *T. dicoccoides*, and wild barley, *H. spontaneum*, that has been conducted at the Institute of Evolution, University of Haifa, Israel, continuously since 1979 as a long-term project. The project is aimed at studying the ecological-genetic, genomic, and domestication evolution of wild cereals, the progenitors of wheat and barley, across their Near East Fertile Crescent center of origin and diversity. The multidisciplinary program has focused on the patterning and maintenance of genetic diversity and divergence, genome organization, and agronomically important traits in natural populations, attempting to highlight their adaptive evolutionary processes and potential genetic resources for crop improvement (Fig. 3.1). The program has involved macrogeographic population genetics and ecology studies in the Near East Fertile Crescent in two countries: Israel, involving 33 central, peripheral, and marginal populations (Nevo et al. 1982; Flavell et al. 1986; Nevo and Payne 1987; Nevo and Beiles 1989; Nevo et al. 1993a, 1995; Pagnotta et al. 1995; Fahima et al. 1998, 1999, 2001), and Turkey, involving four peripheral populations (Nevo et al. 1988a). In all, it has revealed a broad view of wild emmer in a substantial part of the Fertile Crescent (Nevo and Beiles 1989; Nevo 2001b). We have also conducted a series of microgeographic population genetics and ecology studies in Israel on allozymes (Golenberg 1986a,b, 1987; Golenberg and Nevo 1987; Nevo et al. 1988b,c, 1991c; Li et al. 2000d, 2001a), and DNA polymorphisms (Li et al. 1999, 2000a–d, 2001b–d). In addition, we have studied agronomic traits (Nevo et al. 1984b), disease resistance polymorphisms (Moseman et al. 1984, 1985; Nevo et al. 1985, 1986b, 1991b; Hunger et al. 1992; The et al. 1993), protein content (Nevo et al. 1986a), amino acid composition (Nevo and Beiles 1992), herbicide resistance (Snape et al. 1991a,b; Nevo et al. 1992c; Krugman et al. 1997), photosynthetic yield (Carver and Nevo 1990; Nevo et al. 1991a), and drought and salt tolerance (Nevo et al. 1984b, 1993c). A methodology has been developed, based on ecological factors and allozyme markers, for assessing wild-cereal genetic resources, enabling us to screen single- and multiple-trait elite genotypes in nature (Nevo 1987). Moreover, we initiated genetic mapping: first, we mapped the chlorotolurone resistance gene (Krugman et al. 1994, 1997), followed by two yellow-rust resistance genes (Sun et al. 1997; Chagué et al. 1999; Peng et al. 1999, 2000a,c), we also produced a genome-wide molecular genetic map (Peng et al. 2000b) with a few dozen QTLs across most chromosomes (Peng et al. 2001).

Part II

Population Genetics of Wild Emmer Wheat

Macrogeographic Population Genetic Studies of *Triticum dicoccoides* in the Fertile Crescent, Israel and Turkey: Allozyme and DNA Polymorphisms

4.1 Allozymic Diversity

4.1.1 General Overview of Molecular Evolution

Modern evolutionary genetics derives its strength from the spectacular progress made during the last three decades in understanding molecular evolution and diversity at the protein and DNA levels. To a large extent this progress has been due to protein electrophoresis and recombinant DNA technology (Nei 1987). Multiple molecular forms of enzymes, called isozymes, recognized in 1959 by Clement Markert, have served as powerful probes of molecular, genetic, developmental, physiological, and evolutionary mechanisms. Isozymes have been studied, or used as markers, in a wide range of biochemical and biological investigations. In population genetics, isozymes, and their allozyme variants, opened up a new era of evolutionary genetics studies. Genetic diversity within and among species is central to evolutionary biology and its two major processes, speciation and adaptation. Molecular biology permitted the characterization of genetic diversity among individuals, populations and species. It did so first, by exposing the relationship between genes and proteins (Lewontin 1974), and second, by elucidating, at the nuclear and extranuclear coding and noncoding DNA regions, the structure, expression, function, and evolution of genes, intergenic spacers, and multigene families.

The genetic structure and differentiation of natural populations of plants and animals are now understandable at the molecular level, primarily in a dynamic, ecological context over space and time. Evidence of protein variation was critically reviewed for 1,111 species of plants and animals by Nevo et al. (1984b), though the representation of inbreeders was small. The results indicated that genetic polymorphism and heterozygosity in nature are structured on a massive scale and are primarily determined by diverse physical and biotic ecological factors.

The evidence of genetic diversity found in natural populations of *T. dicoccoides,* the wild progenitor of all cultivated wheats, is important for two major reasons. First, theoretically, for understanding the evolutionary history and dynamics of this unique progenitor of wheats, and second, practically, for evaluating the potential genetic resources for conservation and utilization for crop improvement. Modern genetic techniques permit the transference of genetic material between ancestors and their derived domesticates. Very few ancestors have been studied across their geographic range at the interface of population genetics and ecology. This interface is important for displaying the actual genetic differentiation of the progenitor and its ecological determinants, thereby yielding clues about the potential genetic resources it harbors for breeding (Nevo 1983a, 1987, 1989a, 1995, 2001b; Tanksley and Orton 1983; Peng et al. 2000a–c, 2001). We had previously analyzed, in a similar way, the genetic differentiation of wild barley, *Hordeum spontaneum*, in the Near East (Nevo et al. 1986b; Nevo 1992). Here, we do the same, though on a smaller scale (Iran's wild emmer is not represented), for wild emmer wheat (Nevo and Beiles 1989).

4.1.2
Patterns of Allozyme Diversity of Wild Emmer

Triticum dicoccoides, was studied for allozyme diversity of proteins encoded by 42 gene loci in 1,815 plants (1,658 and 157 from Israel and Turkey, respectively), representing 37 populations, 33 from Israel and four from Turkey, sampled in 33 localities from Israel (Fig. 4.1, Table 4.1) and studied from 1979 to 1987 (Nevo and Beiles 1989). The results indicated that: (1) six loci (14%) were monomorphic in all populations, 15 loci (36%) were locally polymorphic, and 21 loci (50%) were regionally polymorphic, similar to the first studies of 12 populations in Israel (Nevo et al. 1982) and in Turkey (Nevo et al. 1988a). All polymorphic loci (except four) displayed high local levels of polymorphism (≥10%). (2) The number of alleles per locus, *A*, was 1.252 (range: 1.050–1.634). The proportion of polymorphic loci per population, *P*, averaged 0.220 (range: 0.050–0.415), and genic diversity, *He*, averaged 0.059 (range: 0.002–0.119). (3) Altogether there were 119 alleles at the 42 putative loci tested, 114 of them in Israel. (4) Genetic differentiation was primarily regional and local, not clinal; 70% of the variant alleles were common (≥10%) and not widespread, but localized or sporadic, displaying an "archipelago" population genetics and ecology structure.

The coefficients of genetic distance between populations were high and averaged $D = 0.134$, range, 0.018–0.297, indicating sharp genetic differentiation over short distances. (5) Discriminant analyses differentiated the Israeli from the Turkish populations, and within Israel between central and three marginal regions, as well as different soil-type populations (Fig. 4.2; Table 4.2) (6) Allozymic diversity comprised 40% within and 60% between populations (Table 4.3). (7) Linkage disequilibria were abundant, and their number positively correlated ($r_s = 0.60$, $p < 0.01$) with the humidity (Table 4.4). (8) Multilocus organization (Brown et al. 1980) was substantial, and also positively correlated with humidity. (9) Allozyme diversity, overall and at single and multilocus structures, was significantly correlated with and partly predictable

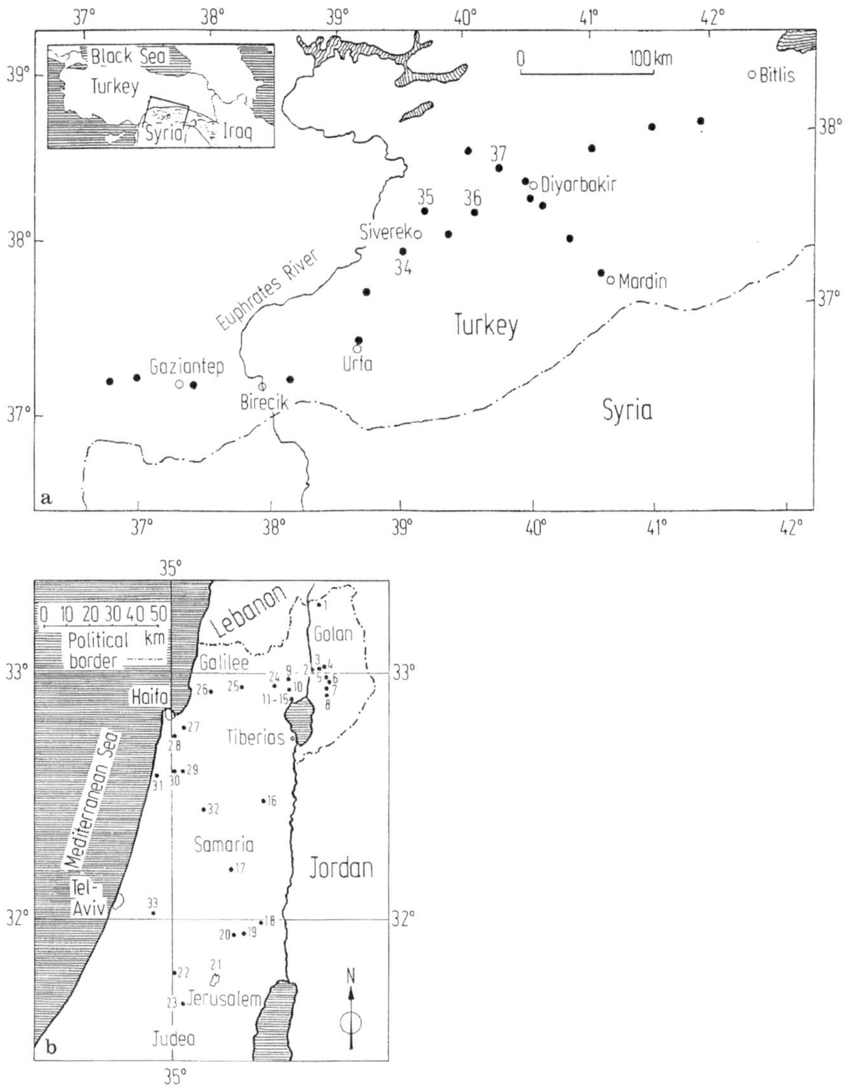

Fig. 4.1a, b. Geographic distributions of 37 tested populations of wild emmer wheat in the Fertile Crescent. **a** Four populations at four sites in Turkish Kurdistan; **b** 33 populations at 29 sites in Israel; for names of numbered populations see list in Table 4.1, The *black, unnumbered solid circles* in Turkey indicate sites at which wild barley was collected but wild emmer was not found. (Nevo and Beiles 1989)

Table 4.1. Summary of genetic variation, based on 42 allozyme loci, and genome organization, X(2), based on 29 polymorphic allozyme loci, in 37 populations of *Triticum dicoccoides* in Israel and Turkey. (Nevo and Beiles 1989)

	Sample size (*N*)	Mean no. of alleles per locus (*A*)	Mean proportion of loci				Genetic diversity (*He*)	Multilocus organization [X(2)]	No. of	
			Polymorphic per population (*P*)		Heterozygous per individual (*H*)				PL[b]	LD
			(1%)	(5%)	Mean	S.E.				
Locality:										
1. Mt. Hermon	40	1.262	0.238	0.214	0.001	0.001	0.060	0.498*	9	32
2. Andarta	30	1.410	0.359	0.308	0.0	0.0	0.095	2.539*	13	68
3. Tsomet	30	1.359	0.308	0.256	0.0	0.0	0.089	1.239*	11	55
4. Kedmot-Zvi	30	1.231	0.231	0.179	0.0	0.0	0.055	1.006*	9	36
5. Qazrin	40	1.310	0.286	0.167	0.004	0.002	0.032	0.568	10	43
6. Aniam	30	1.231	0.231	0.231	0.0	0.0	0.065	0.905*	9	36
7. Yehudiyya	39	1.167	0.167	0.119	0.001	0.001	0.047	0.382ns	7	20
8. Gamla	30	1.333	0.308	0.308	0.0	0.0	0.102	1.238*	12	66
9. Rosh-Pinna	40	1.333	0.286	0.238	0.001	0.001	0.093	0.901*	12	63
10. Ammiad	232	1.634	0.366	0.293	0.000	0.000	0.091	1.366*	18	135
11. Tabigha 1979	40	1.262	0.262	0.167	0.005	0.002	0.054	0.347ns	11	53
12. Tabigha, basalt 1984	92	1.452	0.357	0.286	0.002	0.001	0.119	1.015*	13	72
13. Tabigha, basalt 1985	90	1.262	0.238	0.238	0.0	0.0	0.077	0.322	9	34
14. Tabigha, t.r., 1984	78	1.452	0.357	0.262	0.005	0.002	0.089	0.409	13	62
15. Tabigha, t.r., 1985	96	1.310	0.310	0.238	0.0	0.0	0.073	0.910*	13	64
11–15 Tabigha[a]	396	1.643	0.381	0.310	0.002	0.001	0.113	0.475*	18	136
16. Mt. Gilboa	32	1.167	0.167	0.143	0.0	0.0	0.044	0.323ns	6	15
17. Mt. Gerizim	25	1.214	0.190	0.167	0.007	0.003	0.059	0.405ns	8	27
18. Gitit	48	1.512	0.415	0.293	0.013	0.005	0.116	0.490	15	93
19. Kokhav-Hashahar	40	1.238	0.190	0.190	0.004	0.003	0.058	0.277ns	8	28
20. Taiyiba	41	1.238	0.214	0.190	0.0	0.0	0.051	0.905*	9	36
21. Sanhedriyya	40	1.143	0.143	0.143	0.0	0.0	0.048	1.960*	6	15

22. Bet-Meir	40	1.095	0.095	0.048	0.001	0.001	0.023	0.221ns	4	6
23. J'aba	50	1.463	0.366	0.244	0.003	0.001	0.072	2.021*	11	51
24. Amirim	37	1.366	0.293	0.268	0.002	0.002	0.102	0.883*	11	55
25. Nahef	44	1.171	0.146	0.098	0.001	0.001	0.039	0.179ns	5	9
26. Ahihud	53	1.073	0.073	0.073	0.0	0.0	0.024	1.579*	3	3
27. Nesher	40	1.122	0.122	0.122	0.0	0.0	0.053	1.113*	3	3
28. Bet-Oren	40	1.050	0.050	0.0	0.0	0.0	0.002	–	1	0
29. Daliyya	37	1.098	0.073	0.049	0.004	0.004	0.009	–0.023ns	3	1
30. Bat-Shelomo	40	1.333	0.310	0.214	0.003	0.002	0.085	0.622*	13	78
31. Kabara	30	1.075	0.075	0.050	0.0	0.0	0.015	0.164ns	3	3
32. Yabad	44	1.073	0.073	0.049	0.001	0.001	0.010	–0.073ns	2	1
33. Givat-Koach	40	1.100	0.100	0.100	0.0	0.0	0.026	3.010*	4	6
34. West Siverek, 9 km	39	1.190	0.190	0.143	0.001	0.001	0.058	1.040*	7	21
35. East Siverek, 20 km	45	1.167	0.167	0.167	0.001	0.001	0.047	0.007ns	6	15
36. W. Diyarbakir, 22 km	48	1.238	0.214	0.167	0.001	0.001	0.055	0.988*	8	26
37. N. Diyarbakir, 20 km	25	1.189	0.162	0.108	0.0	0.0	0.048	1.131*	6	12
Means:										
Total (37 populations)	1815	1.252	0.220	0.176	0.002	0.000	0.059	0.857		
Range		1.050–1.634	0.050–0.415	0.0–0.308	0.0–0.013	0.0–0.005	0.002–0.119	–0.073–3.010		
Israel (33 pop.)	1658	1.259	0.224	0.180	0.002	0.001	0.060	0.866		
Range		1.050–1.634	0.050–0.415	0.0–0.308	0.0–0.013	0.0–0.005	0.002–0.119	–0.073–3.010		
Israel (29 pop.[a])	1658	1.257	0.213	0.175	0.002	0.001	0.057	0.899		
Range		1.050–1.643	0.050–0.415	0.0–0.308	0.0–0.013	0.0–0.005	0.002–0.116	–0.073–3.010		
Central (14 pop.)	897	1.339	0.290	0.235	0.001	0.00	0.077	0.939		
(= large) Range		1.167–1.634	0.167–0.366	0.119–0.308	0.0–0.005	0.0–0.002	0.032–0.119	0.322–2.539		
Central (10 pop.[a])	897	1.365	0.292	0.241	0.001	0.000	0.078	1.062		
(= large) Range		1.167–1.643	0.167–0.381	0.119–0.310	0.0–0.004	0.0–0.002	0.032–0.113	0.322–2.539		
Margins (23 pop.)	918	1.199	0.177	0.141	0.002	0.000	0.048	0.805		
(+ Turkey) Range		1.050–1.512	0.050–0.415	0.0–0.293	0.0–0.013	0.0–0.005	0.002–0.116	–0.073–3.010		
Margins (19 pop.)	761	1.199	0.175	0.140	0.002	0.001	0.047	0.809		
(Israel) Range		1.050–1.512	0.050–0.415	0.0–0.293	0.0–0.013	0.0–0.005	0.002–0.116	–0.073–3.010		
North margins (1 pop.)	40	1.262	0.238	0.214	0.001	0.001	0.060	0.498		
West margins (10 pop.)	405	1.146	0.132	0.102	0.001	0.001	0.037	0.828		
Range		1.050–1.366	0.050–0.310	0.0–0.268	0.0–0.004	0.0–0.004	0.002–0.102	–0.073–3.010		

Table 4.1. *Continued*

	Sample size (*N*)	Mean no. of alleles per locus (*A*)	Mean proportion of loci				Genetic diversity (*He*)	Multilocus organization [X(2)]	No. of	
			Polymorphic per population (*P*)		Heterozygous per individual (*H*)				PL[b]	LD
			(1%)	(5%)	Mean	S.E.				
S.E. margins (8 pop.)	311	1.259	0.223	0.177	0.004	0.001	0.059	0.825		
Range		1.095–1.512	0.095–0.415	0.048–0.293	0.0–0.013	0.0–0.005	0.023–0.116	0.221–2.021		
Turkey (4 pop.)	157	1.196	0.183	0.146	0.001	0.000	0.052	0.792		
Range		1.167–1.238	0.162–0.214	0.108–0.167	0.0–0.001	0.0–0.001	0.047–0.058	0.007–1.131		
Terra rossa (20 pop.)	1086	1.250	0.209	0.169	0.002	0.000	0.057	0.927		
Range		1.050–1.634	0.050–0.415	0.0–0.293	0.0–0.013	0.0–0.005	0.002–0.116	0.164–3.010		
Rendzina (3 pop.)	121	1.168	0.152	0.104	0.002	0.001	0.035	0.175		
Range		1.073–1.333	0.073–0.310	0.049–0.214	0.001–0.004	0.001–0.004	0.009–0.085	–0.073–0.622		
Basalt (14 pop.)	608	1.271	0.254	0.198	0.001	0.000	0.067	0.910		
(+ Turkey) Range		1.167–1.452	0.162–0.359	0.108–0.308	0.0–0.005	0.0–0.002	0.032–0.119	0.007–2.539		
Basalt (10 pop.)	451	1.302	0.282	0.219	0.001	0.000	0.073	0.957		
(Israel) Range		1.167–1.452	0.167–0.359	0.119–0.308	0.0–0.005	0.0–0.002	0.032–0.119	0.322–2.539		
Large pop. (14 pop.)	897	1.339	0.290	0.235	0.001	0.000	0.077	0.939		
(= central) Range		1.167–1.634	0.167–0.366	0.119–0.308	0.0–0.005	0.0–0.002	0.032–0.119	0.322–2.539		
Large pop. (10 pop.[a])	897	1.365	0.292	0.241	0.001	0.000	0.078	1.062		
(= central) Range		1.167–1.643	0.167–0.381	0.119–0.310	0.0–0.004	0.0–0.002	0.032–0.113	0.322–2.539		
Medium pop. size (12 pop.)	477	1.259	0.224	0.180	0.003	0.001	0.061	0.711		
Range		1.073–1.512	0.073–0.415	0.048–0.293	0.0–0.013	0.0–0.005	0.010–0.116	–0.073–2.021		
Small pop. size (11 pop.)	441	1.132	0.123	0.098	0.001	0.000	0.034	0.919		
(+ Turkey) Range		1.050–1.238	0.050–0.214	0.0–0.167	0.0–0.004	0.0–0.004	0.002–0.058	–0.023–3.010		
Small pop. size (7 pop.)	284	1.095	0.088	0.070	0.001	0.001	0.024	1.004		
(Israel) Range		1.050–1.171	0.050–0.146	0.0–0.122	0.0–0.004	0.0–0.004	0.002–0.053	–0.023–3.010		

[a] The five Tabigha subpopulations are considered to be one population
[b] PL, Number of polymorphic loci out of the 29 loci, included in the X(2) analysis, followed by the number of LDs
$*p < 0.05$, *ns* $p > 0.05$

Table 4.2. Stepwise discriminant analysis of allozyme allele frequencies of *T. dicoccoides* in 37 populations, between marginal-central populations in Israel and Turkey, based on 22 loci and 34 alleles. The analysis includes: one cold steppic, ten western marginal, eight eastern and southern marginal, four Turkish, and 14 central populations. (Nevo and Beiles 1989)

Actual group	No. of cases	Predicted group membership				
		1	2	3	4	5
1. Cold steppe	1	1	0	0	0	0
		100.0%	0.0%	0.0%	0.0%	0.0%
2. Western margins	10	0	8	2	0	0
		0.0%	80.0%	20.0%	0.0%	0.0%
3. Eastern and southern margins	8	0	1	7	0	0
		0.0%	12.5%	87.5%	0.0%	0.0%
4. Turkey	4	0	0	0	4	0
		0.0%	0.0%	0.0%	100.0%	0.0%
5. Central populations	14	0	1	2	0	11
		0.0%	7.1%	14.3%	0.0%	78.6%

83.78% of cases correctly classified

by climatic and edaphic factors (Table 4.5). (10) The distribution of the significant positive and negative autocorrelations, and their absence in the correlogram, revealed no similar geographic patterns across loci, eliminating migration as a prime factor of population genetic differentiation.

These results suggest: (1) during the evolutionary history of wild emmer, diversifying and balancing natural selections, through climatic, edaphic, and biotic factors, were a major agent of genetic structure and differentiation at both the single and multilocus levels; (2) wild emmer harbors large amounts of genetic diversity exploitable as genetic markers in sampling, and abundant adaptive genetic resources utilizable for wheat improvement.

4.1.3 Population Genetic Structure of Wild Emmer Wheat

4.1.3.1 "Archipelago" Population Genetic Structure of Wild Emmer Wheat

Wild emmer grows in lush and extensive stands in the catchment area of the Upper Jordan Valley (in northern Israel, in the eastern upper Galilee Mountains and the Golan Heights). However, elsewhere in the Fertile Crescent (Fig. 2.4), populations of wild emmer are semi-isolated and isolated and mostly display a patchy structure. At least in Israel, but possibly also elsewhere across the range of wild emmer in the Fertile Crescent, populations are subdivided into demes or clumps of varying sizes, including large, medium, and small patches. This archipelago structure (Table 4.4) was originally described for a smaller sample of wild emmer in Israel (Nevo et al. 1982) and

Table 4.3. Partition of genetic diversity of *T. dicoccoides* within and between 37 populations in Israel and Turkey, based on allozyme polymorphic loci (*Gst* analysis). (Nevo and Beiles 1989)

Locus	Alleles	Sample	*Ht*	*Hs*	*Dst*	*Gst*	*Dm*	*Rst*
Aat-1A	4	1,754	0.0136	0.0129	0.0007	0.0526	0.0007	0.0570
Aat-1B	3	1,795	0.0155	0.0146	0.0009	0.0559	0.0009	0.0609
Aat-2A	2	1,788	0.1179	0.0895	0.0284	0.2407	0.0292	0.3257
Aat-2B	1	1,799	0.0	0.0000	0.0000	–	0.0000	–
Aat-3A	2	1,756	0.0445	0.0022	0.0423	0.9496	0.0435	19.3587
Aat-3B	1	1,796	0.0	0.0000	0.0000	–	0.0000	–
Acph-3	3	1,561	0.2505	0.1362	0.1143	0.4561	0.1179	0.8656
Acph-x	2	1,169	0.3744	0.1604	0.2139	0.5715	0.2208	1.3767
Adh-1A	2	1,295	0.2644	0.0813	0.1830	0.6924	0.1881	2.3130
Adh-1B	2	1,298	0.0840	0.0307	0.0532	0.6338	0.0547	1.7791
Adh-2A	2	1,607	0.0124	0.0092	0.0032	0.2585	0.0033	0.3584
Adh-2B	2	1,608	0.0318	0.0096	0.0222	0.6978	0.0228	2.3734
Cat-A	1	1,424	0.0	0.0000	0.0000	–	0.0000	–
Cat-B	1	1,425	0.0	0.0000	0.0000	–	0.0000	–
Est-4A	3	1,426	0.0373	0.0315	0.0058	0.1557	0.0060	0.1896
Est-4B	5	1,426	0.6803	0.1995	0.4807	0.7067	0.4945	2.4780
Est-5A	6	1,747	0.3996	0.1475	0.2521	0.6309	0.2591	1.7564
Est-5B	6	1,689	0.5353	0.2632	0.2722	0.5084	0.2797	1.0628
Gdh-A	4	1,790	0.0385	0.0288	0.0097	0.2524	0.0100	0.3469
Gdh-B	4	1,790	0.0644	0.0525	0.0119	0.1847	0.0122	0.2328
Gluc-B	4	1,405	0.4572	0.1391	0.3180	0.6956	0.3286	2.3617
Hk	4	1,135	0.3587	0.1680	0.1906	0.5315	0.1970	1.1723
Ipol	4	1,180	0.2985	0.1103	0.1882	0.6306	0.1941	1.7604
Ipor-A	3	1,780	0.0090	0.0079	0.0010	0.1123	0.0010	0.1301
Ipor-B	2	1,789	0.1295	0.0572	0.0723	0.5580	0.0743	1.2974
Mdh-1A	2	1,735	0.3545	0.0941	0.2603	0.7344	0.2676	2.8420
Mdh-1B	2	1,736	0.0012	0.0011	0.0000	0.0339	0.0000	0.0361
Mdh-2	4	1,778	0.0679	0.0180	0.0499	0.7352	0.0513	2.8533
Nadhd-1A	3	1,539	0.2224	0.0677	0.1547	0.6954	0.1589	2.3468
Nadhd-1B	3	1,540	0.1016	0.0309	0.0707	0.6956	0.0727	2.3483

Nadhd-2A	3	1,576	0.0553	0.0438	0.0115	0.2072	0.0118	0.2698
Pepc	3	1,502	0.1281	0.0117	0.1165	0.9088	0.1197	10.2471
Pept-1A	2	1,719	0.0023	0.0022	0.0001	0.0389	0.0001	0.0416
Pept-1B	2	1,717	0.4123	0.1363	0.2760	0.6695	0.2837	2.0816
Pept-2	3	969	0.2126	0.979	0.1147	0.5397	0.1181	1.2068
Pgi-A	5	1,782	0.4254	0.2492	0.1763	0.4143	0.1811	0.7270
Pgi-B	2	1,784	0.2426	0.0853	0.1573	0.6482	0.1616	1.8939
Pgm-A	3	1,779	0.0316	0.0273	0.0043	0.1350	0.0044	0.1605
Pgm-B	1	1,779	0.0	0.0000	0.0000	–	0.0000	–
6Pgdh-1A	2	1,766	0.0113	0.0085	0.0028	0.2457	0.0028	0.3348
6Pgdh-1B	3	1,766	0.0202	0.0166	0.0036	0.1788	0.0037	0.2237
6Pgdh-2	3	1,605	0.4174	0.1265	0.2909	0.6969	0.2990	2.3627
Mean			0.1649	0.0659	0.0989	0.6000	0.1018	1.5434
Central populations ($N = 14$)								
Mean			0.1600	0.0811	0.0789	0.4931	0.0853	1.0516
Israeli marginal populations ($N = 19$)								
Mean			0.1298	0.0487	0.0811	0.6250	0.0857	1.7613
Western marginal populations ($N = 10$)								
Mean			0.1363	0.0377	0.0987	0.7238	0.1103	2.9300
South-eastern marginal populations ($N = 8$)								
Mean			0.1078	0.0604	0.0474	0.4400	0.0542	0.8983
Terra rossa populations ($N = 20$)								
Mean			0.1575	0.0658	0.0918	0.5825	0.0968	1.4712
Israeli basalt populations ($N = 10$)								
Mean			0.1577	0.0777	0.0800	0.5074	0.0899	1.1575
Large populations ($N = 14$)								
Mean			0.1600	0.0811	0.0789	0.4931	0.0853	1.0516
Medium populations ($N = 12$)								
Mean			0.1111	0.0623	0.0487	0.4388	0.0532	0.8537
Israeli small populations ($N = 7$)								
Mean			0.1414	0.0244	0.1169	0.8271	0.1384	5.6607

Abbreviations: *Ht* total gene diversity, *Hs* average gene diversity within populations, *Dst* average gene diversity between populations, *Dm* average of interpopulational diversity only, *Gst* gene diversity between populations relative to *Ht*, *Rst* interpopulational diversity relative to *Hs*

Table 4.4. Examples of linkage disequilibria (LD) between pairs of 11 allozyme loci of *T. dicoccoides* in each of the 37 populations in Israel and Turkey. All values of LD are × 1000. (Nevo and Beiles 1989)

Locus and allele	Population														
	Hermon	Central populations													
	1	2	3	4	5	6	7	8	9	10	11	12	13	14	15
Adh-1A b		c						c			c	c	c	c*	
× *Est*-5B a	42	94	.	.	.	.	.	27	.	.	−25	−15	30	112	.
Adh-1A b		a													
× *Est*-4B null	.	250	.	.	.	.	.	.	.	.	.	15	.	.	.
Adh-1A b							c							c	
× *Pgi*-A d	.	.	.	.	.	.	6	−55	.	.	−27	67	23	42	7
Adh-1A b		c													
× *Est*-5A a	.	−188	.	.	.	.	.	.	.	.	.	.	.	.	.
Adh-1A b							*				c		**	c	
× *Acph*-3 b	.	.	.	.	.	.	92	.	.	.	23	63	52	−31	8
Adh-1A b												c*		c	
× *Acph*-x a	.	.	.	.	.	.	.	.	.	.	.	−142	20	−14	.
Adh-1A b	c**	a						c							
× *Gluc*-B b	79	188	.	.	.	.	.	24	.	.	.	.	.	.	−9
Adh-1A b		c						c					c		
× *Pept*-1B a	.	−63	.	.	.	.	.	−24	.	.	−8	88	15	.	4
Pept-1B a		c*	c					c						c	
× *Est*-5B a	.	89	27	6	.	−60	.	8	.	−11	9	−11	21	22	12
Pept-1B a										c					
× *Est*-4B null	.	−19	.	.	.	.	.	.	.	1	.	.	.	.	.
Pept-1B a		c				c		c			c		***		*
× *Pgi*-A d	.	11	33	.	.	−31	.	27	.	4	15	32	78	1	46
Pept-1B a				c		c		c							
× *Est*-5A a	.	35	.	12	.	−31	.	−66	.	6	.	.	.	.	.
Pept-1B a										**	c				
× *Acph*-3 b	.	.	.	.	.	.	.	.	.	−13	8	19	−14	0	−19
Pept-1B a										c		*			**
× *Acph*-x a	.	.	.	.	.	.	.	.	.	−1	.	−95	55	−5	96
Pept-1B a		c	**	c*				c**						c	
× *Cluc*-B b	.	−58	−121	−57	.	8	.	−87	.	−3	.	.	.	4	2
Pgi-A d		c	c		c	c		c			c	*		c*	
× *Est*-5B a	.	−8	−44	.	−1	20	.	−16	.	−16	−5	50	−7	43	4
Pgi-A d		c								**		c			
× *Est*-4B null	.	−17	.	.	.	.	.	.	.	30	.	4	.	.	.
Pgi-A d		c			c	c			c	**					
× *Est*-5A a	.	−10	.	.	2	−4	.	−53	−54	48	.	.	.	.	.
Pgi-A d					c		c				c	**	*	**	c**
× *Acph*-3 b	.	.	.	.	−1	.	10	.	9	−3	5	81	42	64	−51
Pgi-A d									c					c	***
× *Acph*-x a	.	.	.	.	.	.	.	.	19	−10	.	−54	45	57	136
Pgi-A d		c			c	c		c	c					c	c*
× *Gluc*-B b	.	5	−80	.	−2	−12	.	−36	15	3	.	.	.	−14	35
Est-4B null												c			
× *Est*-5B a	.	.	.	.	.	.	.	.	.	2	.	16	.	.	.
Est-4B null		c													
× *Est*-5A a	.	−20	.	.	.	.	.	.	.	−15	.	.	.	.	.
Est-4B null										c					
× *Acph*-3 b	.	.	.	.	.	.	.	.	.	−3	.	−6	.	.	.
Est-4B null		c								*					
× *Gluc*-B b	.	17	.	.	.	.	.	.	.	19	.	.	.	.	.
Est-5A a		c		c*	c	c		c		**					
× *Est*-5B a	.	−15	.	−101	−24	−47	.	−30	.	−34	.	.	.	.	
Est-5A a					c				c						
× *Acph*-3 b	.	.	.	.	4	.	.	.	−15	−3	.	.	.	.	.
Est-5A a															
× *Acph*-x a	.	.	.	.	.	.	.	.	.	9	.	.	.	.	.

Southeastern margins								Western margins[$]				Turkey[$]		
16	17	18	19	20	21	22	23	24	26	30	33	34	35	37
.	.	.	.	.	.	c 24	.	c 22	.	.	.	.	.	41
.	.	.	.	.	.	.	.	.	.	.	.	.	.	.
.	.	.	.	c 9	.	.	.	c −24	.	c 1	.	.	.	.
.	.	.	.	.	.	.	.	c −7	c*** −151	c −15	.	.	.	.
.	.	.	.	.	.	.	.	c −44	.	c 8	.	.	.	.
.	.	.	.	.	.	.	.	.	c*** −145	c 15	.	.	.	.
.	.	.	.	.	.	.	.	.	.	c −15	.	.	.	.
.	.	.	.	.	.	.	.	.	.	c 16	.	.	.	.
.	.	0	.	.	.	.	.	.	.	.	.	.	.	.
.	.	c 18	.	.	.	.	.	.	.	.	.	.	.	.
.	c −39	.	.	.	.	.	.	.	.	c 10	.	.	.	.
.	.	.	.	.	.	.	.	.	.	* −85	.	.	.	.
.	.	c 7	.	.	.	.	.	.	.	32	.	.	.	.
c −8	c −14	−24	.	.	.	.	.	.	.	*** 163	.	.	.	.
.	c 4	.	.	.	.	.	.	.	.	−74	.	.	.	.
.	.	c 24	c 32	.	.	.	c 17	51	.	.	.	.	c 35	.
.	.	c −10	.	.	.	.	c −28	.	.	.	.	.	.	.
.	.	.	−19	.	.	.	c 0	c 63	.	c −11	a*** 130	.	.	.
.	.	c 9	.	.	.	.	c −22	25		c −7	.	.	.	.
.	48	c −21	.	.	.	.	.	.	.	c 10	.	.	c** −86	.
.	−5	.	−3	.	.	.	.	.	.	c 15	.	.	c −18	.
.	.	55	.	.	.	.	** −85	.	.	.	.	.	.	.
.	.	.	.	.	.	.	c 0	.	.	.	.	.	.	.
.	.	45	.	.	.	.	26	.	.	.	.	.	.	.
.	.	.	.	.	.	.	.	.	.	.	.	.	.	.
.	.	.	46	.	c*** −127	.	2	74	.	.	.	c −4	.	.
.	.	.	.	.	.	.	c −1	−1	.	c*** −117	.	.	.	.
.	.	.	.	.	*** 138	.	c −2	.	*** 129	*** −142	.	.	.	.

Table 4.4. *Continued*

Locus and allele	Population														
	Hermon	Central populations													
	1	2	3	4	5	6	7	8	9	10	11	12	13	14	15
Est-5A a		c*		c	c	c*				**					
× *Gluc*-B b	.	−61	.	−30	−41	57	.	51	.	−37	.	.	.	.	.
Est-5B a					c						c			*	
× *Acph*-3 b	.	.	.	.	1	.	.	.	.	12	−27	22	21	63	.
Est-5B a															
× *Acph*-x a	.	.	.	.	.	.	.	.	.	2	.	8	.	−43	7
Est-5B a	c	c												c	
× *Gluc*-B b	28	8	3	2	.	−50	.	26	.	11	.	.	.	−6	.
Acph-3 b									c	c		c			
× *Acph*-x a	.	.	.	.	.	.	.	.	5	2	.	6	32	0	.
Acph-3 b					c										
× *Gluc*-B b	.	.	.	.	−2	.	.	.	24	7	.	.	.	.	−13
Acph-x a															
× *Gluc*-B b	.	.	.	.	.	.	.	.	.	−13	.	.	.	.	.
Nadhd-1A b									***						
× *6Pgd*-2 a	.	.	.	.	.	.	.	.	148	.	.	.	.	.	.
6Pgd-2 a		c*	c	c				c			c		c		c
× *Est*-5B a	.	−80	−27	−61	.	−60	.	−8	.	−1	−19	−14	−18	32	1
6Pgd-2 a		c										c			
× *Est*-4B null	.	17	.	.	.	.	.	.	.	4	.	−4	.	.	.
6Pgd-2 a		c	c			c		c	***						
× *Pgi*-A d	.	17	−67	.	.	−9	.	−27	−158	12	−6	17	29	2	−12
6Pgd-2 a				c		c*		c	c						
× *Est*-5A a	.	−36	.	−18	.	58	.	66	55	−16	.	.	.	.	.
6Pgd-2 a										**	c				
× *Acph*-3 b	.	.	.	.	.	.	.	.	10	16	18	−12	1	10	−19
6Pgd-2 a									c	***					c
× *Acph*-x a	.	.	.	.	.	.	.	.	11	−35	.	32	−5	−26	−5
6Pgd-2 a		c	**			c***		c***	c	***				c	
× *Gluc*-B b	.	52	121	51	.	114	.	87	−16	46	.	.	.	−5	22
6Pgd-2 a		a***	***					a***		**					
× *Pept*-1B a	.	−202	−127	−27	.	−29	.	−90	.	−16	51	−14	2	−4	−23
6Pgd-2 a		c						c						c	
× *Adh*-1A b	.	63	.	.	.	.	.	24	.	.	1	2	−17	54	12
Nadhd-1A b	c				c										
× *Est*-5B a	52	.	.	.	−20	.	.	.	.	.	.	.	.	.	.
Nadhd-1A b															
× *Est*-4B null	.	.	.	.	.	.	.	.	.	.	.	.	.	.	.
Nadhd-A b					c				**						
× *Pgi*-A d	.	.	.	.	6	.	.	.	−111	.	.	.	.	.	.
Nadhd-1A b									c						
× *Est*-5A a	.	.	.	.	22	.	.	.	50	.	.	.	.	.	.
Nadhd-1A b															
× *Acph*-3 b	.	.	.	.	−14	.	.	.	12	.	.	.	.	.	.
Nadhd-1A b									c						
× *Acph*-x a	.	.	.	.	.	.	.	.	10	.	.	.	.	.	.
Nadhd-1A b					c				c						
× *Gluc*-B b	33	.	.	.	4	.	.	.	−17	.	.	.	.	.	.
Nadhd-1A b															
× *Pept*-1B a	.	.	.	.	.	.	.	.	.	.	.	.	.	.	.
Nadhd-1A b	*														
× *Adh*-1A b	78	.	.	.	.	.	.	.	.	.	.	.	.	.	.

$ Populations 27, 28, and 29 had no LD with these 11 loci; in population 25, the only LD is *Est*-4B null × *Acph*-3 b = 0.002; in population 31, the only one is *Est*-5B a × *Acph*-3 b = 0.008c; in population 32, the only one is *Est*-4B null × *Acph*-3 b = −0.012c; and in population 36, the only one is *Est*-5B a × *Acph*-x a = −0.013c. These seven populations are omitted from the table

$^{*}p < 0.05$; $^{**}p < 0.01$; $^{***}p < 0.001$; p by χ^2 with Yates correction, or by Fisher's exact test

c, Complete association: one gametic type is missing (Clegg et al. 1976); *a*, absolute association: only two gametic types are present (Clegg et al. 1976); *dot* one or both alleles of the pair were not in a polymorphic state, or missing data

Southeastern margins								Western margins[$]				Turkey[$]		
16	17	18	19	20	21	22	23	24	26	30	33	34	35	37
.	.	.	−13	.	.	.	.	.	.	80*	.	−4^{c}	.	.
.	.	16	.	.	.	.	−31	2	.	.	.	.	.	4^{c}
.	.	0	.	.	−111***	.	−106^{c***}	.	.	.	.	.	−30	−7^{c}
.	.	.	19	.	.	.	.	.	.	.	.	28	23^{c}	.
.	.	.	.	.	.	.	22^{c}	.	.	60	.	.	.	7^{c}
.	.	.	.	.	.	.	.	.	.	−69	.	.	.	.
.	9^{c}	.	.	.	.	.	.	.	.	−86*	.	.	−86^{c**}	.
.	.	.	.	5^{c}	.	.	.	.	.	.	.	.	.	.
.	.	.	.	.	.	.	.	.	.	.	.	.	.	.
.	.	.	.	.	.	.	.	.	.	.	.	.	.	.
	.	.	.	−14^{c}	.	.	.	.	.	.	.	.	.	.
.	.	.	.	.	.	.	.	.	.	.	.	.	.	.
.	.	.	.	.	.	.	.	.	.	.	.	.	.	.
.	.	.	.	.	.	.	.	.	.	.	.	.	.	.
.	.	.	.	.	.	.	.	.	.	.	.	.	.	.
.	.	.	.	.	.	.	.	.	.	.	.	.	.	.
.	.	.	.	−31^{c}	.	.	.	.	.	.	.	.	.	.
.	.	32^{c}	.	.	−124^{c***}	.	.	−37	.	.	.	.	−15	.
.	.	76^{c*}	.	.	.	.	.	.	.	.	.	.	.	.
.	.	20	.	1	.	.	.	−40^{c}	.	.	.	.	−15^{c}	.
.	.	.	.	.	170***	.	.	−86**	.	.	.	.	.	.
.	.	18	.	.	.	.	.	−9	.	.	.	.	.	.
−48	.	−34^{c}	.	.	134***	.	.	.	.	.	.	.	−72^{c*}	.
.	.	.	.	.	.	.	.	.	.	.	.	.	−15^{c}	.
3^{c}	.	8^{c}	.	.	.	.	.	.	.	.	.	.	.	.
.	.	.	.	3^{c}	.	.	.	3^{c}	.	.	.	.	.	.

Turkey (Nevo et al. 1988a), and for wild barley, *H. spontaneum,* in the Near East (Nevo 1992; Nevo et al. 1986b).

The highly subdivided, archipelago-type ecological population structure of wild emmer is even more distinct than that of wild barley and is matched by its genetic population structure. We found substantially more gene differentiation within and between populations that were sometimes geographically very close in Israel than between wild emmer in Israel and Turkey (Table 3 in Nevo and Beiles 1989). Of the total genetic diversity of *T. dicoccoides,* 40% was within populations and 60% was between populations, whereas only 5% was found between Israel and Turkey, as metapopulations (see Gst analysis, Table 4.3). This conclusion is reinforced by the microgeographic allozyme analysis based on edaphic, topographic, and temporal differentiation (Nevo et al. 1988b), on microclimatic local differentiation (Nevo et al. 1988c), and on the extreme case of local differentiation in the Golan Heights (Nevo et al. 1982; Golenberg and Nevo 1987). Similar patterns to those of the allozyme analysis were obtained with DNA analyses at macro- (Fahima et al. 1999, 2001) and microlevels (Li et al. 1999, 2000a–d, 2001a–d).

The high genetic differentiation within and between populations of *T. dicoccoides* is also reflected by the analysis of allele distribution. The latter revealed sharp local and regional differentiation in overall genetic indices (Table 4.1) as well as in individual allele frequencies (Appendix in Nevo and Beiles 1989). Out of the 119 alleles, 61 occurred in Turkey and 114 were found in Israel. Of the 61 alleles at 42 shared loci across Israel and Turkey, five alleles (8.2%) were *unique* for Turkey. Likewise, allele *uniqueness* was also found for the ecogeographic subdivisions in Israel as follows: (1) out of 93 alleles occurring in *central* populations, 17 alleles (18.3%) were *unique.* In *marginal* populations: (2) the southeastern margins had 84 alleles, six *unique* (7.1%). (3) The western margins had 79 alleles, six *unique* (7.6%). (4) The northeast (Mt. Hermon) had 53 alleles, one *unique* (1.9%). (5) Based on *soil types*: basalt had 89 alleles, 11 *unique* (12.4%). And finally, (6) regarding *population size*: "large" populations had 93 alleles, 17 *unique* (18.3%); "medium" populations had 78 alleles, 11 *unique* (14.1%); and "small" populations had 82 alleles, 10 *unique* (12.2%) (see discriminant analysis in Fig. 4.2, and Table 3 in Nevo and Beiles 1989).

Moreover, 70% of all variant alleles were not widespread, but revealed *localized* and *sporadic* distribution. Likewise, the analysis of genetic distances between populations supports the conclusion, based on genetic differentiation and allele distribution, that sharp local differentiation over short geographic distances is the rule, and the frequency of some common alleles (>10%) is *localized* and high. The population genetic structure of wild emmer is *mosaic.* This genetic mosaicism appears to reflect the underlying ecological heterogeneity which derives from local and regional geological, edaphic, climatic, and biotic differentiations. The resulting structure, as in wild barley, is an ecological genetic "archipelago", where the genetic structures are in accordance with the ecological ones.

Table 4.5. Coefficient of multiple regressions (R^2) of allozyme genic diversity (*He*) and allele frequencies as the dependent variables, and climatic and edaphic variables in 33 populations of *T. dicoccoides* in Israel as independent variables, and in the two subsets of 14 central and 19 marginal populations. (Nevo and Beiles 1989)

Stepwise model									
	Overall (33 populations)			Central (14 populations)			Marginal (19 populations)		
Genetic indices									
A	Huan 0.353***	Huan EV 0.424***	Huan Ev Ren 0.481***	Hu14 0.418*	Hu14 Rr 0.659**	Hu14 Rr Tr 0.718**	Huan 0.285*	Huan Rr 0.352*	
P-1%	Huan 0.433***	Huan Ev 0.504***	Huan Ev Ren 0.568***	Hu14 0.368*	Hu14 Rr 0.563**		Huan 0.284*	Huan Rr 0.349*	
P-5%	Td 0.496***			Hu14 0.279@	Hu14 Rr 0.435*	Hu14 Rr Td 0.517@	Huan 0.367**	Huan Trd 0.478**	Huan Trd Tr 0.528**
He	Huan 0.368***	Huan Ev 0.453***	Huan Ev Ren 0.491***	Hu14 0.246@	Hu14 Td 0.338 ns		Huan 0.285*	Huan Trd 0.399*	Huan Trd Tj 0.465*
Alleles									
Aat-1A b	Tdd 0.063 ns	Tdd Rv 0.140 ns		Huan 0.275@			Rv 0.078 ns	Rv Rd 0.260@	Rv Rd Rn 0.347@
Aat-1B a	Tdd 0.048 ns	Tdd Ev 0.106 ns		Huan 0.298*					
Aat-2A a	Hu14 0.069 ns	Hu14 Rr 0.128 ns	Hu14 Rr Al 0.207@	Rr 0.150 ns	Rr Hu14 0.549*	Rr Hu14 Al 0.657*	Hu14 0.093 ns		
Acph-3 a	Ba 0.353***			Ev 0.189 ns			Tr 0.414**		
Acph-x a	Tj 0.112@	Tj Tr 0.207@		Td 0.482*	Td Tr 0.617@		Rd 0.086 ns	Rd Ev 0.352*	
Adh-1A b	Td 0.295***	Td Huan 0.444***	Td Huan Tm 0.523***	Tm 0.422*			Rn 0.675***	Rn Rd 0.791***	Rn Rd Ta 0.850***
Adh-1B a	Dw 0.298***	Dw Huan 0.382***	Dw Huan Ta 0.430***				Dw 0.345**	Dw Ta 0.455**	Dw Ta Al 0.559**

Table 4.5. *Continued*

Stepwise model									
	Overall (33 populations)			Central (14 populations)			Marginal (19 populations)		
Est-4A a	Al 0.203**			Rr 0.508**	Rr Hu14 0.842***	Rr Hu14 Rn 0.942***	Al 0.232*	Al Rr 0.317*	Al Rr Ta 0.421*
Est-4B c	Ta 0.334***	Ta Rn 0.487***	Ta Rn Rr 0.529***	Huan 0.653***	Huan Rr 0.702***	Huan Rr Ev 0.868***	Td 0.112 ns	Td Al 0.339*	Td Al Ev 0.603**
Est-5A a	Huan 0.277**	Huan Rd 0.341**	Huan Rd Rn 0.390**	Tm $0.241^{@}$			Hu14 0.274*		
Est-5B e	Ta 0.402***	Ta Tdd 0.505***		Huan 0.512**			–		
Gdh-A b	Rd 0.151*	Rd Dw 0.268**	Rd Dw Rn 0.375**	Td $0.222^{@}$	Td Rr $0.379^{@}$	Td Rr Rn $0.476^{@}$	–		
Gdh-B b	Rd $0.101^{@}$	Rd Rr 0.244*	Rd Rr Dw 0.328**	Tdd $0.237^{@}$			Rr $0.161^{@}$	Rr Rd 0.460**	Rr Rd Trd 0.570**
Gluc-B b	Ta 0.634***	Ta Tr 0.705***	Ta Tr Tdd 0.781***	Trd 0.925***	Trd Rv 0.947***	Trd Rv Td 0.960***	Tr 0.817***	Tr Td 0.868***	
Hk b	Ba 0.361***	Ba Al 0.429***	Ba Al Td 0.486***	Rn 0.190 ns			Hu14 $0.258^{@}$	Hu14 Ta $0.376^{@}$	Hu14 Ta Rv $0.465^{@}$
Ipol b	Ev 0.202*	Ev Tdd 0.347**		Tdd 0.314*			Ev 0.443**	Ev Tdd 0.753***	Ev Tdd Hu14 0.844***
Ipor-B b	Rd 0.073 ns	Rd Ev 0.280**	Rd Ev Rv 0.344**	Dw 0.531**	Dw Rn 0.798***	Dw Rn Hu14 0.908***	Rn 0.096 ns	Rn Ev 0.187 ns	Rn Ev Dw 0.306 ns
Mdh-1A a	Ta 0.545***	Ta Ba 0.608***	Ta Ba Tdd 0.690***	Trd 0.834***	Trd Td 0.859***		Al 0.469***	Al Rv 0.614***	Al Rv Dw 0.773***
Mdh-2 b	Ren 0.299***			–			Tr 0.278*		

Nadhd-1A b	Tm 0.166*	Tm Ta 0.217*	Tm Ta Trd 0.253*	Rn 0.295*	Rn Tr 0.401@		Trd 0.165@		
Nadhd-1B b	Ev 0.105@			Rn 0.532**	Rn Td 0.769***	Rn Td Al 0.896***	Ev 0.167@	Ev Ta 0.275@	Ev Ta Tdd 0.348@
Nadhd-2A c	Huan 0.089 ns	Huan Ev 0.198@	Huan Ev Td 0.383**	Al 0.899***	Al Tdd 0.991***	Al Tdd Rn 0.999***	Hu14 0.229*	Hu14 Tj 0.360*	Hu14 Tj Tdd 0.543**
Pept-1B a	Ta 0.477***	Ta Ba 0.578***	Ta Ba Rn 0.637***	Trd 0.735***	Trd Tj 0.828***	Trd Tj Rn 0.894***	Tr 0.234*	Tr Trd 0.308@	Tr Trd Rr 0.401*
Pept-2 c	Dw 0.358***	Dw Rv 0.447***	Dw Rv Ta 0.542***	Dw 0.698***	Dw Rv 0.934***	Dw Rv Hu14 0.951***	Ta 0.469***	Ta Al 0.702***	Ta Al Tm 0.736***
Pgi-A d	Td 0.140*	Td Ren 0.372***		Rr 0.339*			Td 0.169@	Td Tr 0.467**	Td Tr Al 0.561**
Pgi-A e	Td 0.158*	Td Ren 0.271**	Td Ren Rd 0.330**	Dw 0.452**	Dw Rn 0.758***	Dw Rn Tr 0.852***	Rd 0.106 ns	Rd Huan 0.243 ns	Rd Huan Rv 0.382@
Pgi-B b	Tj 0.094@	Tj Al 0.299**	Tj Al Rv 0.350**	Trd 0.175 ns	Trd Al 0.813***	Trd Al Rr 0.925***	Tm 0.181@	Tm Al 0.408*	Tm Al Rv 0.692***
Pgm-A b	Ba 0.217**	Ba Ta 0.302**	Ba Ta Tdd 0.364**	Huan 0.171 ns	Huan Tr 0.335 ns		–		
6Pgd-1B c	Rn 0.384***	Rn Td 0.478***	Rn Td Rv 0.559***	–			Rn 0.409**	Rn Td 0.605***	Rn Td Trd 0.679***
6Pgd-2 a	Ta 0.401***	Ta Ev 0.533***	Ta Ev Trd 0.570***	Trd 0.754***	Trd Rr 0.879***	Trd Rr Hu14 0.921***	Ev 0.200@	Ev Tdd 0.319*	

Al Altitude, *Tm* mean annual temperature, *Ta* mean August temperature, *Tj* mean January temperature, *Td* mean seasonal temperature difference, *Tdd* mean day-night temperature difference, *Trd* mean number of tropical days, *Ev* mean annual evaporation, *Dw* mean number of dew nights in summer, *Rn* mean annual rainfall (in mm), *Rd* mean number of rainy days, *Rr* coefficient of variation in rainfall, *Hu14* mean humidity at 14:00 h, *Rv* interannual rainfall variation, *Huan* mean annual humidity, *Tr* terra rossa, *Ren* rendzina soil, *Ba* basaltic soil

***$p < 0.001$, **$p < 0.01$, *$p < 0.05$, @$p < 0.10$, *ns* $p > 0.10$

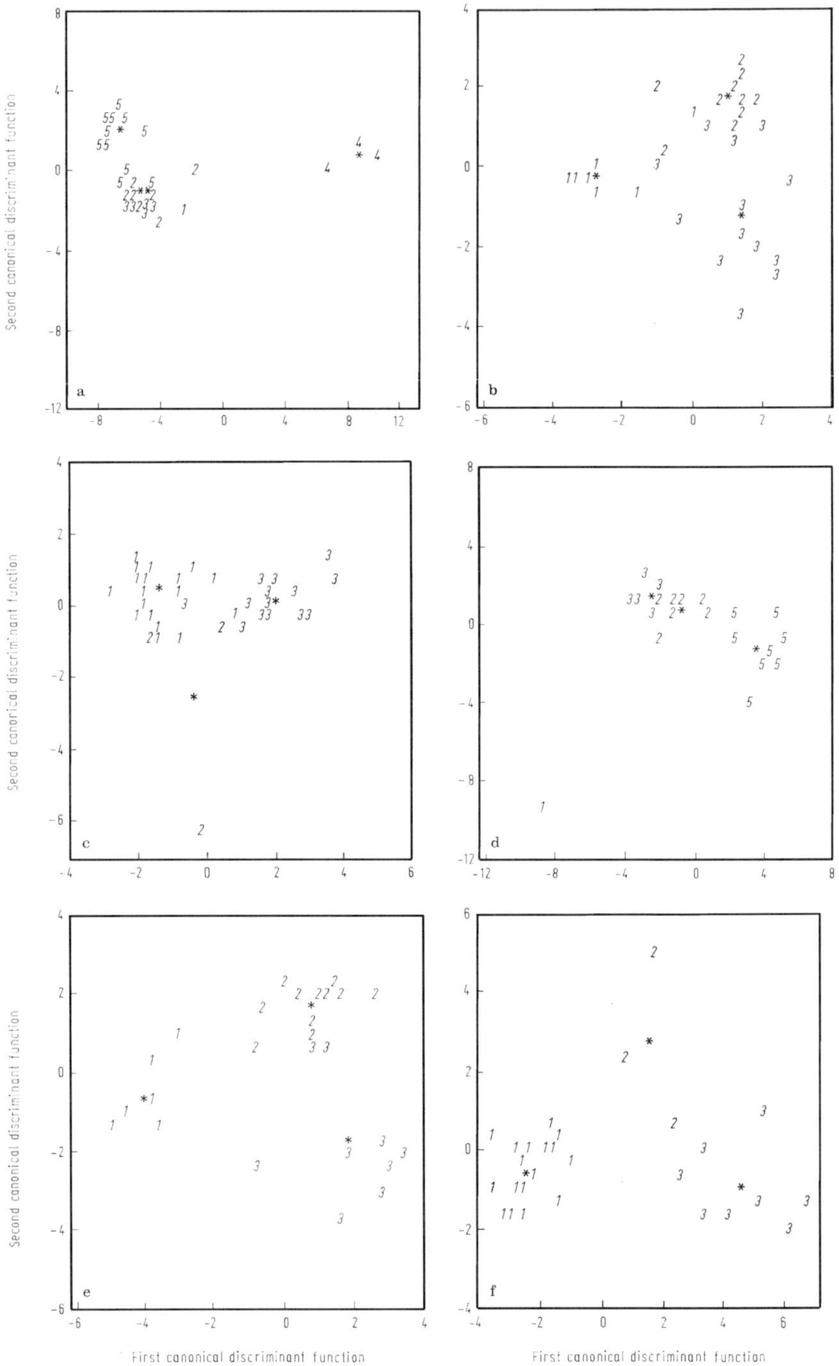

Fig. 4.2a–f. Discriminant analysis of allozyme allele frequencies between ecogeographic, edaphic, and population size categories in *Triticum dicoccoides* in Israel and Turkey (see Table 4.1 and details in Nevo and Beiles 1989). **a–c** Pattern of 37 populations in Israel and Turkey; **d–f** pattern of 29 populations from Israel (excluding the four micropopulations of Tabigha). *Left* **a** and **d** discriminate between four types of marginal and one type of central populations: *1* cold steppe, *2* western margin, *3* eastern and southern margin, *4* Turkey, *5* central populations, * group centroid. *Middle* **b** and **e** discriminate between small-, medium-, and large-size populations; population size: *1* small, *2* medium, *3* large, * group centroid. *Right* **c** and **f** discriminate between populations growing on three soil types: *1* terra rossa. *2* rendzina, and *3* basalt, * group centroid. (Nevo and Beiles 1989)

4.1.3.2
Spatial Autocorrelation of Allozymes

Spatial autocorrelation analysis (Sokal and Wartenberg 1983; Sokal and Oden 1978a,b) predicts that migration will cause high positive correlations (similarity) in the low-order distant groups (i.e., between neighbors), starting from the first distant group in the transect. Migration is expected to cause similarity between loci and alleles. This prediction is not realized in our results. By contrast, loci and alleles of wild emmer differ drastically in their autocorrelation pattern between loci. In addition, positive autocorrelations emerge at intermediate and distant groups, thus negating migration as an important evolutionary factor (Table 6 in Nevo and Beiles 1989).

The autocorrelation analysis also predicts that genetic drift will not create any autocorrelative pattern. Our data are structured, thus negating randomness. However, despite the between-loci variation, a general tendency of more positive correlations on the left-hand side of the table, and more negative correlations on the right-hand side are evident, suggesting that neither migration nor genetic drift can generate the observed patterns, and selection remains a vital explanatory model.

Environmental selection is also partly autocorrelated and affects loci differentially, but differently from migration. This is supported by our data for the following three reasons: (1) variation among loci; (2) positive correlation in different distant groups, and not necessarily the first one as would be expected if migration determined the interpopulation genetic structure; (3) the predominance of negative correlations in the larger distant groups is expected due to decreasing ecological similarity often with increasing distance (see Table 6 for details and discussion in Nevo and Beiles 1989).

4.1.3.3
Multilocus Structure of Allozymes

The multilocus organization estimates expressed by X(2) in Table 4.1 need elucidation. The measure was first suggested by Brown et al. (1980) as an estimate of multilocus associations for summing up multiple-linkage disequilibria. The behavior of this statistic, including its strengths and drawbacks, was discussed by Brown et al. (1980). The merit of this estimate, compared to a set of two-locus linkage disequilibria, is that it effectively reduces multilocus association to a few parameters for comparative studies of populations and species. This measure is an inverse function of two-locus gametic frequencies in inverse analogy with the single-locus Simpson diversity measure. The drawbacks include, first, a severe loss of information and ignorance of the behavior of particular allelic combinations. Second, it is assumed that all loci are of equal interest, which is biologically unjustified. Third, in some cases, the summation combines positive and negative contributions, thus existing multilocus organizations may be obscured. Finally, this measure is partially dependent on single-loci diversity and/or the number of loci scored, which is important in comparing estimates that are based on radically different single-locus data bases.

We acknowledge all the above-mentioned four drawbacks. In our present case, there was no one locus that was polymorphic across all 37 populations (Appendix in Nevo and Beiles 1989). Therefore, we based our analysis on 29 loci that were polymorphic to different degrees in different populations. This explains the wide variation in the number of polymorphic loci and the number of pairwise LDs on which X(2) is based in each population (last two columns of Table 4.1). We ran several analyses of X(2), some excluding very small populations, but nevertheless we obtained similar average estimates for the different subdivisions (Table 4.1, last two columns) as well as similar correlations, as described below. Our results indicated a significant negative Spearman rank correlation of X(2) with evaporation across the range ($r_s = -0.40$, $p < 0.05$), reinforced in the large, highly polymorphic central populations. We speculate that the intensity of X(2) as a summary statistic may vary in content, though it may be similar in level because of different linkage disequilibria, LDs, in diverse ecologies. In other words, the same level of intensity of genome organization may be caused by different ecological stresses.

4.1.4 Adaptive Nature of Allozyme Polymorphisms

The multiple lines of evidence discussed here support the hypothesis that allozyme polymorphisms are at least partly adaptive and are determined by natural diversifying and balancing selection. The evidence includes: (1) the association of single loci and multilocus structures with climatic and soil parameters (Table 4.5). Genetic differentiation appears to match, to a large extent, the ecologically heterogeneous background. (2) Genetic differentiation is not correlated with geographic distance, but with local ecological conditions of soil and climate (Tables 1, 3, and 6 of Nevo and Beiles 1989). (3) Macrogeographic genetic differentiation is partially paralleled by microgeographic soil fluctuations, as well as topographical and yearly climatic fluctuations (compare Nevo et al. 1988b, c, 1991c). (4) The genetic differentiation of wild emmer is far from random at the level of single-, two-, and multilocus organization. Significant linkage disequilibria abound, displaying allele association at the two-locus level (Table 4.4). (5) Strong allele associations, above those expected by chance, occur at the multilocus level (Table 4.1). (6) Spatial autocorrelation analysis reveals different significant spatial patterns for different alleles and loci (Table 6 in Nevo and Beiles 1989), ruling out migration as a differentiating factor. Likewise, the general order of negative and positive correlations rules out random drift as a major agent of genetic differentiation in wild emmer.

Our overall results indicate that genetic diversity among localities and over time (see Appendix in Nevo and Beiles 1989, and Table 4.1 for microsite spatial and temporal variation at Tabigha, and Table 6 for autocorrelation analysis in Nevo and Beiles 1989) display weak correlations with geography and stronger ones with ecology for single loci and for multilocus organization. This evidence permits us to eliminate genetic drift and migration as basic factors of genetic differentiation in wild emmer. By contrast, the structures described suggest that climatic and edaphic diversifying and balancing selection regimes at single- and multilocus levels, operate at different spatial scales, from the micro- to the macrogeographic. These results appear largely to match those found in wild barley locally (Nevo et al. 1981, 1983, 1986a) and regionally (Nevo 1992; Nevo et al. 1979, 1986b). Similar patterns described in wild emmer are demonstrated by DNA diversity, as shown with random amplified polymorphic DNA (RAPD) and short sequence repeats (SSRs) (Fahima et al. 1999, 2001; Li et al. 1999, 2000a–d, 2001b–d).

Stochastic factors certainly interact with natural selection in varying environments over space and time. Some of the very small populations of wild emmer described in Nevo and Beiles 1989 (e.g., nos. 28, 29, 31, and 33 in Table 4.1) may have derived their low levels of *P* and *He* from either founder effects and/or genetic drift. The latter can be a potent force for removing variation, particularly in small populations and when the random environment tends to occasionally push alleles down to low frequencies (Nei 1980; Gillespie 1985). However, most of the populations described here are medium and large, and they appear to derive and maintain their genetic polymorphisms and structure from climatically and edaphically varying environments. Moreover, in subterranean mole rats we have shown that long-term genetic polymorphisms persist even in *small* and *isolated* populations and are maintained by balancing natural selection (Nevo et al. 1997).

The maintenance of polymorphisms in wild emmer may be explicable by both spatial and temporal variations in selection. Theory indicates that selection, acting differentially in space, coupled with limited migration (which is typical of wild emmer, mean dispersal of 1.25 m per generation at Yehudiyya, Israel; Golenberg 1987), will maintain a substantial amount of polymorphism (Karlin and McGregor 1972; Hedrick 1986; Nevo et al. 2000). Two-niche models (Levene 1953) can explain genetic differentiation in wild emmer. Different homozygotes are favored in different climatic and edaphic niches. These operate on all levels, from regional to local, and at a minuscule level at a locality.

Recently, in the case of an additive two-locus panmictic model, we found in both haploids and diploids that stabilizing selection with a cyclically moving optimum may be an efficient factor in protecting polymorphisms for linked loci additively affecting the selected trait (Korol et al. 1994, 1996a; Kirzhner et al. 1995, 1999). We found that, within the same class of fitness functions, nonequal gene action and/or a dominance effect for one or both loci may lead to local polymorphism stability with substantial polymorphism-attracting domains. These important mechanisms may be relevant also to situations with different violations of panmixia (e.g. Ryndin et al. 2001), including inbreeders like wild emmer, but they need additional, critical theoretical studies.

The limited migration of seeds between niches over many generations and the accumulated seed banks in the soil provide rich reservoirs of genetic diversity for natural selection to operate on. Such accumulated seed pools provide a memory of past selection regimes. These seed pools can greatly reduce the fitness uncertainty generated by cyclical or random environments and thus free the plant population from having to respond genetically to the fitness conditions occurring from year to year (Templeton and Levin 1979). The temporal differentiation in allele frequency, presumably realized by the richness of seed pools, is exemplified in our microsite emmer wheat study in Tabigha (Appendix in Nevo and Beiles 1989, and details in Nevo 1989a). We found higher correlations of spatiotemporal allele frequency changes within, rather than between, allozymic groups, a strong indicator of ecological selection (Li et al. 2000a–d, 2001a–d).

4.2 RAPD Genetic Diversity

Random amplified polymorphic DNA (RAPD) markers represent amplification products from a polymerase chain reaction (PCR) utilizing arbitrary primers and genomic DNA (Williams et al. 1990). Most variation among individuals for RAPDs probably arises from base pair substitutions or insertions/deletions that modify (or eliminate) the primer site, or insertions in the genomic sequence that separate the primer sites to a distance that will not permit amplification (Williams et al. 1990). RAPD markers have been useful in studies of taxonomic identities, systematic relationships, population genetic structures, species hybridizations, and parentage identifications (for review see McClelland and Welsh 1994; Penner 1996). RAPDs have contributed to the estimation of genetic diversity among diploid (Vierling and Nguyen 1992), tetraploid (Joshi and Nguyen 1993a; Fahima et al. 1999), and hexaploid (Devos and Gale 1992; He et al. 1992; Joshi and Nguyen 1993b) wheats.

4.2.1 Patterns of RAPD Variation and Genetic Diversity

Genetic diversity of RAPDs was studied at the Institute of Evolution in 110 genotypes of *T. dicoccoides* from 11 populations sampled in Israel and Turkey (Fahima et al. 1999) that were tested allozymically (Nevo et al. 1982; Nevo and Beiles 1989). Our results showed high levels of diversity of RAPD markers in wild wheat populations in Israel. The ten primers used amplified 59 scorable RAPD loci, of which 48 (81.4%) were polymorphic and 11 monomorphic. A typical gel demonstrating RAPD polymorphism among *T. dicoccoides* populations is presented in Fig. 4.3.

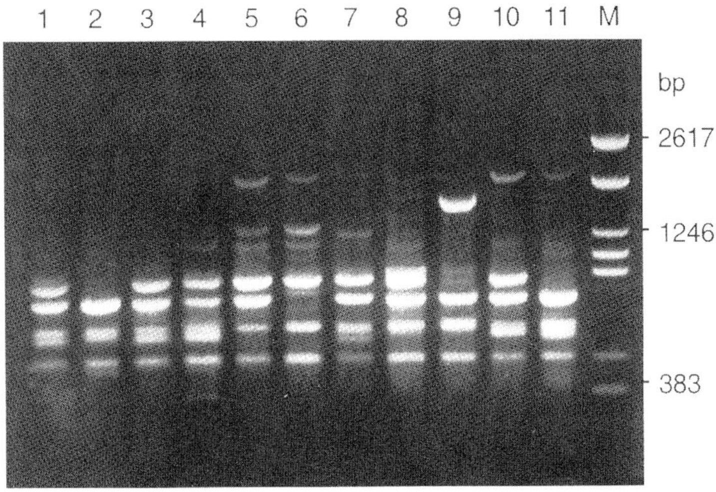

Fig. 4.3. Amplification products generated from genomic DNA of 11 genotypes representing 11 populations of *T. dicoccoides* with RAPD primer F15. Each lane shows the RAPD profile of one representative genotype from one wild wheat population. *Lane 1* Gitit 18–35, *lane 2* Givat-Koach 33–31, *lane 3* Kokhav-Hashahar 19–36,*lane 4* Gerizim 17–35, *lane 5* Mt. Hermon 1–49, *lane 6* Tabigha 11–54, *lane 7* Rosh-Pinna 9–6, *lane 8* Gamla 8–35, *lane 9* Bat-Shlomo 30–30, *lane 10* Diyarbakir, Turkey 36–67, *lane 11* Gilboa 16–61. *Lane M* molecular weight marker, pBR322 digested with *Alw441*/*Mval*. (From Fahima et al. 1999)

The mean proportion of polymorphic loci, *P* (5%), and gene diversity, *He*, of the 11 populations of wild emmer wheat were 0.524 and 0.188, respectively (Table 4.6). The range of *He* between the wild wheat populations was large, 0.138–0.213. The highest value of *He* (0.213) was obtained in the high-altitude and arid-cold steppic population of Mt. Hermon, which represents one of the most stressful xeric-cold habitats in Israel. High *He* values (0.212) were also obtained for Rosh Pinna and Tabigha populations, which experience xeric-warm habitats. By contrast, the populations from cool Mt. Gerizim in Israel and Diyarbakir in Turkey were characterized by lower levels of diversity, with *He* values of 0.138 and 0.149, respectively.

4.2.2 Genetic Structure

4.2.2.1 RAPD Discriminant Analysis

RAPD analysis was found to be highly effective, as were allozymes (Nevo and Beiles 1989; Fig. 4.2), in distinguishing genotypes of *T. dicoccoides* that originated from diverse ecogeographic sites in Israel and Turkey, with 95.5% of the 110 genotypes correctly classified into sites of origin by discriminant analysis based on RAPD genotyping (Fahima et al. 1999). A graphical illustration

Table 4.6. Summary of genetic variation, based on 59 RAPD loci in 11 populations of *T. dicoccoides* in Israel and Turkey. (Fahima et al. 1999)

No[a]	Population[b]	RAPD proportion of polymorphic loci (*P*)	Allozyme[c] proportion of polymorphic loci (*P*)	RAPD gene diversity *He* (SE)	Allozyme[d] gene diversity *He*
1	Mt. Hermon	0.593	0.214	0.213 (0.027)	0.060
8	Gamla	0.627	0.308	0.212 (0.025)	0.102
9	Rosh-Pinna	0.593	0.238	0.212 (0.026)	0.093
11	Tabigha	0.525	0.167	0.203 (0.028)	0.054
16	Mt. Gilboa	0.542	0.143	0.192 (0.026)	0.044
17	Mt. Gerizim	0.407	0.167	0.138 (0.024)	0.059
18	Gitit	0.509	0.293	0.189 (0.027)	0.116
19	Kokhav-Hashahar	0.509	0.190	0.187 (0.027)	0.058
30	Bat-Shelomo	0.525	0.214	0.201 (0.028)	0.085
33	Givat-Koach	0.492	0.100	0.178 (0.026)	0.026
36	W. Diyarbakir, 22 km	0.441	0.167	0.149 (0.025)	0.055
Average		0.524	0.200	0.188	0.068
Range		0.407–0.627	0.100–0.308	0.138–0.213	0.026–0.116

P Proportion of polymorphic loci (5%), *He* genetic diversity
[a] Population numbers are according to Nevo and Beiles (1989)
[b] The RAPD study included a sample size of ten accessions per population, while the allozyme study included 25–48 (average of 38) accessions per population
[c] Allozyme data are according to Nevo and Beiles (1989)

is displayed in Fig. 4.4 and is based on the presence/absence of the bands at 48 polymorphic RAPD loci. It shows that the genotypes of most of the populations are clustered together. The clear separation of isolated marginal populations (e.g., Hermon, Givat Koach, Gitit, and Gerizim) from the other populations by discriminant analysis demonstrates the influence of unique ecogeographic conditions on the genetic structure of these populations. The soil type at each location also plays an important role in genetic differentiation, as can be seen by the separation of the populations located on basalt soil (e.g., Tabigha, Gamla, and Diyarbakir) from those located on terra-rossa-type soil (e.g., Hermon, Rosh Pinna, Mt. Gilboa, Mt. Gerizim, Kokhav Hashahar), or rendzina-type (Bat Shelomo) soil. This is also true in a microscale analysis (Li et al. 2000d).

4.2.2.2
RAPD Genetic Distance

Coefficients of genetic distance, *D*, calculated for paired comparisons of the 11 populations according to Nei (1978), showed a mean value of 0.089 with a range of 0.017–0.165. The highest genetic distance ($D = 0.165$) was obtained

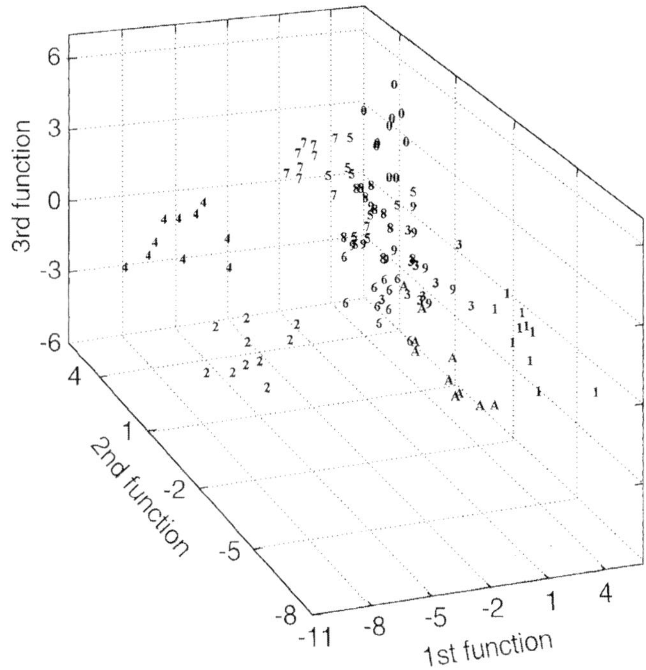

Fig. 4.4. Plot of canonical discriminant functions 1, 2, and 3 based on the 48 polymorphic RAPD loci: *1* Mt. Hermon, *2* Gamla, *3* Rosh-Pinna, *4* Tabigha, *5* Mt. Gilboa, *6* Mt. Gerizim, *7* Gitit, *8* Kokhav-Hashahar, *9* Bat-Shelomo, *0* Givat-Koach. *A* Diyarbakir. The proportion of the eigenvalues of functions 1, 2, and 3 from the sum of the eigenvalues of the ten discriminant functions are 36.9%, 16% and 12%, respectively. (Fahima et al. 1999)

between the populations of Mt. Gerizim and Tabigha, while the lowest (D = 0.017) was obtained between Bat-Shelomo and Kokhav-Hashahar. The estimates of D were geographically independent. Large Ds, i.e., sharp genetic differentiation, were recorded over very short geographic distances, compared to low Ds between geographically distant populations. For example, the genetic distance obtained between Mt. Gilboa and Tabigha populations, located only about 50 km apart, was 0.155, whereas the genetic distance between the Bat-Shelomo population in Israel and the Diyarbakir population in Turkey, separated by 750 km, was only 0.065. These results negate a simple isolation-by-distance model and reinforce the importance of shared ecological factors regardless of geographic distance.

4.2.2.3 RAPD Spatial Autocorrelation

Spatial autocorrelation (Sokal and Wartenberg 1983) of RAPD frequencies suggests that migration is not influential. Our RAPD results (Fahima et al.

1999) did not show correlation predictions, i.e., high correlations in the lower-order distant groups. Likewise, we found great differences across loci and alleles in their autocorrelation pattern. The autocorrelation analysis predicts that migration will cause high positive correlations (similarity) in the low-order distant groups (i.e., between neighbors), starting from the first distant group in the transect. Migration is expected to cause similarity between loci and alleles. This prediction is not realized in our results. Instead, RAPD loci of wild emmer differ drastically in their autocorrelation pattern between loci. In addition, positive autocorrelations emerge at intermediate distance groups, thus negating migration as an important evolutionary factor. The autocorrelation analysis also predicts that genetic drift will not create any autocorrelative pattern. Our data are structured, hence they negate randomness. Despite the between-loci variation, a general tendency is evident of more positive correlations on the left-hand side of Table 6 in Fahima et al. (1999) and more negative correlations on the right-hand side.

Environmental selection is also partly autocorrelated and affects loci differentially. This is supported by our data because of the following three reasons: (1) variation among loci; (2) a positive correlation in different distance groups, and not necessarily the first distant group in the transect; (3) the predominance of negative correlations in the larger distance groups, which is expected due to the decreasing ecological similarity that often occurs with increasing distance.

4.2.3 Multiple Regression Analysis of Environmental Variables and RAPD Polymorphisms

A test of the best predictors of *P*, *He*, and representative allele frequencies at the 48 polymorphic loci in the ten Israeli populations was conducted by stepwise multiple regression (MR) analysis employing these characters as dependent variables, and geographic, climatic and edaphic factors as independent variables (Fahima et al. 1999). The following environmental variables were included in the analysis: (1) *geographic*: longitude (Ln), latitude (Lt), altitude (Al), (2) *climatic*: *temperature*, annual (Tm), January (Tj), August (Ta), seasonal temperature difference (Td), daily temperature difference (Tdd), number of tropical days (Trd), evaporation (Ev), (3) *moisture;* annual rainfall (Rn), number of rainy days (Rd), number of dewy nights in summer (Dw), annual humidity (Huan), humidity at 14:00 (Hu14), interannual rainfall variation (Rv), coefficient of variation in rainfall (Rr); and (4) three *edaphic* dummy variables, one per each soil type: basalt (Ba), rendzina (Ren), and terra rossa (Tr). The number of variables is in fact much lower because many are highly intercorrelated, for example, temperature variables or the variables of water availability. The test involved the ten Israeli populations. (The Turkish population was not included in this test due to missing ecological data.) The results are given in Table 4.7.

Table 4.7. Coefficient of multiple regressions (R^2) of RAPD genetic indices and allele frequencies as the dependent variables, and environmental variables in ten populations of *T. dicoccoides* in Israel as independent variables. (Fahima et al. 1999)

	Stepwise model					
	Ecogeographic variables			Ecological variables		
Genetic indices						
P	Lt	Lt Tdd	Lt Tdd Tm	Td		
	0.526*	0.781**	0.936***	0.277^{ns}		
He	Lt	Lt Tdd	Lt Tdd Rr			
	0.485*	0.698*	0.876**			
Alleles						
486-1100	Trd	Trd Tr		Trd	Trd Tr	
	0.752**	0.848**		0.752**	0.848**	
486-750	Ln	Ln Trd	Ln Trd Td	Td	Td Trd	
	0.491*	0.745**	0.835**	0.374$^{@}$	0.561$^{@}$	
486-600	Trd	Trd Rr	Trd Rr Al	Trd	Trd Rr	Trd Rr Hul4
	0.608**	0.799**	0.887**	0.608**	0.799**	0.872**
551-790	Ba			Ba		
	0.444*			0.444*		
551-500	Ba			Ba		
	0.929***			0.929***		
494-580	Ba			Ba		
	0.279^{ns}			0.279^{ns}		
494-530	Ba	Ba Ev	Ba Ev Rr	Ba	Ba Ev	Ba Ev Rr
	0.898***	0.945***	0.971***	0.898***	0.945***	0.971***
F15-830	Ln	Ln Ren	Ln Ren Dw			
	0.359$^{@}$	0.602*	0.824*			
F15-740	Ba			Ba		
	0.444*			0.444*		
F15-550	Lt	Lt Tr		Tr	Tr Ti	
	0.499*	0.638*		0.375$^{@}$	0.785**	
537-800	Hu14	Hu14 Rv		Hu14	Hu14 Rv	
	0.355$^{@}$	0.566$^{@}$		0.355$^{@}$	0.566$^{@}$	
537-700	Trd	Trd Rr	Trd Rr Ba	Trd	Trd Rr	Trd Rr Ba
	0.400$^{@}$	0.722*	0.840**	0.400$^{@}$	0.722*	0.840**
537-560	Rn	Rn Lt		Rn		
	0.489*	0.666*		0.489*		
561-1690	Ev	Ev Dw	Ev Dw Tr	Ev	Ev Dw	Ev Dw Tr
	0.507*	0.747**	0.863**	0.507*	0.747**	0.863**
561-1020	Rv			Rv		
	0.296^{ns}			0.296^{ns}		
561-650	Ren			Ren		
	0.464*			0.464*		
A20-1600	Al			Tj		
	0.342$^{@}$			0.202^{ns}		
A20-1000	Ln	Ln Huan		Tj	Tj Ba	Tj Ba Tdd
	0.407*	0.589*		0.323$^{@}$	0.690*	0.872**
A20-600	Trd	Trd Rr		Trd	Trd Rr	
	0.334$^{@}$	0.533$^{@}$		0.334$^{@}$	0.533$^{@}$	

Table 4.7. *Continued*

	Stepwise model					
	Ecogeographic variables			Ecological variables		
A19-850	Ba 0.603**			Ba 0.603**		
A19-530	Ba 0.878***			Ba 0.878***		
A19-500	Rd 0.585**			Rd 0.585**		
T17-800	Trd 0.398$^{@}$	Trd Tr 0.611*		Trd 0.398$^{@}$	Trd Tr 0.611*	
T17-650	Rd 0.898**	Rd Ba 0.934***		Rd 0.898**	Rd Ba 0.934***	
T17-470	Tm 0.853***	Tm Trd 0.952***	Tm Trd Rr 0.975***	Tm 0.853***	Tm Trd 0.952***	Tm Trd Rr 0.975***
T17-400	Ln 0.525*	Ln Trd 0.673*		Td 0.413*	Td Trd 0.621*	Td Trd Ta 0.785*
A12-720	Al 0.356$^{@}$			Trd 0.313$^{@}$	Trd Hu14 0.562$^{@}$	
A12-690	Lt 0.473*			Tdd 0.291^{ns}		
A12-490	Ln 0.313$^{@}$			Td 0.309$^{@}$		
A12-430	Lt 0.284^{ns}					

Lt Latitude, *Ln* longitude, *Al* Altitude, *Tm* mean annual temperature, *Ta* mean August temperature difference, *Tj* mean January temperature, *Td* mean seasonal temperature difference, *Trd* mean number of tropical days, *Ev* mean annual evaporation, *Dw* mean number of dew nights in summer, *Rn* mean annual rainfall (in mm), *Rd* mean number of rainy days, *Rr* coefficient of variation in rainfall *Hu14* mean humidity at 14:00 h, *Rv* interannual rainfall variation, *Huan* mean annual humidity, *Tr* terra rossa, *Ren* rendzina soil, *Ba* basalt soil
***$p < 0.001$, **$p < 0.01$, *$p < 0.05$, $^{@}p < 0.10$, *ns* $p > 0.10$

Latitude, temperature, and water availability factors, singly or in combination, explained a significant proportion of the variation in polymorphism of RAPDs. The best two variable-predictors of *P* and *He*, explaining significantly 0.7–0.8 of their variance, were Lt and Tdd (latitude and day-night temperature difference). A three-variable combination involving both geographic (latitude) and climatic factors (day-night temperature difference and mean annual temperature, Lt, Tdd, and Tm), accounted significantly ($p = 0.001$) for 0.94 of the variance in *P*, while the combination LtTddRv (latitude, day-night temperature difference and mean relative variability of rainfall) accounted significantly ($p = 0.01$) for 0.88 of the variance in *He*.

RAPD loci can cautiously be classified into several categories in terms of their prime ecological predictors: (a) water factors (Rn, Hu, Ev, Rd): 537–800, 537–560, 561–1690, A19–500, T17–650; (b) soil type (Ba, Tr, Ren): 551–790,

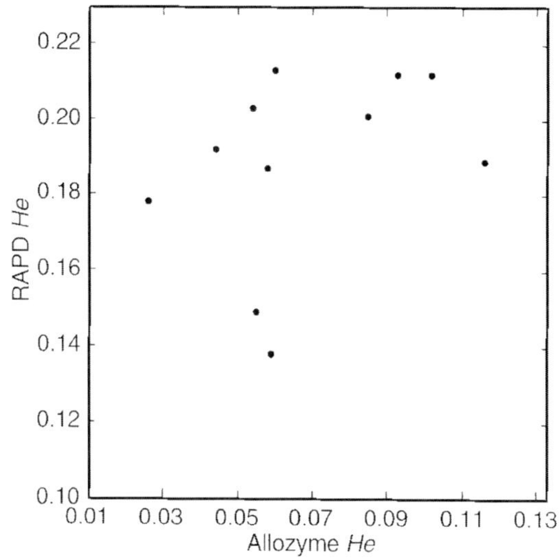

Fig. 4.5. Correlation between RAPD and allozyme gene diversity profiles. The Spearman rank correlation was tested between the *He* values produced by RAPD and allozyme profiles in 11 populations of *T. dicoccoides* $r_s = 0.414$. (Fahima et al. 1999)

551–500, 494–530, F15–740, 561–650, A19–850, A19–530; (c) temperature (Ta, Td, Tm, Trd): 486–750, 486–600, 537–700, A12–490, A20–600, T17–800, T17–470, T17–400; (d) water+temperature (Hu, Rr, Trd, Tm): 486–1100, 486–600, 486–750, T17–470, A12–720, 537–700; (e) soil+temperature (Ba, Ren, Tr, Tj, Tdd, Trd, Ev): T17–800, 486–1100, 494–530, F15–550, A20–1000; (f) soil+water (Ba, Ren, Rd, Dw); T17–650, F15–830. For verification, this classification needs critical testing by permutation tests and by testing repetitive samples. The RAPD results seem to be in agreement with the allozyme patterns (Nevo et al. 1982; Nevo and Beiles 1989; Fig. 4.5), although the genetic diversity values obtained with RAPDs are much higher than the allozyme values. Significant correlates of RAPD markers with various climatic and soil factors suggest that, as in allozymes, natural selection causes adaptive RAPD ecogeographic differentiation, at least partly in *noncoding* genomic DNA (Table 4.7).

4.2.4 RAPD versus Allozyme Genetic Diversity Profiles

The genetic diversity profiles obtained with RAPD markers (Fahima et al. 1999) were compared with those obtained previously with allozyme markers (Nevo et al. 1982; Nevo and Beiles 1989). Our results showed agreement between RAPD and allozyme genetic diversity profiles, at both the regional (Fig. 4.5) and local (Li 2000; Li et al. 1999, 2000a–d, 2001b–d) levels. However, RAPDs yielded much higher values of diversities than allozymes, both regionally and locally. The average regional RAPD values for *P* and *He* were 0.524 (range: 0.407–0.627) and 0.188 (range: 0.138–0.213), as compared with allozyme values of 0.200 (range: 0.100–0.308) and 0.068 range: (0.026–0.116), respectively. These results showed that RAPD loci were more polymorphic than allozyme loci among wild wheat populations from Israel and Turkey, both regionally and locally (Li et al. 2000d).

There was a positive, albeit small Spearman rank correlation ($r_s = 0.414$) between RAPD gene diversity (*He*) values and allozyme *He* values (Fig. 4.5). All the populations showing high allozyme *He* values also showed high RAPD *He* values. However, in many cases low allozyme *He* values were accompanied by high RAPD values, suggesting that these two genetic systems are selected independently and do not display a clonal genotypic "hitchhiking" due to the high inbreeding level of *T. dicoccoides*. There were no opposite cases, in which high allozyme *He* values were accompanied by low RAPD *He* values. It is noteworthy that the allozyme analysis (Nevo and Beiles 1989) included 423 *T. dicoccoides* individuals (on average 38 genotypes per population) belonging to the same 11 populations tested for RAPD, while the RAPD analysis included only 110 genotypes (10 genotypes per population) (Fahima et al. 1999). We would expect a higher correlation level between RAPDs and allozymes if the analysis was based on identical plant material with the same sample size. Other studies have also compared the patterns of variation detected at allozyme vs RAPD loci in several nonbreeders (Puterka et al. 1993; Isabel et al. 1995; Peakall et al. 1995; Lanner-Herrera et al. 1996). As in our case, the overall patterns were generally in good agreement, although the levels of genetic variation and fine-scale genetic structures sometimes differed with the two kinds of markers. The parallels between RAPD and allozyme patterns indicate that similar, primarily deterministic (selection) evolutionary forces are involved in shaping the genomic structure of both RAPD and allozyme loci, i.e., *noncoding* and *coding* genomic regions.

4.3
Microsatellite (SSR) Diversity

Microsatellites are a relatively new class of molecular markers based on tandem repeats of DNA sequences called short sequence repeats (SSRs) (Litt and Luty 1989). These 2- to 6-bp repeats are highly polymorphic, even among closely related cultivars, due to mutations causing variation in the number of repeating units. This kind of polymorphism at specific loci is easily detected by the use of specific primers in the flanking regions of such loci and subsequent amplification by PCR (Litt and Luty 1989; Weber and May 1989). The high level of polymorphism, combined with the high interspersion rates, makes microsatellites an abundant source of genetic markers. Extremely high-density molecular maps based on microsatellite DNA markers have been constructed in both the human (Dib et al. 1996) and mouse (Dietrich et al.

1996) genomes. Microsatellites appear to serve as a major source of "tuning knobs" for molecular adaptive evolution, including the evolutionary control of the mutation process itself (reviewed and critically analyzed in Kashi et al. 1997; King et al. 1997; Li et al. 2001d).

The usefulness of microsatellites as genetic markers in plants was demonstrated for many species, including soybean (Akkaya et al. 1995), rice (Wu and Tanksley 1993), maize (Senior and Heun 1993) *Arabidopsis* (Bell and Ecker 1994), and tropical trees (e.g., Aldrich et al. 1998). These studies indicated that microsatellites in plants can be up to tenfold more variable than other marker systems, such as restriction fragment length polymorphism (RFLPs). Furthermore, the efficiency of microsatellite markers was demonstrated also for self-pollinating species with relatively low levels of intraspecific polymorphism, such as hexaploid wheat, *Triticum aestivum* (Plaschke et al. 1995; Röder et al. 1995), and cultivated barley, *Hordeum vulgare* (Saghai-Maroof et al. 1994; Liu et al. 1996). The studies conducted at the Institute of Evolution, University of Haifa, demonstrate the application of wheat microsatellite (WMS) markers for the differentiation and estimation of genetic diversity among different accessions of populations and subpopulations of *T. dicoccoides* (Fahima et al. 1998, 2001; Li et al. 2000a–d, 2001b–d) and for tagging of agronomically important genes (Chagué et al. 1999; Peng et al. 1999, 2000a–c, 2001).

4.3.1 SSR Genetic Diversity

We examined the diversity of wild emmer wheat, *T. dicoccoides*, in 20 microsatellite loci (SSRs) in 15 populations (135 genotypes), representing a wide range of ecological conditions of soil, temperature, and water availability, in Israel and Turkey (Fahima et al. 2001). A large amount of diversity at microsatellite loci was observed despite the predominantly selfing nature of this plant species. The 20 Gatersleben wheat microsatellites (GWM), representing 13 chromosomes of genomes A and B of wheat, revealed a total of 363 alleles, with an average of 18 alleles per GWM marker (range: 5–26). The proportion of polymorphic loci per population averaged 0.90 (range: 0.45–1.00); genic diversity, *He*, averaged 0.50 (range 0.094–0.736); Shannon's information index averaged 0.84 (range 0.166–1.307). The coefficients of genetic distance between populations were high and averaged D=1.903 (range: 0.923–3.348), an indication of sharp genetic divergence over short distances. Interpopulation genetic distances showed no association with geographic distance between the population sites of origin, negating a simple isolation-by-distance model. Genetic dissimilarity values between genotypes, calculated from the GWM-derived data, were used to produce a dendrogram of the relationships among wild wheat populations using the UPGMA unweighted pair-group method with arithmetic averages (Fig. 4.6). The results showed that all of the wild emmer wheat populations could be distinguished. Microsatellite

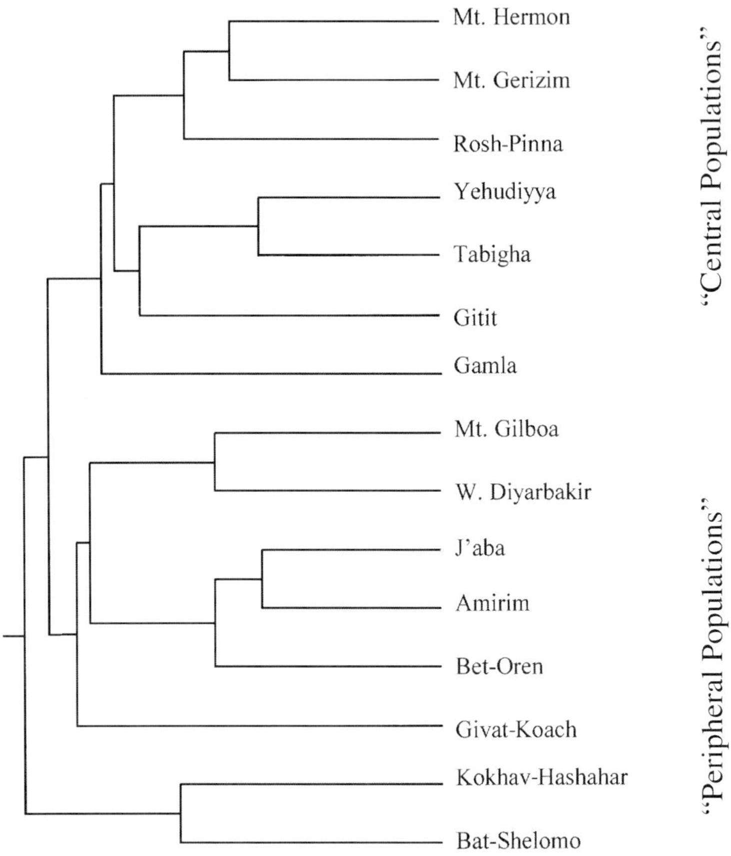

Fig. 4.6. Dendrogram of 15 wild emmer wheat populations from Israel and Turkey, based on the genetic distance calculated from data of 20 wheat microsatellites, using the UPGMA as the clustering method. (Fahima et al. 2001)

analysis was found to be highly effective in distinguishing genotypes of *T. dicoccoides* that originated from diverse ecogeographic sites in Israel and Turkey, with 88% of the 135 genotypes correctly classified into sites of origin by discriminant analysis. Our present microsatellite results were nonrandom and in agreement with the previously obtained allozyme (Nevo and Beiles 1989) and RAPD (Fahima et al. 1999) patterns, although the genetic diversity values obtained with microsatellites were much higher. Significant correlates of microsatellite markers with various climatic and soil factors suggest that, as in allozymes and RAPDs, natural selection causes adaptive microsatellite ecogeographic differentiation, not only in *coding* but most importantly in *noncoding* genomic regions. Hence, the concept of "junk DNA" needs to be replaced by one that takes into account, at least partly, a DNA regulatory function. The results obtained suggest that microsatellite markers are not only adaptive markers, but are useful for the estimation of genetic diversity in natural populations of *T. dicoccoides* and for tagging agronomically important traits derived from wild emmer wheat (see chapter 8).

4.3.1.1
Patterns of SSR Diversity

Wild emmer wheat was analyzed for diversity in 15 populations totaling 135 genotypes, including 126 from Israel and nine from Turkey (22 km west of Diyarbakir). The collections from Israel covered 14 locations across the ecogeographic range of *T. dicoccoides* in Israel. The geographic distribution of 14 wild emmer wheat populations in Israel is described in Fig. 4.1 (the ecogeography is presented in Table 1 of Nevo and Beiles 1989). In total, 363 microsatellite alleles were detected with 20 GWM markers, representing 20 microsatellite loci; the alleles were located on 13 different chromosomes and 20 different chromosome arms of the A and B genomes of wheat (Table 4.8). All 20 GWM markers yielded polymorphic fragments among the 135 *T. dicoccoides* accessions used in this study. The number of alleles per GWM ranged from 5 to 26

Table 4.8. Wheat microsatellite, chromosomal-arm locations, number and size range of alleles for the GWM employed. (Fahima et al. 2001)

Designation	Chromosomal location	Number of alleles	Fragment size in Chinese Spring (bp)	Range of allele size (bp)
GWM18	1BS	17	186	181–217
GWM60	7AS	17	211	171–221
GWM95	2AS	14	121	107–135
GWM99	1AL	20	119	93–179
GWM120	2BL	20	139	119–171
GWM154	5AS	20	102	99–153
GWM155	3AL	22	141	134–196
GWM169	6AL	21	196	186–230
GWM186	5AL	19	140	102–186
GWM218	3AS	26	153	130–216
GWM219	6BL	15	181	142–190
GWM234	5BS	21	241	208–256
GWM251	4BL	18	103	57–119
GWM265	2AL	5	200	166–184
GWM340	3BL	21	132	97–167
GWM361	6BS	10	126	135–153
GWM389	3BS	24	130	84–172
GWM400	7BS	13	139	128–164
GWM408	5BL	22	176	144–202
GWM577	7BL	18	133	136–182

GWM marker designation and chromosomal location are according to Röder et al. (1995, 1998) and Plaschke et al. (1995, 1996)

(Table 4.8). On average, 18.2 alleles were detected per locus. The most polymorphic microsatellite was GWM218, with 26 alleles. The sizes of the 26 alleles observed ranged from 130 to 216 bp. GWM218 is a perfect microsatellite with a motif (CT) in Chinese Spring (CS) (Korzun et al. 1997). Hence, since the size of the amplified fragment in CS obtained in this study was 156 bp, it is assumed that the number of dinucleotide repeats of the repetitive motif of GWM218 ranged from 5 to 48 in the *T. dicoccoides* collection used in this study.

4.3.1.2
Summary of SSR Genetic Diversity

A summary of the genetic data for each of the 15 populations of *T. dicoccoides* is given in Table 4.9. Mean levels of the proportion of polymorphic loci, *P* (5%), genetic diversity, *He* (Nei 1978), and Shannon information index (Lewontin 1972) of the 15 populations of wild emmer wheat were 0.90, 0.50, and 0.84, respectively. The range of diversity in *He* between the wild wheat populations was large, 0.094–0.736. The highest value of *He* (0.736) was obtained for the high-altitude and arid-cold steppic population of Mt.

Table 4.9. Summary of genetic diversity based on 20 microsatellite loci in 15 populations of wild emmer wheat, *T. dicoccoides*, in Israel and Turkey. (Fahima et al. 2001)

No.[a]	Population[b]	Polymorphic per population *P*	Genetic diversity *He* (SE)	Shannon's information index[c] (SE)
1	Mt. Hermon	1.00	0.736 (0.142)	1.307 (0.362)
7	Yehudiyya (sun/shade)	0.90	0.506 (0.202)	0.761 (0.317)
8	Gamla	0.70	0.288 (0.262)	0.439 (0.412)
9	Rosh-Pinna	0.90	0.473 (0.206)	0.723 (0.348)
11	Tabigha	1.00	0.636 (0.124)	1.059 (0.277)
16	Mt. Gilboa	0.80	0.381 (0.241)	0.632 (0.410)
17	Mt. Gerizim	1.00	0.634 (0.153)	1.086 (0.332)
18	Gitit	0.95	0.648 (0.194)	1.129 (0.395)
19	Kokhav-Hashahar	0.95	0.494 (0.205)	0.801 (0.375)
23	J'aba	1.00	0.569 (0.154)	0.968 (0.304)
24	Amirim	1.00	0.618 (0.193)	1.097 (0.402)
28	Bet-Oren	0.45	0.094 (0.155)	0.166 (0.270)
30	Bat-Shelomo	0.95	0.573 (0.197)	0.960 (0.376)
33	Givat-Koach	0.85	0.249 (0.185)	0.421 (0.309)
36	W. Diyarbakir, 22 km	1.00	0.603 (0.225)	1.049 (0.449)
Mean		0.90	0.500 (0.198)	0.840 (0.355)

P Proportion of polymorphic loci, *He* genetic diversity
[a] Population numbers are according to Nevo and Beiles (1989)
[b] Sample size of nine accessions per population
[c] Shannon's information index (Lewontin 1972)

Hermon, which, as noted above, represents one of the most stressful xeric-cold habitats in Israel. In contrast, low levels of diversity, with *He* values of 0.094 and 0.249, characterized the isolated small populations from Bet-Oren on Mt. Carmel and Givat-Koach near Tel Aviv on the coastal plains, respectively.

4.3.1.3
SSR versus RAPD and Allozyme Genetic Diversity Profiles

The genetic diversity profiles obtained in this study with microsatellite markers were compared with those obtained previously with allozyme (Nevo et al. 1982; Nevo and Beiles 1989) and RAPD (Fahima et al. 1999) markers (Fig. 4.7). All populations studied here showed a higher proportion of polymorphic loci (*P*) and higher genetic diversity (*He*) values for microsatellite loci than for allozyme and RAPD loci. The average microsatellite values for *P* and *He* were 0.90 (range: 0.45–1.00) and 0.50 (range: 0.094–0.736) compared with allozyme (Nevo and Beiles 1989) values of 0.200 (range: 0.100–0.308) and 0.068 (range: 0.026–0.116) and RAPD (Fahima et al. 1999) values of 0.524 (range: 0.407–0.627) and 0.188 (range: 0.138–0.213), respectively. These results showed that microsatellite loci were far more polymorphic than allozyme and RAPD loci among wild wheat populations from Israel and Turkey. Similar results were obtained in the microscale analysis at Ammiad (Li et al. 2000d).

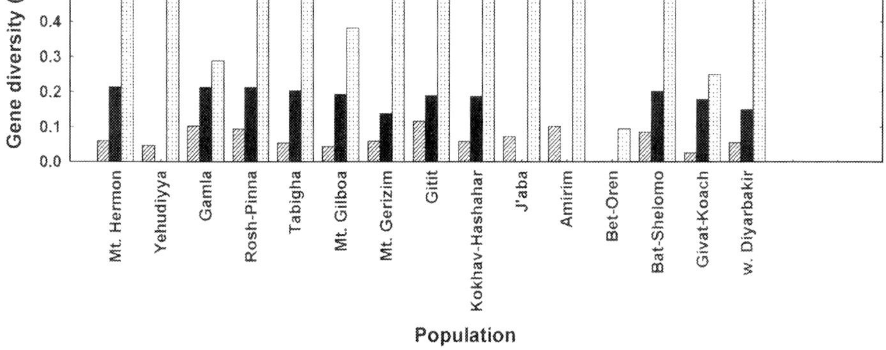

Fig. 4.7. Gene diversity (*He*) profiles of microsatellite, RAPD and allozyme loci in 15 wild emmer wheat populations from Israel and Turkey. (Fahima et al. 2001)

4.3.2 Population Genetic Structure

4.3.2.1 SSR Genetic Distance

Coefficients of genetic distance, *D*, were calculated for paired comparisons of the 15 populations based on the normalized identity of all loci between each of the populations (Nei 1978). The genetic distances for all 105 pairs ranged from 0.923 to 3.348 and averaged 1.903. The highest obtained genetic distance (3.348) was between the population of Bet-Oren and the population of Tabigha, while the most similar populations were Amirim and Jaba, with a genetic distance of 0.923. In many cases, the estimates of *D* were geographically independent. They displayed large *D*s, i.e., sharp genetic divergence, over very short geographic distances, compared to low *D*s between geographically distant populations. For example, the genetic distance between the Bat-Shelomo and Bet-Oren population, both on Mt. Carmel and 15 km apart, was 2.797, whereas the genetic distance between the Mt. Gilboa population in Israel and the Diyarbakir population in Turkey, 750 km apart, was only 1.159. Many other such examples occur and are given in Table 4 of Fahima et al. (2001). This suggests that geographic distance alone may not explain inter-population genetic divergence and negates an isolation-by-distance model (Wright 1943).

Cluster analysis was then carried out by the UPGMA method and resulted in the dendrogram shown in Fig. 4.6. The dendrogram differentiates all wild wheat populations tested. Two major and several minor groups can be distinguished. The dendrogram shows that the pattern of clustering of most of the *T. dicoccoides* populations is related to the ecogeographic distribution of the wild wheat populations. The two major groups differentiate the largely central populations located near the Sea of Galilee (Tabigha, and Rosh-Pinna) and the Golan Heights (Gamla, Yehudiyya, and Mt. Hermon), from the peripheral populations located in Judea and Samaria (Jaba, Kokhav-Hashahar, Gilboa), Mt. Carmel (Bat-Shelomo, Bet-Oren), and the coastal plains (Givat Koach). However, some of the clustering is between populations that are located in different geographic regions but share similar ecological conditions. For example, the Mt. Hermon and Mt. Gerizim populations are clustered together, although they are located 120 km apart. This clustering may be explained by the similarity in the ecological conditions, both being located on high mountains with large amounts of precipitation (altitude 1300 m and 800 m; and rainfall 1400 and 800 mm for Mt. Hermon and Mt. Gerizim, respectively). Furthermore, both of these populations are located on terra rossa-soil type and both are peripheral rather than central populations.

4.3.2.2
SSR Discriminant Analysis

The discriminant analysis was able to differentiate significantly, on the basis of allele frequencies, between almost all 15 populations, based on ten loci (Fig. 4.8). The correct classification of individual genotypes into their respective population, according to their geographic distribution, ranged between 67 and 100%, with an average of 88%. These results indicate that microsatellite markers can be used to differentiate *T. dicoccoides* genotypes and classify them according to their respective populations based on multilocus analysis of predominantly *noncoding genomic regions*. Figure 4.8 shows that, in most cases, the genotypes of each population are clustered together. In addition, most of the central populations from the eastern Galilee (Rosh Pinna and Tabigha) and the Golan Heights (Gamla, Yehudiyya, and Mt. Hermon), are clustered together with extensive overlapping between them. Terra rossa soil in the first group and basalt in the second may be a major edaphic diversifying factor. By contrast, the peripheral small populations from Judea (Jaba and

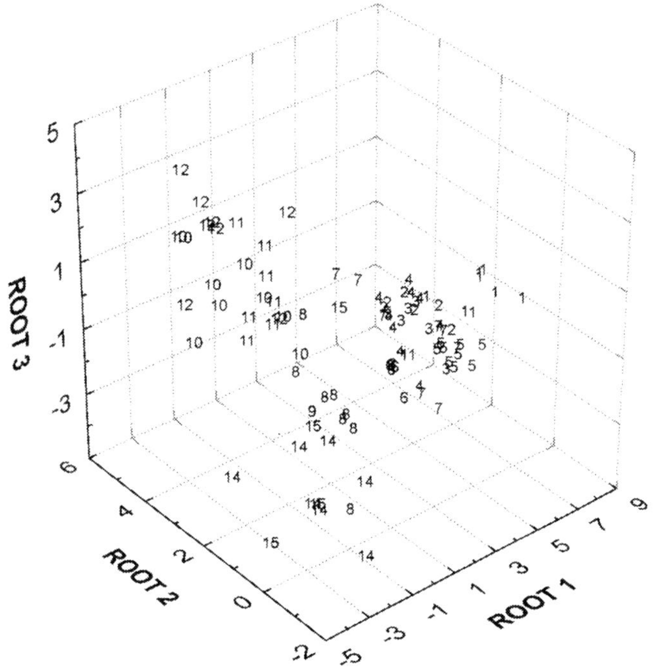

Fig. 4.8. Plot of canonical discriminant functions 1, 2 and 3 based on 10 polymorphic microsatellite loci of 15 populations of wild emmer wheat, *T. dicoccoides*, from Israel and Turkey. *1* Hermon, *2* Yehudiyya, *3* Gamla, *4* Rosh-Pinna, *5* Tabigha, *6* Mt. Gilboa, *7* Mt. Gerizim, *8* Gitit, *9* Kokhav-Hashahar, *10* Jaba, *11* Amirim, *12* Bet-Oren, *13* Bat-Shelomo, *14* Givat-Koach, *15* Diyarbakir. (Fahima et al. 2001)

Gitit), the coastal plains (Givat Koach) and Mt. Carmel (Bet-Oren) in Israel, and Diyarbakir in Turkey, show a different pattern of clustering. The peripheral populations show a clear separation from each other, with almost no overlap between them, and can easily be distinguished from the major cluster of the central populations. Most of these populations are isolated marginal populations that can also be distinguished by their unique ecogeographic conditions.

4.3.2.3
Genetic Structure of Wild Emmer Wheat Populations

Wild emmer wheat grows in lush and extensive stands in the catchment area of the Upper Jordan Valley, in the eastern upper Galilee Mts. and the Golan Heights in Israel. However, as previously stated, elsewhere in the Fertile Crescent, populations of wild emmer are semi-isolated and isolated, and largely display a patchy structure (Nevo et al. 1982, 1995; Nevo and Beiles 1989). Microsatellite analysis was found to be highly effective in distinguishing genotypes and populations of *T. dicoccoides* originating from these diverse ecogeographic sites in Israel and Turkey, as was demonstrated by the high level of polymorphic loci, genic diversity, and Shannon's Information Index (Table 4.9), by high genetic distance values between populations (Fig. 4.6), and by discriminant analysis (Fig. 4.8). Remarkably, 88% of the 135 genotypes were correctly classified into sites of origin by discriminant analysis on the basis of microsatellite genotyping (Fig. 4.8). Genotypes chosen for this study spanned most of the ecological range of emmer wheat in Israel. Central populations (Yehudiyya, Gamla, Rosh-Pinna and Tabigha) were collected in warm, humid environments of the Golan Plateat and near the Sea of Galilee. Marginal steppic populations were collected across a wide geographic area in the northern (Mt. Hermon), southeastern (Mt. Gilboa, Gitit, Kokhav-Hashahar), and southern (Jaba) borders of wild emmer distribution, ranging from xeric-cold to xeric-warm regions. Marginal mesic (Mediterranean) populations (Amirim, Bet-Oren, Bat-Shlomo, Givat-Koach) were collected from the western border of the wild emmer range (Nevo et al. 1982; Nevo and Beiles 1989). The dendrogram (Fig. 4.6) and the discriminant analysis (Fig. 4.8) show a clustering together of the central populations and a clear separation of most of the isolated marginal populations (e.g., Hermon, Givat Koach, Gitit, Bet-Oren, and Turkey) from the other populations. These results demonstrate the influence of unique ecogeographic conditions on the genetic structure of these populations based on the *noncoding* genomic regions. The *noncoding* DNA genome also outlines an archipelago-type genetic structure of wild emmer, as earlier demonstrated by the *coding* DNA genome expressed by allozymes (Nevo and Beiles 1989).

4.3.3 Ecological Correlates: Multiple Regression Analysis of SSR Polymorphism on Environmental Variables

A test of the best predictors of *P*, *He*, Shannon's information index, allele number per locus, and representative allele frequencies (alleles that were present in at least four populations) at the 20 polymorphic GWM loci in the 15 wild wheat populations was conducted by stepwise multiple regression (MR) analysis (Fahima et al. 2001). These characters were used as dependent variables, and the geographic, climatic, and edaphic factors listed in Section 4.2.3 were used as independent variables. The results are given in Table 4.10. Altitude, temperature, and water availability factors, singly or in combination, explained a significant proportion of the diversity in the polymorphism of the microsatellites. The best two-variable predictors of *He* and Shannon's Information Index, explaining significantly 0.38–0.4 of their variance, were Al and Ta (altitude and mean temperature in August). A three-variable combination, involving both geographic (altitude) and climatic (mean temperature in August and mean annual rainfall) variables, AlTaRn, accounted significantly ($p = 0.05$) for 0.49 of the variance in *P* and Shannon's Information Index.

Based on the number of microsatellite alleles per locus, microsatellite loci can be classified into several categories in terms of their prime ecogeographic predictors: (a) *water factors* (Rn, Hu, Rv): GWM95, GWM251, GWM265, GWM408; (b) *soil type* (So): GWM155; (c) *temperature* (Td): GWM120; (d) *geographic factors* (Al): GWM154, GWM186, GWM234; (e) *water + temperature* (Rn, Hu, Trd): GWM251; (f) *soil + temperature* (So, Trd, Td, Ta): GWM120, GWM155; (g) *geographic factors + temperature* (Al, Ta): GWM154; (h) *geographic factors + water* (Al, Ln, Dw, Rn, Hu, Rv, Rd): GWM234, GWM186, GWM408, GWM265. Microsatellite alleles can be classified similarly based on allele frequencies, as can be seen in the results presented in Table 4.10. Clearly, ecogeographic factors affect the differentiation pattern of primarily *noncoding genomic* SSRs just as they affect *coding* genomic regions (allozymes). The idea that *noncoding genomic* regions are "junk DNA" again seems unrealistic (Li et al. 2001d). As indicated for the RAPD diversity, although SSR and allozymes are correlated, cases of low SSR and high allozymes exist, possibly suggesting that the SSR ecological correlate may be partly independent from allozymes, so the SSR pattern is not only a result of the inbreeding system but due to direct selection on the SSRs.

4.3.4 Diversity and Utility of Wheat SSRs

The main objective of the SSR study (Fahima et al. 2001) was to assess microsatellite diversity among wild emmer wheat populations originating mainly from Israel and to test for relationships with ecological parameters. The GWM developed for cultivated wheat were used to detect DNA

Table 4.10. Coefficient of multiple regressions (R^2) of microsatellite (SSR) genetic indices, allele number per locus, and allele frequencies as the dependent variables, and environmental variables in 15 populations of wild emmer wheat, *T. dicoccoides*, in Israel and Turkey as independent variables. (Fahima et al. 2001)

Genetic indices	Stepwise model by ecogeographic variables		
	Step 1	Step 2	Step 3
P	Huan 0.196$^{@}$		
He	Al 0.199$^{@}$	Al Ta 0.382*	
Shannon's information index	Al 0.232$^{@}$	Al Ta 0.389*	Al Ta Rn 0.492*
Allele number / locus			
GWM95	Rn 0.283*		
GWM120	Td 0.242$^{@}$	Td So 0.481*	Td So Ln 0.579*
GWM154	Al 0.431**	Al Ta 0.607**	Al Ta Rn 0.744**
GWM155	So 0.327*	So Trd 0.502*	So Trd Ta 0.555*
GWM186	Al 0.318*	Al Huan 0.389$^{@}$	Al Huan Rn 0.512*
GWM234	Al 0.252$^{@}$	Al Dw 0.506*	Al Dw Rd 0.574*
GWM251	Rn 0.553**	Rn Huan 0.696***	Rn Huan Trd 0.750**
GWM265	Huan^{ns} 0.188	Huan Rv 0.470*	Huan Rv Ln 0.588*
GWM408	Rv 0.293*	Rv Ln 0.399*	Rv Ln Trd 0.481$^{@}$
Allele frequency			
GWM18-B	Ta 0.304*		
GWM18-F	Huan 0.301*	Huan So 0.429*	Huan So Rn 0.589*
GWM60-K	Rr 0.171^{ns}	Rr Ln 0.344$^{@}$	Rr Ln Rn 0.540*
GWM60-L	So 0.465**	SoTj 0.517**	
GWM95-C	Hu14 0.321*		
GWM95-L	So 0.296*	So Rr 0.483*	
GWM99-F	So 0.298*		
GWM99-Q	Huan 0.285*	Huan Rv 0.406*	
GWM120-K	Rn 0.322*		
GWM120-M	Huan 0.273*	Huan Td 0.448*	
GWM154-F	Ta 0.208$^{@}$	Ta Trd 0.525*	Ta Trd Al 0.624**

Table 4.10. *Continued*

Genetic indices	Stepwise model by ecogeographic variables		
	Step 1	Step 2	Step 3
GWM154-G	Huan 0.327*	Huan So 0.475*	Huan So Rn 0.535*
GWM155-C	Trd 0.441**	Trd Ta 0.541**	
GWM155-K	Rn 0.369*	Rn Ta 0.648**	Rn Ta Rd 0.708**
GWM186-O	So 0.300*	So Hu14 0.481*	
GWM218-B	Trd 0.257*		
GWM218-J	So 0.341*	So Tdd 0.538**	So Tdd Rv 0.583*
GWM219-D	So 0.468**		
GWM219-E	Tm 0.297*	Tm Huan 0.353@	Tm Huan Tj 0.562@
GWM219-G	Trd 0.292*	Trd Rd 0.522*	Trd Rd Ev 0.646**
GWM234-D	Al 0.344*	Al Rn 0.410*	
GWM234-F	So 0.296*	So Hu14 0.504*	So Hu14 Ta 0.690**
GWM251-B	Ta 0.143ns	TaRn 0.424*	
GWM251-K	So 0.283*		
GWM265-C	Hu14 0.205@	Hu14 Lt 0.579**	Hu14 Lt Ev 0.651**
GWM340-P	Trd 0.280*		
GWM389-Q	Al 0.368*	Al Tm 0.442*	Al Tm Ta 0.557*
GWM400-D	Rd 0.258*	Rd Tm 0.471*	Rd Tm Trd 0.518*
GWM400-F	Ta 0.126ns	Ta Rn 0.426*	
GWM400-H	So 0.104ns	So Rn 0.248ns	So Rn Rd 0.642**
GWM400-J	Huan 0.279*	Huan Ev 0.397*	
GWM408-P	Ta 0.118ns	Ta Rd 0.289ns	Ta Rd Rd 0.797*
GWM577-K	Al 0.240*		

Lt Latitude, *Ln* longitude, *Al* Altitude, *Tm* mean annual temperature, *Ta* mean August temperature, *Tj* mean January temperature, *Td* mean seasonal temperature difference, *Trd* mean number of tropical days, *Ev* mean annual evaporation, *Dw* mean number of dew nights in summer, *Rn* mean annual rainfall (in mm), *Rd* mean number of rainy days, *Rr* coefficient of variation in rainfall *Hu14* mean humidity at 14:00 h, *Rv* interannual rainfall variation, *Huan* mean annual humidity, *So* soil type

$^{***}p < 0.001$, $^{**}p < 0.01$, $^{*}p < 0.05$, $^{@}p < 0.10$, *ns* $p > 0.10$

polymorphism among 15 populations of the wild progenitor of wheat *T. dicoccoides*.

The utility of microsatellite markers in the study of genetic diversity within and between wild and cultivated wheat germplasm was already demonstrated *regionally* (Fahima et al. 1998) and *locally* (Li et al. 2000a–d, 2001b–d). Regions flanking microsatellite loci are often conserved between closely related species (Moore et al. 1991), allowing the use of primers to amplify loci in closely related species. Although the *T. dicoccoides* accessions used in this study originated from a wide range of habitats and show a high level of polymorphism, only a small proportion of marker-accession combinations failed to amplify a product in *T. dicoccoides*. These missing amplification products (or "null alleles") are most likely due to sequence alterations, such as point mutations, deletions, or insertions, within the priming sites (Devos et al. 1995a, b). Although the GWM markers were derived from hexaploid wheat (AABBDD), *T. aestivum* (Röder et al. 1995, 1998), they were successfully used to study genetic diversity in several cultivated and wild *Triticum* species, including tetraploid wheats (AABB), such as *T. durum*, *T. aethiopicum* (Plaschke et al. 1995), and *T. dicoccoides* (Fahima et al. 1998, 2001; Li et al. 2000a–d, 2001b–d), and diploid *Triticum* species, such as *Aegilops tauschii* (genome DD, Pestsova et al. 2000).

4.3.5 Genetic Diversity of Wild Emmer versus Cultivated Wheat

In the regional wild wheat study (Fahima et al. 2001), using 20 GWM, 363 alleles were revealed among 135 wild wheat accessions, an average of 18 alleles per GWM. In a previous study, also using GWM markers, Plaschke et al. (1995) revealed a total of 142 alleles among 40 cultivated wheat lines, an average of 6.2 alleles per GWM. The results obtained in the wild emmer study demonstrate the high diversity in microsatellite sequences among *T. dicoccoides* accessions compared with the cultivated germplasm, as was already demonstrated by RAPD (Fahima et al. 1999) and allozyme markers (Nevo and Beiles 1989). Clearly, genetic diversity was eroded across the *coding* and *noncoding* genome during the domestication of major cereal crops, as was displayed by allozymes (Nevo 1986a), RAPD, and SSR loci (Fahima et al. 1998, 1999, 2001). This contributed to the susceptibility and vulnerability of the wheat cultivars to abiotic and biotic stresses.

The dendrogram presented in Fig. 4.6 clearly demonstrates the ability of microsatellites developed from *T. aestivum* sequences to detect a large amount of genetic diversity in wild emmer wheat and to identify intergroup differences. All the examined *T. dicoccoides* populations were distinguishable by the GWM markers used, even within closely related populations originating from close geographic locations. Our results demonstrated that the SSR DNA polymorphism of wild emmer wheat was correlated with *macro-* and *microscale* ecogeographic factors. In particular, the Israeli populations exhibited high interpopulation and interregional polymorphism. These observations are consistent with previous results obtained with isozymes and different DNA markers for different collections of wild emmer wheat covering a much wider geographic range. They demonstrate geographic and genomic congruence and continuity from *macro-* to *microscales* and from *coding* (protein) to *non-coding* (presumably partly regulatory) DNA (Nevo et al. 1982; Nevo and Beiles 1989; Li et al. 1999, 2000a–d, 2001a–d).

4.4 Ribosomal DNA Diversity

Ribosomal RNAs are involved in translation and interact directly with translation initiation, hence their cardinal importance in protein synthesis. The rRNA genes are separated by nontranscribed intergenic spacers. What could be the function of these spacers? We explored the variation in rRNA gene structures in wild emmer and its relationship to allozymic diversity in wild emmer (Flavell et al. 1986). The rRNA genes, the infrastructure of protein synthesis, lie in tandem arrays at the nucleolus organizers (NOR) on chromosomes 1B and 6B of *T. dicoccoides*. In hexaploid wheat and in *T. dicoccoides*, there is variation in the amount of DNA residing between the transcribed regions specifying the rRNAs. This variation often involves the number of 135-bp repeats that are clustered in this intergenic region (Appels and Dvorak 1982). A map of a typical rDNA repeat unit from chromosomes 1B or 6B from hexaploid wheat is shown in Fig. 4.9. The variable segment containing the 135-bp repeats is marked. To understand the incidence and possible significance of the spacer-DNA variation in hexaploid wheat, it was considered useful to study the equivalent variation in the Israeli populations of *T. dicoccoides* that had already been studied genetically and ecologically.

The variation in the intergenic spacer of ribosomal DNA (rDNA) of tetraploid wild emmer wheat, *T. dicoccoides*, in Israel was examined in 112

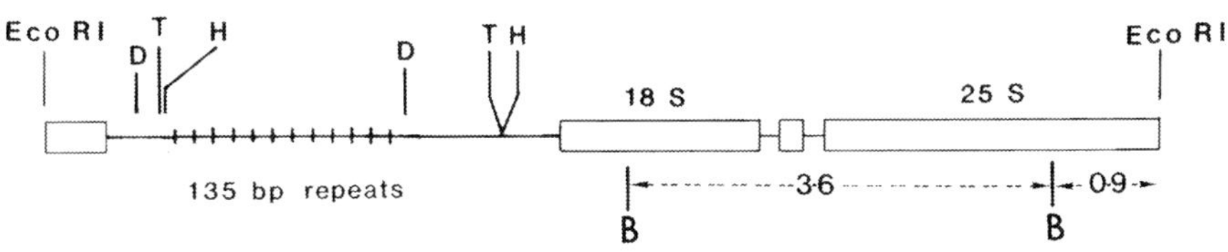

Fig. 4.9. Reference restriction-endonuclease map of a rRNA gene from the hexaploid wheat variety Chinese Spring. The map of the *Eco*RI repeat unit, taken from a tandem array of rRNA genes in *Triticum aestivum*, is redrawn from information published by Appels and Dvorak (1982) and from new sequence information (R.F. Barker and R.B. Flavell, unpubl. data). Only the *Dde*I (*D*), *Taq*I (*T*), and *Hinf*I (*H*) sites relevant to the spacer-variation assays are shown. Many other cleavage sites for these endonucleases exist outside this region. *B Bam*HI cleavage sites. The sizes of the *Bam*HI–BamHI and *Bam*HI–EcoRI fragments are shown in kilobase pairs. (Flavel et al. 1986)

individual plants representing 12 populations (Flavell et al. 1986) that had been tested earlier for allozymic diversity encoded by 50 gene loci (Nevo et al. 1982). The variation, detected by means of restriction endonucleases, results in part from variation in the number of 135-bp repeats in tandem array in the intergenic DNA (Fig. 4.9). Intergenic spacers of genes within a single locus, such as rDNA locus, are relatively homogeneous, but populations of *T. dicoccoides* display a wide spectrum of rDNA-spacer-length variation. Some populations are very homogeneous, whereas others are heterogeneous. The most homogeneous population (Bet Meir in Table 4.11) consists of plants with spacer-rDNA lengths that are the most common in *T. dicoccoides* in Israel.

We showed that: (1) natural populations of tetraploid *T. dicoccoides* in Israel display a wide spectrum of rDNA diversity. (2) The allozymic and rDNA diversities are highly and significantly intercorrelated. (3) Both allozymic and rDNA diversities are significantly correlated with climatic variables. It thus appears that both allozyme and rDNA genic markers may be molded, at least partly, by diversifying natural selection in accordance with extreme xeric ecological stresses, such as the Samaria eastern-xeric marginal-biotic environment.

Of prime importance for evolutionary theory of protein and DNA diversities, is the fact that the genetic landscape is nonrandom between loci, populations, and habitats, displaying possibly adapted patterns, since the allozyme and DNA variation is predictable on the basis of climatic factors. Allozymic and rDNA diversities are not only positively intercorrelated (Table 4.12), but, most importantly, both polymorphic systems are significantly correlated with and predictable by climatic factors (Table 4.12; Flavel et al. 1986). The extent of fragment-length polymorphism was found to be high in the steppic populations 9 and 10 (Taiyiba, Kokhav Hashahar; Table 4.11). These populations occupy a climatically highly fluctuating region situated between the Mediterranean and Samaria Desert climatic regimes. Could this polymorphism represent a regulatory adaptation to a climatically fluctuating environment, characterized by interchanging dry-hot and wet-cool years? And complementarily, an adaptation to shorten the life cycle to avoid the early summer drought?

If selection is responsible for the varying degrees of rDNA heterogeneity in these populations, it is necessary to consider whether selection acts on the rDNA loci or on alleles tightly linked to them. Few data are available on this point. In a study of 54 generations of a barley population created by a composite cross, Saghai-Maroof et al. (1984) showed that a class of ribosomal spacer DNA that was originally infrequent became predominant, whereas the originally most frequent phenotype decreased drastically and all other variants disappeared from the population. They concluded that selection was operating on the particular rDNA loci that became predominant and/or on associated loci. Again, it was not possible to decide unequivocally whether selection of rDNA variants alone was responsible, but their study clearly

Table 4.11. Natural populations of *T. dicoccoides* in Israel, a summary of the large *Taq*I rDNA intergenic fragments observed in individuals from the 12 populations, and the indices of allozyme diversities. (Flavell et al. 1986)

Population (site)[a]	Designation of population in Nevo et al. (1982)[c]		No. of individuals scored for rDNA	Total no. of different fragments	Mean no. of fragments per individual	No. of fragments occurring independently	Allozyme diversities based on 50 loci[b]		
							Alleles per locus	Mean proportion of loci polymorphic per population at 1%	Diversity index (*He*)
Qazrin	2	5	13	2	1.23	2	1.32	0.26	0.031
Bet-Meir	12	22	16	1	1.0	1	1.20	0.16	0.043
Sanhedriyya	11	21	7	3	1.85	2	1.22	0.16	0.057
Mt. Gerizim	8	17	2	3	1.50	2	1.22	0.20	0.059
Mt. Gilboa	7	16	3	3	2.33	2	1.24	0.22	0.060
Yehudiyya	3	7	9	5	2.11	3	1.28	0.22	0.061
Taiyiba	10	20	10	9	2.1	8	1.36	0.26	0.075
Kokhav-Hashahar	9	19	12	9	3.25	8	1.38	0.26	0.079
Tabigha	5	11	10	10	3.3	9	1.40	0.29	0.080
Mt. Hermon (Mt. Dov)	1	1	9	5	1.44	3	1.42	0.28	0.084
Rosh-Pinna	4	9	10	4	2.90	3	1.42	0.30	0.101
Bat-Shelomo	6	30	11	5	1.9	5	1.46	0.38	0.118

[a] See Fig. 4.1b and Table 4–1 for geographic distribution of population sites

[b] From Table 3 in Nevo et al. (1982) these numbers are employed in the rDNA text and Fig. 4.10

[c] From Nevo and Beiles 1989, and in Fig. 4.1

Table 4.12. Spearman rank correlations within and between allozymic and rRNA gene variables. (Flavell et al. 1986)

	A	*P*1%	*He*	TF	FI
A	–				
*P*1%	0.97****	–			
He	0.89****	0.84****	–		
TF	0.65**	0.59**	0.71***	–	
FI	0.41[ns]	0.42[ns]	0.55*	0.70**	–
HF	0.74***	0.70**	0.74***	0.97***	0.70**

A Average number of alleles per population, *P*1% proportion of polymorphic loci per population (criterion of polymorphism = 1%), *He* genetic diversity, *TF* total number of different rDNA long *Taq*I fragments, FI mean number of long *Taq*I fragment types per individual, HF number of long *Taq*I fragment types inherited independently
$^{*}p < 0.1$, $^{**}p < 0.05$, $^{***}p < 0.01$, $^{****}p < 0.001$; *ns*, $p > 0.1$

illustrates that major changes in the frequency of specific rDNA loci can occur in populations over relatively short time periods.

When considering the possibility that selection may operate on the intergenic spacer DNA of an rDNA gene cluster, it is important to consider what function the spacer DNA could have. The spacer DNA measured in these experiments includes tandemly arrayed repeats, and some of the spacer-length variation is likely to arise from variation in the number of these repeats (Appels and Dvorak 1982). These repeats adopt a distinct configuration that facilitates access to DNase I in the chromatin of active, but not of inactive rDNA loci. It can be postulated, therefore, that these sequences are likely to play a special role in rDNA expression that involves protein binding. Equivalent repeats in the intergenic spacer DNA of *Xenopus* rRNA genes have been shown to play an important role in stimulating rDNA transcription, and variation in the number of spacer repeats determines the relative efficiency of transcription of the adjacent gene (Moss 1983; Reeder 1984; Reeder and Roan 1984).

For wheat also there is some evidence that the variation in the number of these sequences may be correlated with a variation in the activity of rDNA in the NOR locus (R. B. Flavell et al., unpubl. data). Thus it is possible that these sequences have a regulatory function and that the particular spacer-DNA length most common in the species, and the only one found in the Bet Meir population, near Jerusalem, is the optimum length and is, therefore, maintained by selection across the wild emmer range.

If selection operates on rDNA loci and their intergenic spacers, how does this occur given: (1) the large number of genes in a locus, and (2) the probable redundancy of many of the genes? It would seem likely that selective forces would not recognize an rDNA variant until it constituted a certain proportion of the genes at the locus (Dover and Flavell 1984). The stochastic turnover mechanisms behind concerted evolution will either eliminate a new variant (the most frequent event) or increase its frequency at a locus and in some cases spread it to other loci. If the new variant increases in frequency by means of these mechanisms, it can become the predominant form at a locus (loci) and then presumably will be tested by natural selection. If the locus is acceptable, the variant can spread more extensively. If the locus reduces fitness, the individuals carrying the locus will presumably decrease in frequency. The turnover mechanisms behind homogenization of rRNA genes are, therefore, the means of: (l) propagating variants to the point at which they can be tested in evolution, and (2) maintaining loci homogeneous with adapted alleles.

Reassortment, by out-crossing in a population, of loci carrying different numbers of rRNA genes leads to individuals with different numbers of rRNA genes. From results such as those shown in Fig. 4.10, it can be concluded that individuals in populations such as 5 (Tabigha), 9 (Kokhav-Hashahar) and 10 (Taiyiba) differ in their total rRNA gene number. Individuals in population 12 (Bet-Meir), however, are likely to have the same rRNA gene number. The consequences, if any, of the variation in total rRNA gene number are unknown. Tetraploid wheat plants are likely to have a large surplus of genes that exceeds the minimum number required to maintain a supply of ribosomal RNAs, as has been discussed with regard to hexaploid wheat (Flavell and Martini 1982). However, in the model quoted above, which advocates a role for spacer repeats in regulating rRNA transcription, genes that are not transcribed but are able to bind regulatory proteins would influence the number of proteins available to stimulate transcription of active genes. If this occurs - as it has in certain circumstances in experiments with *Xenopus* oocytes (Reeder and Roan 1984) - then the total rRNA gene number might be a highly regulated parameter that would aid the regulation of a subset of rRNA genes by acting as a sink for regulatory proteins. If this aspect of the model or something akin to it is correct, then selection might also operate on the total gene number. This might be another reason for the absence of NOR heterogeneity in populations such as 12 (Bet-Meir in Table 4.11).

4.5 Structure, Function, and Evolution of Ribosomal DNA

The structure and evolution of the intergenic region in an rDNA repeat unit of wheat has been further examined (Barker et al. 1988). Processes of unequal crossing over and gene conversion may be involved in increasing a few ancestral sequences. Methylation patterns correlate with the transcription potential of an rRNA gene (Flavell et al. 1988). Repeated sequences in wheat intergenic rDNA are presumed to function as upstream promoters of those of *Xenopus* (Thompson and Flavell 1988). Sardana et al. (1993) suggested a model in which the 135-bp repeats play a part in the control of gene activity

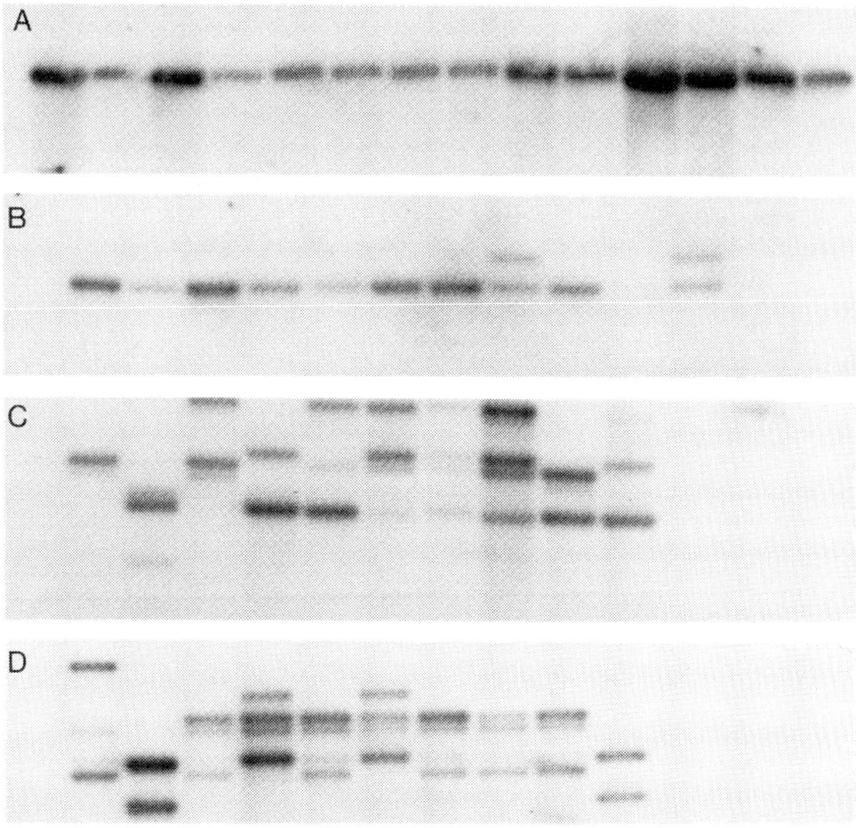

Fig. 4.10A–D. The long *Taq*I fragments found in the intergenic spacer regions of some of the individuals from four populations. **A** population 12, **B** population 2, **C** population 5, **D** population 9. DNAs were prepared from individual seedlings, digested with *Taq*I, fractionated by electrophoresis, transferred to nitrocellulose, and hybridized with ^{32}P-labeled rDNA. The rDNA fragments were revealed by autoradiography. (Flavel et al. 1986)

by binding (a) protein(s) that is in limiting supply, thereby predisposing the neighboring gene to become active preferentially. Molecular cloning and characterization of an unusually large intergenic spacer from the NOR-B2 locus of hexaploid wheat may be responsible for the higher activity of the NOR-B2 loci (Sardana and Flavell 1996).

4.6 Growth Characteristics in Wild Barley, *Hordeum spontaneum*, Populations from Different Habitats in Israel

Can the genotypic variation within and between populations be translated into phenotypic diversity? We initiated this study first in wild barley, which has a wider ecological range than that of wild emmer wheat. We tested the interpopulation, *interindividual*, and *intraindividual* phenotypic divergence of wild barley across Israel (van Rijn et al. 2000). *H. spontaneum* shows a large genetic variation and occupies a wide range of different habitats. Our aim was to quantify the variation in growth characteristics of *H. spontaneum* from different sites in Israel and to relate this variation to different environmental conditions. To this end, 84 accessions of 21 populations were grown in a growth chamber in near-optimal conditions and a range of physiological, morphological, allocation-related, and chemical characteristics were measured. These parameters included rates of photosynthesis, shoot and root respiration, specific leaf area, biomass allocation, and seed mass. The averaged overall traits variation that could be explained by differences between populations was 26%, and that by differences between accessions was 21%, whereas the traits variation explained by differences within accessions was 53%. In contrast to most other genetic studies, we found that variation between populations was larger than that between accessions. The phenotypic growth characteristics were subdivided into physiological, allocation-related, chemical, and morphological traits. The largest between-population variation (46%) was for morphological traits. In particular, seed mass, leaf thickness, and leaf width differed markedly between populations. Variation in growth characteristics between populations was poorly related to mean annual rainfall, mean humidity, or January temperature at the sites of origin. We expect these differences between populations to be *larger* and correlation with environmental parameters to be more notable in plants grown in stressful conditions. According to our study, seed mass is more important than relative growth rate in determining variation in early plant biomass in *H. spontaneum*. It is of utmost importance to conduct a parallel study in wild emmer wheat.

In summary, the population genetics of allozymes and DNA (RAPD and SSR) of wild emmer wheat, *T. dicoccoides*, the progenitor of most wheats, has been reviewed across its regional geographic range in the Near East Fertile Crescent. We found massive nonrandomness and *heavy structure*. Diversity in both the *coding* (allozymes) and *noncoding* (primarily RAPD and SSR DNA) genomes was largely correlated with and predictable by abiotic and biotic factors. Natural selection (both diversifying and balancing selection) seems to be the major evolutionary driving-force responsible in wild emmer for the pattern of primarily adaptive structures to environmental stress and heterogeneity. Do the macroscale patterns reflect a microscale evolutionary ecological theater? We tested genetic polymorphisms at four microsites in northern Israel, in the Golan Heights (Yehudiyya, Qazrin) and Upper Galilee (Tabigha, Ammiad), displaying microclimatic, edaphic, and topographic ecological contrasts at a microsite.

Chapter 5

Microgeographic Studies of Allozyme and DNA Polymorphisms in *Triticum dicoccoides*

Microgeographic studies of allozyme and DNA differentiation could contribute to a better understanding of the nature and dynamics of genetic polymorphisms at the protein and DNA levels and the evolutionary forces shaping their genomic structure (Nevo 1983b).

During the last two decades we have conducted extensive studies on allozyme and DNA differentiation in many species of plants and animals in Israel, both at the macrogeographic (reviewed in Nevo 1983b,c, 1990, 1998a, Nevo et al. 1994, 1996) and microgeographic level (reviewed in Nevo 1983c, 1989a, 1990, 1995, 1997, 1998b; Nevo et al. 1981, 1982, 1983, 1986c, 1988a–c, 1991c; Li et al. 1999, 2000a–d, 2001a–d). These studies suggest that both allozyme and DNA polymorphisms are at least partly adaptive across the *coding* and *noncoding* genomic regions. Our microsite studies, incorporating microclimatic, rock, soil, and topographical variation, indicate that microniche ecological selection is the primary cause of genetic differentiation in the plants and animals studied, rather than stochastic processes and the neutrality of allozymic and DNA variants. Here, we provide an overview of four different microgeographic studies in wild emmer wheat related to microclimate (Yehudiyya), soil (Tabigha) and topography (Qazrin-Yehudiyya, Ammiad).

5.1 Divergence of Multilocus Allozyme Structure in Wild Emmer Wheat Across a Dipping Basalt Plateau in the Golan: Qazrin Versus Yehudiyya

Our first microgeographic study in wild emmer wheat was conducted on a dipping basalt slope between Qazrin and Yehudiyya (Golenberg 1986a, b, 1987; Golenberg and Nevo 1987; Fig. 5.1). The importance of multilocus units in the evolution of inbreeding plant species was investigated in an ecological-genetic study of wild emmer wheat, *Triticum dicoccoides*. Two multilocus genotypes, as defined by eight loci, *Mdh-1, Ipol, β-Glu, Pept-1, Pept-3, 6Pgdh-2, Hk,* and *Rc,* were identified in the area of the Yehudiyya Nature Preserve in northern Israel and were found to be distributed in a step cline along a 10 km transect (Fig. 5.2).

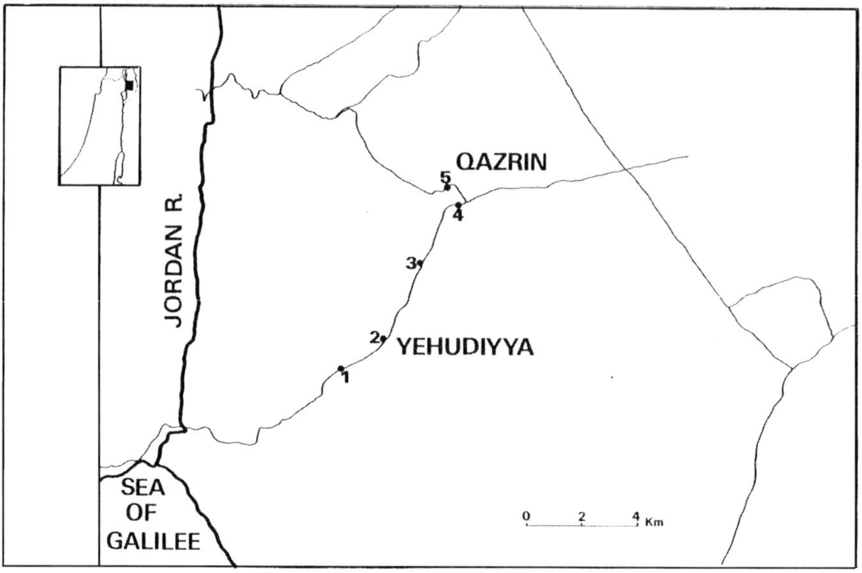

Fig. 5.1. Locations of the sample sites in the Yehuddiya-Qazrin transect. *1–5* Populations analyzed across the transect. (Golenberg 1986a)

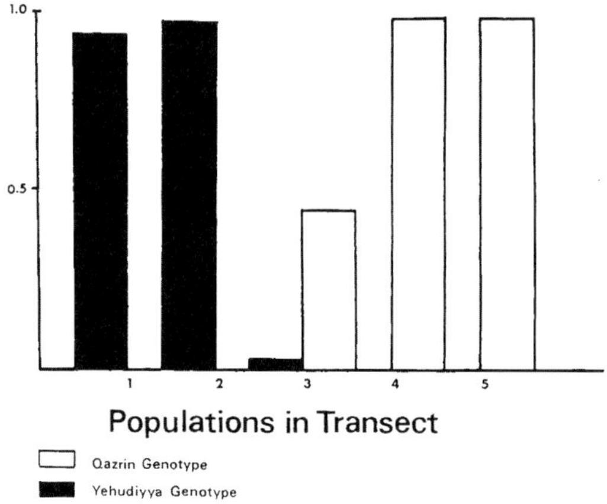

Fig. 5.2. Histogram of the proportion of multilocus genotypes by site. (Golenberg 1986a)

The linkage relationships and chromosomal locations of the eight loci were determined by analysis of crosses of Yehudiyya and Qazrin genotypes and by comparison with isozyme patterns of ditelocentric accessions of *Triticum aestivum* cultivar Chinese Spring. It was determined that *Mdh-1* and *Hk* are linked and probably located on chromosome 1B. *Pept-1* and *Rc* are also linked and probably located on chromosomal arm 6B1. For updated information of chromosomal localization of wheat allozymic loci see http://wheat.pw.usda.gov/ggpages/wgc/98/Protein2b.htm. Linkage disequilibria between loci were tested over four generations and found to increase. Two cases of probable epistatic selection were noted; however, no evidence suggesting the regeneration of the original multilocus structures was found (Golenberg 1986b).

Significant differences were found among populations of Yehudiyya and Qazrin for morphological, germination, and phenological characters when grown under standardized greenhouse conditions. Reciprocal transplantation and replacement series competition experiments did not, however, reveal any local adaptations in the years studied, possibly because of the short duration of the experiment and technical disturbances.

Outcrossing rates were estimated by direct methods from field and experimental garden material and were well below 1%. Gene flow rates were estimated by two indirect methods and were, in general, found to be low. Genetic neighborhood area was estimated to be about 5 m in diameter, implying that the mean gene flow distance per generation was about 1.25 m.

In conclusion, there was no strong evidence indicating that multilocus structures were the units of selection in *T. dicoccoides* in this area and in this experiment. Our results suggested that sporadic selection pressures, high selfing rates, and limited gene flow are of major importance in the evolution of multilocus associations in natural populations of wild emmer wheat and other inbred plants.

5.2 Microclimatic Stress and Adaptive Allozyme and DNA Differentiation: *Sun* versus *Shade* at the Yehudiyya Microsite

The Yehudiyya microsite is located in an open oak park forest of *Quercus ithaburensis* in the lower foothills of the Golan Heights, northeast of the Sea of Galilee, Israel (Figs. 4.1, 5.1, 5.7). *T. dicoccoides* sampling was conducted in 1985 in an area smaller than 1,000 m^2 and involved 12 repeated experiments (trees and their immediate circumference). Trees were mosaically distributed, 10–20 m in height, with canopy diameters up to 20 m, and they were separated by only 10–20 m from each other. Each experimental tree involved two microclimatic niches separated by a few meters: (1) *shade*, mostly shaded during the daytime under the canopies of oak trees and thus receiving a much lower light intensity than between trees in the sun; and (2) *sun*, exposed in daytime to continuous sun radiation and drying in the circumference around each tree and between trees (Nevo et al. 1988a). The microclimates varied significantly in the two microniches (a single day's measurement is shown in Table 5.1a). The soil temperature in the *sunny* niche was almost 10 °C higher than in the *shady* niche. The relative humidity in soil was similar in the *shady* and *sunny* niches. During the growing season (Oct.–May) of *T. dicoccoides*, the *shady* niche was under less stresses due to lower radiation intensity than that affecting the drought-stressed *sunny* niche. The two microniches vary significantly in plant formations (Nevo et al. 1988b). Wild emmer is sparse in the shade

Table 5.1a. Ecological backgrounds of sunny and shady microniches at Yehudiyya ($N = 10$ trees) measured on a single day. (Nevo et al. 1988b)

Ecological factors	Sun	Shade	Sun-shade differential
	Mean SE	Mean SE	Mean SE
Temperature[a]			
Air	32.45 ± 0.617	30.35 ± 0.269	2.1* ± 0.515
Soil	39.35 ± 0.943	29.37 ± 0.440	9.98* ± 0.922
Soil-air differential	6.9* ± 0.833	−0.98* ± 0.289	
Relative humidity			
Air	36.95 ± 1.184	37.6 ± 1.213	-0.65^{ns} ± 0.460
Soil	37.30 ± 1.265	37.7 ± 1.274	-0.40^{ns} ± 0.476
Soil-air differential	0.35^{ns} ± 0.279	0.1^{ns} ± 0.277	0.25^{ns} ± 0.393
Plant formation	Sun: *Psoralea hirsuta Carthamus glaucus Ami majus, Olchoria pumilum, Eryngium creticum, Gundelia Tournefortii, Lavatera trimestris* with dense wild cereals		Shade: *Ricotia lunaria, Tordylium aegyptiacum* with sparse other species
Wild emmer	Abundant	Very sparse	

*$p < 0.001$; *ns* not significant
[a] Air temperature taken 1 m above ground

under the oak canopy and abundant in the sun. The rarity of wild emmer in the shade suggests that it is primarily stressed for light, whereas in the sun it may be more stressed for water availability. In each pair of niches of each tree, several plants (4–5) were sampled first from the *shade* under the tree canopy and then nearby from the *sunny* niche immediately surrounding the tree circumference. Thus, the 12 trees sampled represent 12 repeated experiments, each involving a *sun-shade* paired comparison, separated by 2–4 m, and extracted from lush and extensive stands of wild emmer involving thousands of individual plants.

5.2.1 Allozymic Diversity: *Sun-Shade* Divergence

Allozymic variation in proteins encoded by 48 loci was analyzed electrophoretically in 1984 and 1985 in 137 individual plants of *T. dicoccoides* from a microsite at Yehudiyya, northeast of the Sea of Galilee, Israel (Nevo et al. 1988b). The test involved two climatic microniches: (1) *sunny* between trees, and (2) *shady* under the trees' canopy, in the open Tabor oak forest, as described above. Significant genetic differentiation at single-, two-, and multilocus structures was found in the neighboring climatic niches, separated by only a few meters (Table 5.1b; Fig. 5.3). Our results suggest that allozyme polymorphisms in wild emmer wheat are partly adaptive and are

Table 5.1b. Allele frequencies at 21 polymorphic loci of four subpopulations of *Triticum dicoccoides* (sunny and shady microniches in two years) at Yehudiyya, Israel. (Adapted from Nevo et al. 1988b)

Year	+ – Allele	Complementary allele	1984		1985			χ^2 between sun and shade
Microniche			Sun	Shade	Sun	Shade	Mean	shade
Size (*N*)			14	16	60	47	137	(1 df)
Locus								
*Aat-2*A	a – +	(b)	0.083	0.393	0.136	0.128	0.155	0.427 ns
	He		0.153	0.477	0.234	0.223	0.272	
*Aat-3*A	a – –	(null)	0.917	1.000	0.932	1.000	0.962	Fisher @
	He		0.153	0.0	0.126	0.0	0.070	
Acph-3	b – +	(c)	0.167	0.727	0.583	0.553	0.546	0.417 ns
	He		0.278	0.397	0.486	0.494	0.414	
Acph-X	a –	(b)	0.0	0.700	–	–	0.389	Fisher**
	He		0.0	0.420	–	–	0.210	
*Adh-1*A	a –	(b)	1.000	1.000	0.310	0.529	0.500	2.656 ns
	He		0.0	0.0	0.428	0.498	0.232	
*Adh-2*A	a –	(b)	1.000	1.000	0.817	0.979	0.902	5.980*
	He		0.0	0.0	0.299	0.042	0.085	
Est-1	b +	(c)	0.429	0.409	–	–	0.417	Fisher ns
	Null –		0.429	0.455	–	–	0.444	Fisher ns
	He		0.612	0.607	–	–	0.610	
*Est-2*A	b +	(c)	0.700	0.0	–	–	0.500	Fisher @
	He		0.420	0.0	–	–	0.210	
*Est-2*B	c –	(a)	0.200	1.000	–	–	0.429	Fisher @
(a = null)	d +		0.700	0.0	–	–	0.500	Fisher*
	He		0.460	0.0	–	–	0.230	
*Est-4*B	c –	(null)	0.400	0.600	1.000	1.000	0.727	0.166 ns
	He		0.480	0.480	0.0	0.0	0.240	
*Est-5*A	a –	(d)	0.167	0.667	1.000	1.000	0.885	0.569 ns
	He		0.278	0.444	0.0	0.0	0.181	
*Est-5*B	a + –	(c)	0.833	0.533	0.136	0.233	0.279	2.766 ns
(a = null)	e – +		0.167	0.467	0.814	0.628	0.651	
	He		0.278	0.498	0.317	0.532	0.406	
Gluc-A	a +	(null)	0.778	0.625	0.0	0.0	0.097	0.0 ns
	He		0.346	0.469	0.0	0.0	0.204	
Gluc-B	a –	(b)	0.333	0.875	1.000	1.000	0.944	Fisher ns
	He		0.444	0.219	0.0	0.0	0.166	
Hk	a +	(b)	0.714	0.100	–	–	0.353	Fisher*
	He		0.408	0.180	–	–	0.294	
Ipol	b –	(c)	0.500	0.643	1.000	1.000	0.954	Fisher ns
	He		0.500	0.459	0.0	0.0	0.240	
*Mdh-1*A	a –	(b)	0.308	0.667	1.000	1.000	0.896	0.277 ns
	He		0.426	0.444	0.0	0.0	0.218	
*Pept-1*B	a – –	(b)	0.300	0.714	0.783	0.864	0.766	1.682 ns
	He		0.420	0.408	0.339	0.236	0.351	
Pet-3	b + –	(c)	0.700	0.385	0.061	0.171	0.196	0.279 ns
	He		0.420	0.473	0.115	0.284	0.323	
Pgi-A	a + +	(b)	0.846	0.615	0.300	0.255	0.368	0.336 ns
	He		0.260	0.473	0.420	0.380	0.383	
6Pgd-2	a + +	(c)	0.818	0.429	0.515	0.381	0.506	2.130 ns
	He		0.298	0.490	0.500	0.472	0.440	

Monomorphic loci: *Aat-1*A, *1*B, *2*B; *Adh-1*B, *2*B; *Cat*-A, B; *Est-4*A; *Gdh*-B; *Ipor*-A; *Mdh-1*B, *2*; *Nadh-1*A, *1*B, *2*; *Pepc*; *Pept-1*A, *2*; *Pgi*-B; *Pgm*-A; *6Pgd-1*A, *1*B
Low polymorphic loci: *Aat-3*B, *Gdh*-A, *Pgm*-B
+ Higher frequency in the sun, – higher frequency in the shade, – in the 1985 column is missing data, *Fisher* Fisher's exact test; $*p < 0.05$, $**p < 0.01$, $***p < 0.001$, $^{@}p < 0.10$, *ns* $p > 0.10$

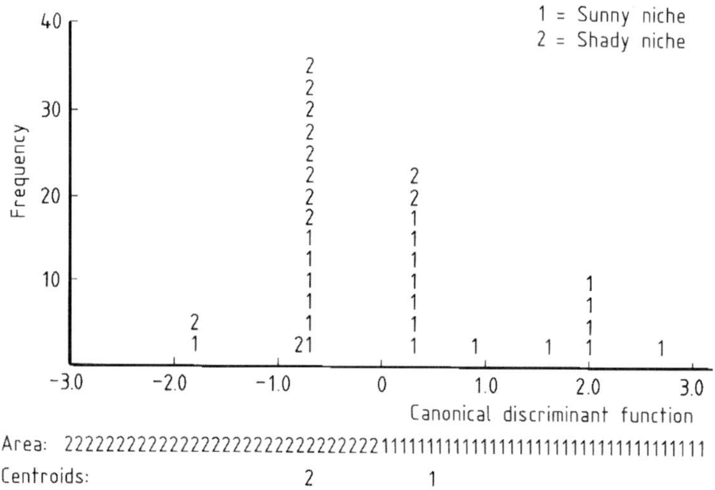

Fig. 5.3. Discriminant analysis of wild emmer wheat in Yehudiyya according to two climatic microniches, sunny and shady, based on seven polymorphic isozyme loci. (Nevo et al. 1988b)

differentiated primarily at the multilocus level by climatic factors, presumably related to temperature, water availability, and light intensity stresses. The following are the detailed results.

5.2.1.1
Distribution of Polymorphism Among Loci

The allele frequencies at 21 polymorphic loci, subdivided according to *shade*, *sun*, and year (1984, 1985), are presented in Table 5.1b. The raw data are followed in the same table by χ^2 or Fisher's exact tests. Of the 48 loci, 24 proved to be totally monomorphic *(Aat-1*A, *1*B, *2*B; *Acph-2; Adh-1*B, *2*B; *Cat*-A, B; *Est-4*A; *Gdh*-B; *Ipor* A, B; *Mdh-1*B, *2; Nadh-l*A, B, 2; *Pgi*-B; *6Pgd-1*A, B; *Pepc; Pgm*-A; *Pept-l*A, *2*). Three were nearly polymorphic *(Aat-3*B; *Gdh*-A; *Pgm*-B*)*. The remaining 21 polymorphic loci (44%) appear in Table 5.1b.

Association of Alleles with the Microclimatic Niches. Several loci varied significantly between the *sun* and *shade* microniches (Table 5.1b). Consistent association between allozymes and microclimatic niches was found in four loci *(6Pgd-2, Pgi-A, Aat-3*A and *Pept-1*B*)* and alleles, first with sun *(6Pgd-1*^{a} *Aat-3*A^{null} *Pgi*-A^{a} *Pept-1*B^{b}*)* and then with shade *(6Pgd-2*^{c}, *Aat-3*A^{a}, *Pgi*-A^{b}, *Pept-1*B^{a}). Interestingly, the *Aat-3*A^{null} allele appeared only in the *sunny* microniche.

Genetic Summary. The genetic summary of the two climatic microniches in each year appears in Table 5.2. The estimates of gene diversity, *He* (Nei 1978), did not vary significantly in the two microniches. However, the levels of poly-

Table 5.2. Summary of genetic variation based on 48 allozyme loci of *Triticum dicoccoides* in two microclimatic niches and 2 years at Yehudiyya. (Nevo et al. 1988b)

Locality	Sample size (*N*)	Mean no. of alleles per locus (*A*)	Mean proportion of loci: Polymorphic per population (*P*)		Heterozygous per individual (*H*)		Genetic diversity (*He*)
			(1%)	(5%)	Mean	SE	
1984							
Sunny niche	14	1.426	0.383	0.383	0.0	0.0	0.141
Shady niche	16	1.383	0.362	0.362	0.006	0.004	0.150
1985							
Sunny niche	60	1.302	0.233	0.233	0.001	0.001	0.077
Shady niche	47	1.233	0.209	0.186	0.0	0.0	0.074
Mean	137	1.336	0.297	0.291	0.002	0.001	0.110
Summarized over 1984 + 1985							
Sunny niche	74	1.521	0.417	0.396	0.001	0.000	0.145
Shady niche	63	1.438	0.396	0.313	0.003	0.002	0.116
Mean	137	1.479	0.406	0.354	0.002	0.001	0.131

morphism, *P*, and number of alleles per locus, *A*, were higher in *sun* than in *shade* in both years.

5.2.1.2 Analysis of Genetic Diversity Within and Between Populations

How is genetic diversity partitioned within and between populations? Gene diversity of a subdivided population can be analyzed according to its components, i.e., the gene diversities within and between subpopulations. The results provide measures of the average *(Dst)*, absolute *(Dm)*, relative *(Gst)* and *inter-* vs *intra*populational *(Rst)* degree of gene differentiation among subpopulations, where *Hs* is the mean gene diversity within a population, *Ht* is the mean gene diversity in the total population, and $Ht = Hs + Dst$ (Nei 1973). *Dm* can be used to compare the degrees of gene differentiation in different organisms.

The *Dst* analysis of gene diversity of the 16 polymorphic loci is given in Table 5.3, considering each of the two climatic microniches as subpopulations. On the average, 77.9% of the total genetic diversity in the subpopulations (2 microniches and 2 years) existed within and 22.1% existed between subpopulations. These results indicate a relatively strong gene differentiation between subpopulations. The contrasting climatic microniches contributed 8.9% to the genetic differentiation, but in certain loci *(Aat-3A, Adh-lA, Adh-*

Table 5.3. Partition of genetic diversity of *T. dicoccoides* in microclimatic niches at Yehudiyya, based on 16 allozyme polymorphic loci: *Gst* analysis. (Nevo et al. 1988b)

Locus	Alleles	Sample	*Ht*	*Hs*	*Dst*@			*Gst*	*Dm*	*Rst*
					Total	Sun-shade	%			
*Aat-2*A	2	132	0.2624	0.2486	0.0138	0.0019	13.8	0.0526	0.0184	0.0740
*Aat-3*A	2	132	0.0729	0.0704	0.0025	0.0025	98.5	0.0343	0.0033	0.0474
Acph-3	2	130	0.4957	0.4623	0.0334	0.0026	7.8	0.0675	0.0446	0.0965
*Adh-1*A	2	56	0.5000	0.3729	0.1271	0.0323	25.4	0.2541	0.1694	0.4543
*Adh-2*A	2	122	0.1774	0.1633	0.0141	0.0110	78.0	0.0793	0.0187	0.1148
*Est-4*B	2	44	0.3967	0.2727	0.1240	0.0051	4.1	0.3125	0.1653	0.6061
*Est-5*A	2	131	0.2028	0.0763	0.1264	0.0016	1.3	0.6236	0.1686	2.2087
*Est-5*B	3	129	0.4932	0.4062	0.0871	0.0052	6.0	0.1765	0.1161	0.2858
Gluc-A	2	124	0.1748	0.0553	0.1195	0.0001	0.1	0.6835	0.1593	2.8793
Gluc-B	2	124	0.1065	0.0464	0.0602	0.0023	3.8	0.5647	0.0802	1.7298
Ipol	2	120	0.0875	0.0518	0.0357	0.0000	0	0.4079	0.0476	0.9186
*Mdh-1*A	2	135	0.1859	0.0904	0.0955	0.0009	0.9	0.5137	0.1273	1.4083
*Pept-1*B	2	128	0.3589	0.3175	0.0414	0.0064	15.5	0.1152	0.0551	0.1736
Pept-3	2	107	0.3155	0.2423	0.0732	0.0018	2.5	0.2319	0.0975	0.4025
Pgi-A	2	133	0.4654	0.3956	0.0698	0.0020	2.9	0.1500	0.0931	0.2353
6Pgd-2	2	79	0.4999	0.4623	0.0376	0.0180	47.9	0.0753	0.0502	0.1086
Mean	2.06	114.1	0.2997	0.2334	0.0663	0.0059	8.9	0.2212	0.0884	0.3787

Ht Total gene diversity, *Hs* average gene diversity within populations, *Dst* average gene diversity between populations, *Dm* average of interpopulation diversity only, *Gst* gene diversity between populations relative to *Ht*, *Rst* interpopulational diversity relative to *Hs*, @ % of sun-shade *Dst* from total *Dst* which also includes the genetic difference between years.

2A, 6Pgd-2) the microclimate contribution was greater. Thus, allozyme differentiation between the two climatic microniches was substantial, as is also displayed in the following discriminant analysis.

Discriminant Analysis. The best differentiating factors were: *Adh*-2A^{b}, *Aat*-^{3}A^{null}, *Pgi*-A^{a}, *Mdh*-1A^{b}, *Est*-5B^{e}, *Aat*-2A^{a}, *Pept*-3^{b}. The results are given in Fig. 5.3. The analysis was able to significantly differentiate plants from the two climatic microniches by pairwise comparison ($F = 3.62$; d.f. 7/74; $p < 0.01$) and by a canonical discriminant function (Fig. 5.3). The correct classification of plants into their two climatic microniches was 73% (62 out of 85 plants analyzed): 68% in the *sunny* microniche and 80% in the *shady* one.

Genetic Distances. Coefficients of genetic distances *(D)* and genetic similarity (*I*) were calculated to compare the four subpopulations, based on the normalized identity of all loci between each pair of subpopulations (Nei 1972). The mean value of *D* was 0.115 (range: 0.004–0.194). All standard errors of individual values of *D* were less than 0.013. The estimates of *D* indicated sharp

Table 5.4. Representative linkage disequilibria (LD) between paired allozyme loci of *T. dicoccoides* in Yehudiyya in the entire sample and separately for each climatic microniche. (Nevo et al. 1988b)

Locus and allele	Locus and allele	Overall *p*	Sun *p*	Shade *p*
*Mdh-1*A a	*Est-4*B c	0.171***	0.185** (c)	0.152*** (c)
*Mdh-1*A a	*Est-5*A a	0.088***	0.109*** (c)	0.062*** (c)
*Est-5*A a	*Est-4*B c	0.152***	0.166** (c)	0.123***
*Pept-1*B a	*Est-4*B c	0.123**	0.133* (c)	0.086*
Pept-3 b	*Pept-1*B b	0.123***	0.112*** (c)	0.142***
Pept-3 c	*Est-5*A a	0.084* (c)	0.101*** (c)	0.065** (c)
Pept-3 c	*Est-4*B c	0.185*** (c)	0.240*** (a)	0.136** (c)
*Pept-1*B b	*Pgi*-A a	0.120***	0.161***	0.064* (c)
Pept-3 b	*Gluc*-A a	0.076***	0.100*** (c)	0.043 NS
*Pept-1*B b	*Gluc*-A a	0.059***	0.072*** (c)	0.041 @
Gluc-A null	*Est-4*B c	0.163***	0.245*** (a)	0.090 @
Gluc-B a	*Est-4*B c	0.123*** (c)	0.204** (c)	0.040 NS (c)
Ipol b	*Est-4*B c	0.098**	0.168** (a)	0.045 NS
Pgi-A b	*Adh-2*A a	0.070*** (c)	0.111*** (c)	0.014 NS (c)
*Pept-1*B a	*Adh-2*A a	0.071***	0.120***	–
6Pgd-2 a	*Adh-1*A a	0.071 NS (c)	0.141* (c)	–
6Pgd-2 a	*Adh-2*A a	0.059** (c)	0.124*** (c)	–
6Pgd-2 a	*Pgi*-A a	–0.009 NS	–0.087*	0.074 @
6Pgd-2 c	*Est-4*B c	0.183*** (c)	0.118 NS (c)	0.228*** (a)
Acph-X a	*Est-4*B c	0.188** (c)	–	0.148 @ (c)
*Aat-2*A a	*Adh-1*A a	0.030 NS	–0.001 NS	0.068 NS (c)
*Aat-2*A a	*Est-4*B c	0.032 NS (c)	–	0.047 NS (c)
Acph-3 b	*Adh-1*A a	0.063 @	0.016 NS	0.112 @

$^*p < 0.05$, $^{**}p < 0.01$, $^{***}p < 0.001$, *NS* $P > 0{,}10$; p by χ^2 with Yates correction, or by Fisher's exact test; *(c)* complete association: one gametic type is missing (Clegg et al. 1976); *(a)* absolute association: only two gametic types are present. (Clegg et al. 1976)

genetic differentiation between the two climatic microniches ($D = 0.05$, range: 0.004–0.096).

5.2.1.3
Genetic Differentiation at the Two-Locus Level

Linkage disequilibria (LD) for the entire sample and for each climatic microniche (sunny and shady) are given in Table 5.4. LDs can be classified into general (occurring in both *sun* and *shade*) and specific (occurring either in *sun* or in *shade*). Examples of general LDs were: *Mdh*-1A^a–*Est*-4B^c (LD = 0.185, 0.152 in *sun* and *shade*, respectively); *Pept*-3^b–*Pept*-1B^b (LD = 0.112, 0.142, respectively). Specific LDs for the *sunny* microclimatic niche were: *Pgi*-A^b-*Adh*-2A^a (LD = 0.111); *Pept*-1B^b–*Adh*-2A^a (LD = 0.120). Specific LDs for the *shady* microclimatic niche were: *6Pgd*-2^c–*Est*-4B^c (LD = 0.228).

Table 5.5. Estimates of allozyme multilocus genetic organization of *Triticum dicoccoides* in two climatic microniches in Yehudiyya. (Nevo et al. 1988b)

Microniche	Polymorphic loci	Expected variance of K	Upper limit of 95%	Estimate of variance of K	X(2)
Sun	16	3.142	<5.553	12.571*	3.001
Shade	15	2.574	<4.295	8.543*	2.319
Sun + shade	16	2.948	<4.464	10.262*	2.481

*$p < 0.05$; K Number of heterozygous loci in two randomly chosen gametes; X(2) Measure of multilocus structure

5.2.1.4
Multilocus Organization: *Sun* versus *Shade*

A multilocus organization index, related to the single-locus Simpson index and based on the observed distribution of the number of heterozygous loci (K) in two randomly chosen gametes was proposed by Brown et al. (1980). It measures multilocus associations when multiple alleles and many loci are analyzed, combining all paired-loci linkage disequilibria. We calculated multilocus organization indices for the entire data set and for the *sunny* and *shady* microniches in this study separately. The results are given in Table 5.5. The differences between the expected and observed variances (S_k^2) were significant in each microniche, and the observed variance of K in the entire data set was 248% above the expected variance. Multilocus organization appeared to be significantly higher in the *sunny* microclimatic niche than in the *shady* one. The standardized index of intensity of multilocus structure X(2) was 3.0 in the *sun* and 2.3 in the *shade*. We concluded that, on average, multilocus organization was higher in the *sunny* microniche.

5.2.1.5
Shade Ecology

The climatic profile of *open shade* as a microclimatic niche was analyzed in Europe by Stoutjesdijk (1974). While different types of open shade may differ in their climatic profiles, the major climatic determinants of the *open shade*, i.e., temperature, humidity, and radiation conditions, may display parallel patterns. Surface temperatures are often 6–8 °C below ambient temperatures, humidity is higher than in the sunny microniche, and dew may persist in the open shade. Light intensity is relatively high, but net radiation may be zero or slightly negative due to the strong long-wave radiation loss to the "cold" bright sky. The specific vegetation in open shade sites reflects habitat conditions. Several angiosperms, mosses, and liverworts prefer this microniche. While the under-oak canopy in our study area at Yehudiyya is not an *open shade*, but rather a covered shade, it appears to share several microclimatic characteristics with the open shade. These involve more relatively constant temperature and humidity conditions in the *shady* microniche than in the *sunny* microniche. In the covered shade, the temperature is lower and the humidity slightly higher (Table 5.1a). The alternative microclimatic niches may represent two different stresses: (1) high solar radiation, temperature, and drought in the *sunny* niche; (2) low light intensity affecting primarily the photosynthetic yield in the *shady niche.*

5.2.1.6
Neighbor Effects, Population Size, and Migration in Wild Emmer

It is important, however, to avoid the pitfall of neighboring effects (Jain 1976) in the attempt to separate *random* (founder effect) from *nonrandom* (niche structure) factors. Our present data, at least partly, overcame this problem, since ours was a series of repetitive experiments, each consisting of an oak tree under-canopy and its surrounding inter-tree space. The mosaic oak tree structure of our experimental design, plus the 18 repetitions in both years, repel genetic neighboring effects, at least partly. The latter was estimated to be about 5 m in diameter in Yehudiyya (Golenberg 1986a). Furthermore, neither small-size population effects nor migration patterns can singly or in combination explain the pattern discovered. The population size of wild emmer in Yehudiyya is large, involving many thousands of individuals. The role of migration in this population was estimated to be low, about 1.25 m per generation (Golenberg 1986a). Nonetheless, it suffices to equalize allele frequencies during relatively short periods. The *sun-shade* genetic differentiation found is, therefore, unlikely to be the result of stochastic processes. Rather, it appears to reflect microclimatic niche divergence in a relatively small area of less than 1,000 m^2 and over a few meters between sun and shade.

5.2.1.7
Multilocus Microclimatic Differentiation in Wild Emmer

The results at the Yehudiyya site of genetic differentiation, according to the microclimatic niche, suggest that genetic diversity and allele frequency differentiation at the single locus level, but mainly at the multilocus level, are subject to selection. This is indicated by: (1) single loci differentiation, but *primarily* by multilocus discrimination (Tables 5.1b–5.5); (2) higher levels of polymorphism in *sun* than in *shade*, according to the prediction of the niche width-variation hypothesis: while *sun* may represent a broader niche than *shade*, the latter may involve a strong limited-light stress; (3) microclimatic specificity of two-locus linkage disequilibria for *sun* as opposed to *shade* (Table 5.4); and (4) multilocus genome organization (Table 5.5). At the multilocus level, there is substantial multilocus organization. On average, the variance of K was 248% higher than expected under random association. Interestingly, the *sunny* multilocus genotype showed more multilocus organization than the *shady* one (300% vs 232%, respectively, Table 5.5). It is worth noting that genetic diversity, *He*, based on our regional analysis (Nevo et al. 1982), is also positively associated with higher ecological unpredictability in climatically fluctuating steppe-like regions (Nevo and Beiles 1989).

5.2.1.8
Aridity and Lightless Stress Selection in Wild Emmer and Other Organisms

The *sun-shade* microniche study at the Yehudiyya microsite supports earlier studies indicating sharp microgeographic differentiation in other plants such as wild oats *Avena barbata* (Hamrick and Allard 1972; Hamrick and Holden 1979) and *Litaris cylindracea* (Schaal 1974, 1975; for reviews on microgeographic genetic differentiation, see Brown et al. 1990; Nevo 1983a, 1995, 1997). The effects of stress as an agent in unravelling the operation of natural selection was demonstrated earlier in microsite differentiation of allozymes in barnacles (Nevo et al. 1977), land snails (Nevo et al. 1982), wild barley, *Hordeum spontaneum* (Nevo et al. 1981, 1983, 1986c), and wild emmer (Nevo et al. 1988c). The effects of stress on allozyme differentiation were also demonstrated in controlled laboratory experiments of pollution biology (reviewed in Nevo 1986b).

Aridity stress was previously implicated in allozyme differentiation of many unrelated species in Israel (Nevo 1983c) and worldwide (Nevo et al. 1984a). In the analysis of genetic parallelism in genetic diversity of 21 species in Israel (Nevo and Beiles 1988), we concluded that natural selection may have operated directly, at least on some of the allozyme loci studied, rather than indirectly through the hitchhiking effect (Hedrick 1982). In the present study, we have no direct evidence implicating ecological selection of specific allozymes as opposed to linked blocks of genes. However, our evidence supports the hypothesis that genetic differentiation in the *sun* at the Yehudiyya microsite is affected, at least partly, by the aridity stress faced by various life cycles of wild emmer. In the *shade*, however, light threshold may stress wild emmer even more than drought, as is reflected by the sparse population of wild emmer in the *shade*.

Our present data are inconsistent with the hypothesis that allozymic diversity is neutral and that its fate is determined by mutation input and random fixation, as suggested by the neutral theory of molecular evolution (Kimura 1983). Allozyme differentiation, particularly at the multilocus level as the unit of selection, corresponds with the two microclimatic niches. The aridity index presumably plays a major differentiating role here and elsewhere (Nevo et al. 1981, 1982, 1983, 1986c, 1988c). The higher levels of polymorphism in the *sunny* microniche may be regarded as an adaptive strategy for increasing fitness in a more climatically fluctuating microniche, in accordance with the niche variation hypothesis (Van Valen 1965). This result supports regional (Nevo 1983c; Nevo and Beiles 1988) and globál (Nevo et al. 1984a) patterns, where allozymic diversity generally increases with increasing spatiotemporal ecological heterogeneity and unpredictability. Our results support the theoretical prediction that the existence of a protected polymorphism is more likely in a more heterogeneous environment (e.g. Karlin 1981, 1982; Karlin and Avni 1981; Karlin and Levikson 1974). We hypothesize that the patterns observed here are, to a large extent, due to diversifying and balancing selection on allozyme polymorphism. Fitness tests through transplant experiments, and biochemical tests of alternative allozyme alleles, should follow to verify the correlative results presented here.

Our results implicate not only theoretical, but, more importantly, practical applications. Since allozymic diversity varies in space and time, at both the *macro*- and *micro*geographic levels, with varying and/or changing ecologies, sampling strategies should be designed accordingly. If *macro*- and *micro*ecologic differentiation are closely followed, sampling strategies, exploitation, conservation, and utilization of wild emmer for breeding could be substantially optimized (Nevo 1983a, 1989a, 1995; Brown and Clegg 1983).

5.2.2 RAPD Diversity: *Sun-Shade* Divergence

An interesting question is whether DNA polymorphism is also subjected to selective forces mirroring allozymic pattern. This is important because the protein polymorphism observed to be correlated to ecology is produced by the *coding* sequences of the genome, while DNA-based marker polymorphism is produced by both *coding* and *noncoding* genomic regions. Moreover, analysis of DNA polymorphism is considered to be a finer tool than analysis of protein polymorphism because it permits much deeper probing into genomic variation (Puterka et al. 1993). The correlation between genomic variation and its environment in wild emmer wheat was addressed at the Yehudiyya microsite by means of RAPD (Li et al. 1999) and microsatellite DNA markers (Li et al. 2001b). The objectives of these studies were to detect DNA polymorphism of wild emmer wheat from stressful microclimatic niches in Yehudiyya, and to analyze the relationship between microscale ecologically divergent microclimates and DNA diversity. In addition, the RAPD polymorphism was also compared and contrasted with the previously studied allozymic differentiation (Nevo et al. 1988b).

Genetic diversity was examined by RAPD-PCR analysis (Li et al. 1999) in 118 registered individuals of wild emmer wheat, *T. dicoccoides,* from the Yehudiyya microsite (Figs. 4.1, 5.1, 5.7) tested earlier for allozymic *sun-shade* divergence (Nevo et al. 1988b). The test involved two climatic microniches in the open oak park forest of *Quercus ithaburensis*: (1) *sunny* between trees and (2) *shady* under-tree canopies. The 118 individuals included 56 plants from the shade and 62 from the sun, involving 12 trees from the 1985 collection (Nevo et al. 1988b). Comparisons were based on 97 loci amplified by 20 oligonucleotide primers (Table 5.6). The 20 RAPD primers amplified 2–8 bands per primer at a size range of 250–1740 bp. Of the total 97 loci, 13% were

Table 5.6. Sequences of RAPD primers used and the amplified fragments produced. (Li et al. 1999)

Primer	Sequence (5′ → 3′)	No. of bands[a]	Range of band size (bp)
OPA12	TCGGCGATAG	8	250–980
OPF2	GAGGATCCCT	8	480–1740
OPM18	CACCATCCGT	4	435–1340
UBC116	TACGATGACG	3	440–890
UBC129	GCGGTATAGT	4	600–1100
UBC143	TCGCAGAACG	4	410–1030
UBC144	AGAGGGTTCT	2	750–1050
UBC269	CCAGTTCGCC	2	410–680
UBC285	GGGCGCCTAG	5	400–950
UBC324	ACAGGGAACG	7	400–1330
UBC416	GTGTTTCCGG	5	510–1150
UBC428	GGCTGCGGTA	5	375–880
UBC465	GGTCAGGGCT	6	375–770
UBC479	CTCATACGCG	7	310–1160
UBC493	CCGAATCACT	6	500–1070
UBC546	CCCGCAGAGT	5	620–1330
UBC611	CCATCGTACC	5	650–1160
UBC621	GTCTGCGCTA	4	625–990
UBC635	CTCAGCTCAG	4	400–1500
UBC636	GGGATATCGC	3	760–1240

OP Primers obtained from Operon, *UBC* primers obtained from University of British Columbia
[a] Number of reproducible and scorable fragments

monomorphic in both *shade* and *sun* subpopulations. The remaining 84 loci (87%) were polymorphic: four of the 84 loci were polymorphic only in the *sun*, while 80 loci were polymorphic in both *shade* and *sun*. Only one band, $UBC324_{480}$, was restricted to one niche. It was found in the *sun* only with allele frequency of 0.1774. Between the *shade* and *sun* microniches, 26 loci varied significantly in allele frequencies at the $p = 0.05$, 0.01, and 0.001 levels, and there were 5 loci at the $p = 0.10$ level. The remaining 53 loci varied non-significantly ($p > 0.10$) between the *shade* and *sun* (Table 5.7).

5.2.2.1 RAPD Discriminant Analysis

Discriminant analysis was conducted to maximize the overall multivariate F ratio of the two climatic microniches based on a multilocus analysis involving all polymorphic loci (Li et al. 1999). According to the most informative 25 loci out of the 80 polymorphic loci in both niches, the discriminant function was highly significant ($p < 0.00005$), and all the individuals (100%) could be correctly classified into the niches from which they were sampled (Fig. 5.4).

Table 5.7. Allele frequency and gene diversity (*He*, Nei 1973) at different RAPD polymorphic loci in the *shady* and *sunny* subpopulations of *T. dicoccoides* from Yehudiyya. (Li et al. 1999)

Locus[a]	Allele frequency (present)		χ^2	*He*	
	Shade	Sun		Shade	Sun
OPA12-**720**	0.4464	0.9032	28.53***	0.4943	0.1748
-500	0.7857	0.9032	3.14[c]	0.3367	0.1748
-470	0.7857	0.7581	0.13	0.3367	0.3668
-370	0.8214	0.8710	0.56	0.2934	0.2448
-300	0.2143	0.2742	0.09	0.3367	0.3980
-250	0.5357	0.8387	12.74***	0.4974	0.2706
OPF2-1740	0.7143	0.8710	4.46*	0.4082	0.2248
-1430	0.2500	0.5806	13.17***	0.3750	0.4870
-1260	0.9286	0.9194	0.04	0.1327	0.1483
-1110	0.8393	0.8871	0.57	0.2698	0.2003
-480	0.9286	0.8548	1.63	0.1327	0.2482
OPM18-1090	1.0000	0.8387	9.87**	0.0000	0.2706
-820	0.1607	0.2742	2.21	0.2695	0.3980
-440	0.2679	0.3226	0.42	0.3922	0.4370
UBC116-890	0.5893	0.6774	0.99	0.4841	0.4370
-670	0.9107	0.8871	0.18	0.1626	0.2003
UBC129-1100	0.1429	0.1935	0.54	0.2449	0.3122
-820	0.4286	0.6129	4.01*	0.4898	0.4745
-750	0.2500	0.2742	0.09	0.3750	0.3980
-600	0.6250	0.6613	0.17	0.4688	0.4480
UBC143-1030	0.6071	0.6613	0.07	0.4770	0.4480
-790	0.7143	0.5645	2.85[c]	0.4082	0.4917
-650	0.4464	0.6290	3.95*	0.4943	0.4667
-410	0.1429	0.4516	13.23***	0.2449	0.4953
UBC144-1050	0.6964	0.7419	0.30	0.4228	0.3829
-750	0.6429	0.5806	0.48	0.4592	0.4870
UBC269-680	0.5357	0.7742	7.47***	0.4974	0.3496
-410	0.2143	0.3871	4.14*	0.3367	0.4745
UBC285-950	0.2679	0.3548	1.03	0.3922	0.4579
-780	0.6607	0.8065	3.23[c]	0.4483	0.3122
-680	0.3929	0.3226	0.63	0.4770	0.4370
-530	0.3214	0.5323	5.33*	0.4362	0.4979
-400	0.8571	0.8710	0.09	0.2449	0.2248
UBC324-**1330**	0.1964	0.4032	5.94*	0.3157	0.4813
-1100	0.6964	0.8065	0.57	0.4228	0.3122
-800	0.3929	0.4677	0.67	0.4770	0.4679
-720	0.7857	0.5000	10.37**	0.3367	0.5000
-580	0.2857	0.1613	1.75	0.4082	0.2706
-480	0.0000	0.1774	10.96***	0.0000	0.2919
-400	0.1429	0.2258	1.33	0.2449	0.3496
UBC416-1150	0.8214	0.8871	1.03	0.2934	0.2003
-1110	1.0000	0.9355	3.74[c]	0.0000	0.1207
-790	0.4464	0.8548	21.90***	0.4943	0.2482
-710	0.7857	0.7581	0.13	0.3367	0.3668

Table 5.7. *Continued*

Locus[a]	Allele frequency (present)		χ^2	*He*	
	Shade	Sun		Shade	Sun
UBC428-710	0.7857	0.7258	0.57	0.3367	0.3980
-580	0.7500	0.7742	0.10	0.3750	0.3496
-500	0.7321	0.4194	11.73***	0.3922	0.4870
-380	0.2143	0.4032	4.88*	0.3367	0.4813
UBC465-770	0.9286	0.7581	6.33*	0.1327	0.3668
-550	1.0000	0.9194	4.72*	0.0000	0.1483
-490	0.5714	0.9355	17.86***	0.4898	0.1207
-400	0.3393	0.4839	2.53	0.4483	0.4995
UBC479-1150	0.5714	0.6935	1.89	0.4898	0.4251
-880	1.0000	0.9194	4.72*	0.0000	0.1483
-820	0.3750	0.3710	0.00	0.4688	0.4667
-710	0.3929	0.5806	4.15*	0.4770	0.4870
-580	0.8393	0.8871	0.57	0.2698	0.2003
-470	0.0893	0.1290	0.47	0.1626	0.2248
-310	0.7143	0.7581	0.29	0.4082	0.3668
UBC493-***1070***	0.6786	0.8548	5.18*	0.4362	0.2482
-970	0.9107	0.9355	0.03	0.1626	0.1207
-830	0.8750	0.9032	0.24	0.2188	0.1748
-760	0.7321	0.8387	2.00	0.3922	0.2706
-650	0.6607	0.6613	0.00	0.4483	0.4480
-500	0.5893	0.3871	4.82*	0.4841	0.4745
UBC546-***1330***	0.3393	0.7581	20.94***	0.4483	0.3668
-1140	0.7143	0.9677	14.62***	0.4082	0.0624
-730	0.2379	0.3387	1.63	0.3922	0.4480
-620	0.1607	0.2581	1.67	0.2698	0.3829
UBC611-1160	0.6964	0.5968	1.27	0.4228	0.4813
-910	0.8929	0.8871	0.01	0.1913	0.2003
-750	0.2857	0.2742	0.02	0.4082	0.3980
-650	0.5893	0.4032	4.08*	0.4841	0.4813
UBC621-**990**	0.6607	0.6129	0.29	0.4483	0.4745
-920	0.3929	0.4839	0.99	0.4770	0.4995
-790	0.8036	0.7581	0.35	0.3157	0.3668
-630	0.5179	0.4839	0.14	0.4994	0.4995
UBC635-1500	0.4286	0.4032	0.08	0.4898	0.4813
-1100	0.4286	0.4194	0.01	0.4898	0.4870
-750	0.1964	0.2742	0.98	0.3157	0.3980
-400	0.7143	0.6290	0.97	0.4082	0.4667
UBC636-1240	0.5179	0.5484	0.11	0.4994	0.4953
-760	0.8571	0.8871	0.24	0.2449	0.2003
Overall 97 loci					
Proportion of loci polymorphic	0.8247	0.8660			
Mean				0.3049	0.3053

13 Monomorphic loci

$*p < 0.05$, $**p < 0.01$, $***p < 0.001$, $cp < 0.10$, df = 1

[a] Loci in bold type were included in the 25-variable model of discriminant analysis; those in italics and bold type were included in the 12-variable model

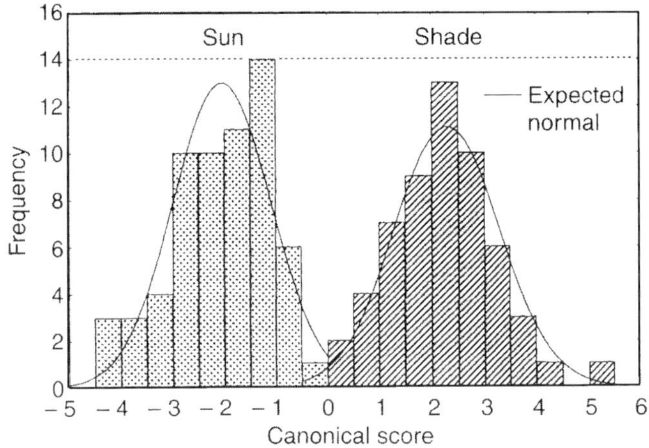

Fig. 5.4. The histogram of frequencies of canonical scores for wild emmer wheat at the shady and sunny niches in Yehudiyya according to 25 polymorphic RAPD loci. (Li et al. 1999)

These results suggest that RAPD discrimination succeeds in correctly characterizing the genotypes according to their neighboring microclimatic niches. *Sun-shade* discrimination was also successful with fewer loci. By choosing 12 of the 25 loci, the discriminant function was also significant ($F = 21.900$, $p < 0.00005$), and 112 (96%) individuals could be correctly assigned to their own niches, whereas only four (7%) plants from the *shade* were incorrectly assigned to the *sun*, and two (3%) individuals from the sun were incorrectly assigned to the *shade*. Even when the discriminant analysis was based on only the two most informative loci (OPA12 and UBC416), with a significant discriminant function ($F = 38.190$, $p < 0.0005$), 95 (80.5%) individuals also could be correctly classified into the two niches, and only eight plants (14.3%) from the *shade* and 15 plants (24.2%) from the *sun* were incorrectly classified into the alternative niche. These results suggest that largely *noncoding* RAPD differentiation identifies individuals characteristic for their respective niches, similar to allozymes.

5.2.2.2
RAPD Linkage Disequilibria

Coefficients (LD) of linkage disequilibrium among polymorphic loci were calculated for the *whole* population and for the *shady* and *sunny* microniches (Table 5.8) using Weir's method (Weir 1979). In the whole population, in the *shady* microniche, and in the *sunny* microniche, 52, 56, and 40 significant linkage disequilibria were found, respectively (Li et al. 1999). Interestingly, the 56 significant two-allele linkage disequilibria in the *shade* were *unique* for the *shady* microclimatic niche, but not for the *sunny* one, whereas the 40 pairs of

Table 5.8. The niche-specific RAPD linkage disequilibrium (LD) and normalized linkage disequilibrium (LD′) with a χ^2 test for wild emmer in shady and sunny niches at Yehudiyya. (Li et al. 1999)

$Locus_{allele}$	LD		LD′		χ^2	
	Shade	Sun	Shade	Sun	Shade	Sun
$OPA12_{720A}$-$OPA12_{250A}$	0.1180	0.0005	0.4774	0.0060	12.68**	0.00
$OPA12_{720A}$-$UBC285_{780A}$	−0.0628	0.0297	−0.093	0.3393	3.99*	4.00*
$OPA12_{500A}$-$UBC116_{670A}$	0.0523	−0.0109	0.6431	−0.0107	11.19**	0.85
$OPA12_{470A}$-$UBC116_{670A}$	0.0523	−0.0273	0.6431	−0.0320	11.19**	2.52
$OPA12_{370A}$-$UBC129_{820A}$	−0.0663	0.0468	−0.0346	0.4167	6.86**	5.10
$OPA12_{300A}$-$UBC116_{670A}$	0.0191	−0.0658	0.2353	−0.2129	1.50	13.47**
$OPA12_{300A}$-$UBC143_{650A}$	0.0651	−0.0918	0.3864	−0.2107	5.70*	11.26**
$OPA12_{250A}$-$UBC546_{1140A}$	−0.0612	0.0271	−0.0816	0.8667	4.14*	10.75**
$OPF2_{1430A}$-$OPF2_{1110A}$	−0.0313	0.0656	−0.1049	0.6545	2.16	10.93**
$UBC116_{890A}$-$UBC621_{990A}$	0.0214	−0.0926	0.0953	−0.0838	0.47	11.42**
$UBC116_{890A}$-$UBC621_{630A}$	0.0877	−0.0536	0.3623	−0.0383	7.13**	3.26
$UBC116_{670A}$-$UBC143_{1030A}$	0.0006	0.0747	0.0078	0.7455	0.00	15.41**
$UBC116_{670A}$-$UBC493_{830A}$	0.0424	−0.0109	0.5216	−0.0111	11.33**	0.85
$UBC129_{820A}$-$UBC636_{700A}$	−0.0867	0.0161	−0.1590	0.0680	10.22**	0.27
$UBC143_{1030A}$-$UBC143_{650A}$	−0.0746	0.0840	−0.0549	0.3751	5.29*	8.38**
$UBC144_{1050A}$-$UBC635_{400A}$	0.0561	−0.0957	0.2750	−0.0812	12.72**	4.09*
$UBC269_{680A}$-$UBC465_{770A}$	−0.0332	0.0583	−0.0575	0.3333	3.73*	6.57**
$UBC285_{530A}$-$UBC621_{790A}$	0.0631	−0.0486	0.4000	−0.0693	6.48**	3.21
$UBC285_{400A}$-$UBC636_{760A}$	0.0689	−0.0146	0.5625	−0.0148	17.72**	1.17
$UBC324_{1330A}$-$UBC465_{770A}$	0.0140	−0.0960	0.2115	−0.1805	1.05	12.95**
$UBC324_{800A}$-$UBC465_{770A}$	0.0102	−0.1126	0.1538	−0.1826	0.37	17.23**
$UBC324_{700A}$-$UBC635_{1100A}$	−0.0867	0.0323	−0.0473	0.1325	10.22**	1.06
$UBC416_{1150A}$-$UBC416_{790A}$	−0.0989	−0.0003	−0.0537	−0.0003	15.10**	0.00
$UBC416_{790A}$-$UBC416_{710A}$	−0.0115	0.0617	−0.0202	0.4969	0.18	10.36**
$UBC428_{580A}$-$UBC465_{770A}$	0.0536	−0.0224	0.8077	−0.0219	12.92**	0.97

$^*p < 0.05$, $^{**}p < 0.01$

linkage disequilibria in the sun were *unique* for the *sunny* microclimatic niche, but not for the *shady* one.

Significant genetic divergence was also found at multilocus structures between neighboring *shady* and *sunny* niches. However, in RAPD the multilocus organization was X(2) = 1.791 and 1.294 in *shade* and *sun*, respectively (Li et al. 1999), in contrast to the allozyme X(2) = 2.319 and 3.001 in *shade* and *sun*, respectively (Nevo et al. 1988a). These DNA polymorphisms appear to be significantly associated with *unique* microclimatic stresses. The *uniqueness* of linkage disequilibrium in the two microniches suggests natural selection effects on two-locus and multilocus DNA associations. The multilocus divergence structures, like the linkage disequilibria, result from natural selection, not by chance from population subdivision or founder effects (Li et al. 1999). These results are largely parallel to the previous allozymic results at the single-locus and multilocus levels (Nevo et al. 1988b). The DNA and allozymic results suggest that microclimatic selection plays an important role in DNA and protein polymorphism maintenance, i.e., on both *noncoding* and *coding* genomic regions.

5.2.3
Microsatellite DNA Differentiation

We also examined the ecological and genetic effects on dinucleotide microsatellite (short sequence repeats, SSR) variation of wild emmer wheat, *T. dicoccoides*, from the same two closely neighboring and contrasting *sunny* and *shady* microclimatic niches at Yehudiyya that were previously analyzed for allozymic (Nevo et al. 1988b) and RAPD diversity (Li et al. 1999). SSR diversity at 28 loci comprising seven types of tandem repeated dinucleotide motifs (Table 5.9) was analyzed in 105 individual plants of *T. dicoccoides* from the Yehudiyya microsite (Li et al. 2001b,c). The experimental area was less than 1,000 m^2 and involved the same 12 repetitive experimental trees distributed in a mosaic pattern, exactly as in the allozyme and RAPD tests. Each experiment involved very close (a few meters apart), but sharply divergent, microclimatic niches in the open park forest of Tabor oak: (1) *sunny*, between trees, and (2) *shady*, under the canopies of the trees. In total, SSR diversity was examined in 105 individual plants (50 from the *shady* and 55 from the *sunny* microniches, 4–5 individuals on the average in the *shade* and 4–5 plants from the *sun* around each tree) from 12 mosaically distributed tree-pairs derived from the 1985 collection of Nevo et al. (1988b).

Of the tested 28 SSR DNA markers, 27 (96.4%) produced fragments. Out of the total 28 loci, two (GWM601 and GWM415) were monomorphic in both microclimatic niches, and the remaining 26 loci (92.9%) were polymorphic. Significant microclimatic divergence characterized many loci displaying *asymmetric* and *nonrandom* distribution of repeat numbers. *Niche-specific* and *unique* alleles and linkage disequilibria abound in the two subpopulations. Microsatellite diversity at both the single- and two-locus levels is affected by microclimatic stress. Genetic factors, involving genome, chromosome, motif, and locus, significantly affected SSR diversity. Genome B had a larger average repeat number (ARN), but a lower variance in repeat number (σ^2) and a smaller number of alleles per locus than genome A. Microsatellites with compound motifs had larger ARNs than those with perfect motifs. The evidence reflects interaction between ecological stresses and genetic factors oriented by natural selection. In the following sections the detailed results are presented. See also the review of Li et al. (2001d).

5.2.3.1
The Distribution of Alleles at SSR Loci

One to 13 alleles were amplified per locus, but the number of alleles (NA) per locus was different at the 28 SSR loci between the two niches (Table 5.10). The

Table 5.9. Repeat types and chromosomal locations of microsatellites in the *Triticum aestivum* cultivar. (Li et al. 2001b)

Marker	Repeat type	Chromosomal	
		Location[a]	Distance[b]
GWM18	$(CA)_{17}GA(TA)_4$	1BS	5.7
GWM60	$(CA)_{30}$	7AS	52.3
GWM95	$(AC)_{16}$	2AS	10.8
GWM99	$(AC)_{21}$	1AL	95.3
GWM120	$(CT)_{11}(CA)_{18}$	2BL	23.7
GWM124	$(CT)_{27}(GT)_{18}$imp	1BL	82.2
GWM136	$(CT)_{58}$	1AS	38.8
GWM162	$(CA)_{14}AA(CA)_4$	3AL	91.0
GWM169	$(GA)_{23}$	6AL	49.6
GWM186	$(GA)_{26}$	5AL	38.4
GWM218	$(CT)_{18}$	3AS	30.4
GWM219	$(GA)_{35}$imp	6BL	40.5
GWM251	$(CA)_{28}$	4BL	28.9
GWM294	$(GA)_9TA(GA)_{15}$	2AL	16.1
GWM332	$(GA)_{36}$	7AL	53.0
GWM340	$(GA)_{26}$	3BL	131.4
GWM361	$(GA)_{20}$imp	6BS	11.4
GWM368	$(AT)_{25}$	4BS	8.7
GWM389	$(CT)_{14}(GT)_{16}$	3BS	118.4
GWM408	$(CA)_{>22}(TA)(CA)_7(TA)_9$	5BL	73.8
GWM415	$(GA)_{25}$imp	5AS	25.4
GWM429	$(CT)_{25}$	2BS	29.0
GWM459	$(GA)_{>28}$	6AS	62.5
GWM537	$(CA)_{18}(TA)_{13}$	7BS	43.6
GWM540	$(CT)_3(CC)(CT)_{16}$	5BS	9.3
GWM577	$(CA)_{14}(TA)_6$	7BL	105.0
GWM601	$(CT)_{17}$	4AS	8.1
GWM637	$(CA)_{18}$	4AL	50.1

imp Imperfect
[a] *A*, *B* genomes A and B; *S*, *L* short and long arm
[b] Distance from the centromere (*D*, in cM) estimated according to the maps in Röder et al. (1998); the distance of GWM218 also refers to the map of Peng et al. (2000b)

total NA was 172 and 179 in the *shady* and *sunny* subpopulations respectively (Li et al. 2001b). The shared NA was 159 in the two subpopulations, and allele distributions were different in the two subpopulations (Table 5.10; Fig. 5.5).

After exclusion of rare alleles, 26 *niche-specific* alleles were observed, 18 of them *shade-specific* and eight *sun-specific*. Among these *niche-specific* alleles, three *shady-unique* alleles were found, two of them under neighboring trees nos. 1 and 2, and another one under trees nos. 6 and 7. Two *sun-unique* alleles

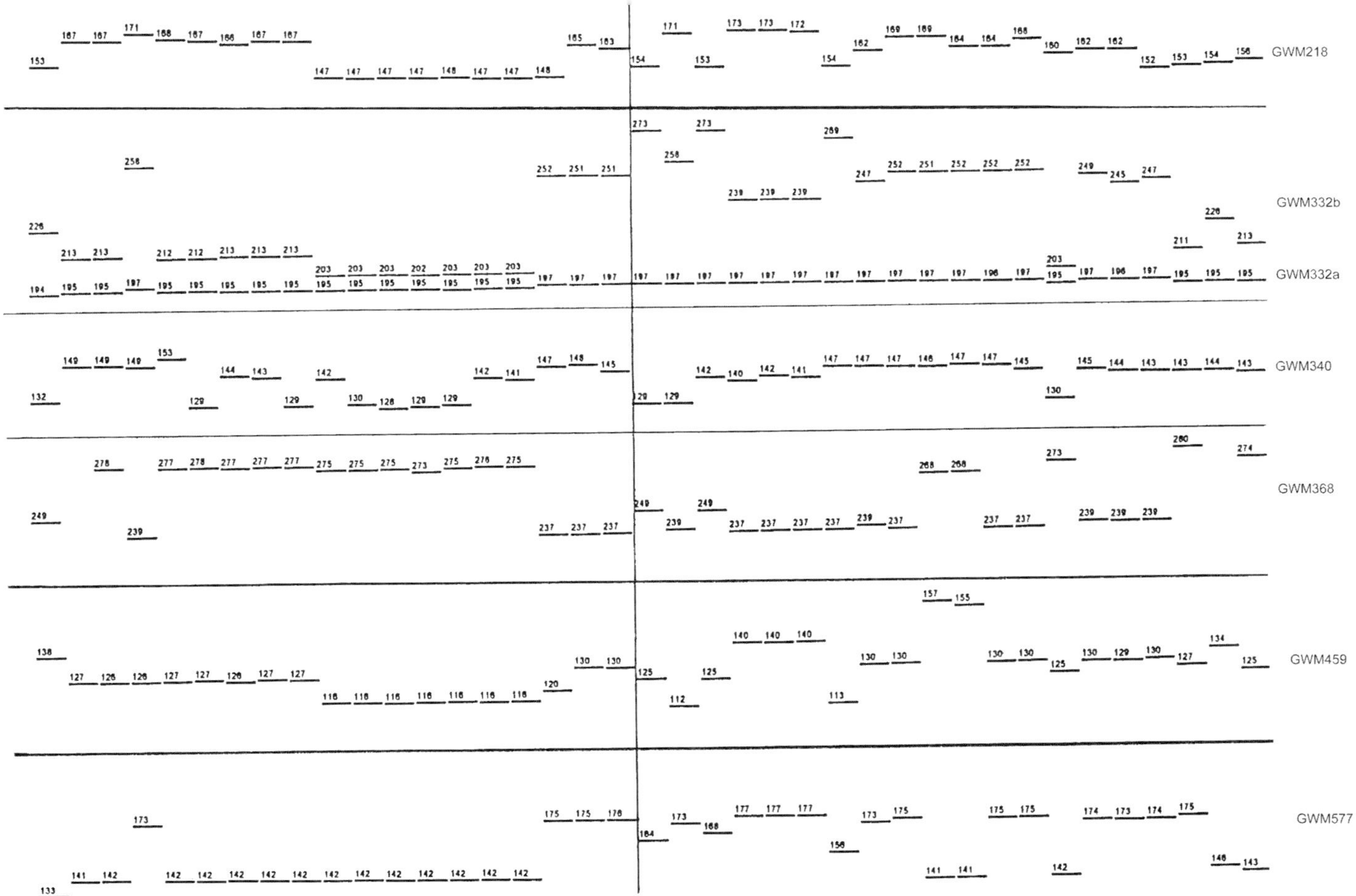

Fig. 5.5. SSR allele distribution in the two microclimatic shady and sunny subpopulations of *Triticum dicoccoides* at Yehudiyya microsite. Allele sizes were labeled by base pairs, including repeated and flanking region. The first *lane on the left* is for the *Triticum aestivum* cultivar Chinese Spring. (Li 2000; Li et al. 2001b)

Table 5.10. Statistics of allele distributions at 27 microsatellite loci in the *shady* and *sunny* subpopulations of *T. dicoccoides* at Yehudiyya, Israel. (Li et al. 2001b)

Locus	Number of alleles[1]			ARN[†]		Sign test z (n)[2]	Variance			Skewness		H_e	
	Shade	Sun	Total	Shade	Sun		Shade	Sun	$F_{(df1,df2)}$	Shade	Sun	Shade	Sun
GWM18	0	1	5	37.3	36.9	1.33 (9)	1.69	2.74	1.62*	–0.14	0.60	0.665	0.737
GWM60	0	0	5	25.2	23.9	2.27* (7)	22.22	20.16	1.10	2.41****	2.27****	0.618	0.630
GWM95	0	1	6	16.9	18.4	–0.00 (11)	2.87	2.07	1.39	–0.07	–0.16	0.748	0.782
GWM99	0	0	5	10.6	11.3	0.00 (5)	22.07	43.19	1.96*	2.40****	1.17***	0.578	0.679
GWM120	0	0	6	33.4	33.2	0.89 (5)	2.12	2.52	1.19	0.93	0.41	0.664	0.732
GWM124	1	2	9	53.4	53.1	1.21 (11)	19.41	34.19	1.76*	0.32	0.37	0.812	0.843
GWM136	3	2	8	68.6	71.1	–	72.15	70.55	1.02	–0.11	–0.44	0.798	0.748
GWM162	0	2	4	23.0	21.5	2.04* (6)	3.80	1.75	2.17**	0.57	2.18****	0.465	0.409
GWM169	1	1	9	20.4	20.1	1.23 (6)	10.81	10.24	1.06	1.24**	1.34****	0.806	0.793
GWM186	0	0	6	23.0	24.2	1.79c (5)	12.75	11.62	1.09	0.55	–0.62	0.664	0.726
GWM218	1	1	9	24.2	23.7	0.35 (8)	20.10	20.52	1.02	–0.21	–0.22	0.818	0.779
GWM219	1	0	8	31.0	23.6	2.67** (9)	83.81	24.48	3.42***	0.53	2.28****	0.799	0.667
GWM251	0	1	6	26.1	25.8	0.76 (7)	2.94	3.54	1.20	0.96	1.19	0.595	0.727
GWM294	1	1	13	24.4	23.2	0.60 (11)	9.56	18.45	1.93*	0.07	0.17	0.862	0.887
GWM332a	0	0	3	18.4	18.2	–0.00 (5)	0.27	0.55	2.03**	0.29	–0.42	0.441	0.632
GWM332b	1	3	13	39.9	39.8	–0.32 (10)	127.01	97.62	1.30	0.60	0.02	0.858	0.889
GWM340	0	0	6	33.8	32.7	2.67** (9)	11.41	8.22	1.39	–1.42	–1.44	0.819	0.806
GWM361	1	0	5	23.0	22.0	2.04* (6)	2.78	0.56	4.96***	1.89***	1.45***	0.550	0.393
GWM368	2	0	11	31.7	23.9	2.67** (9)	65.52	41.23	1.59	–0.01	1.37*	0.876	0.804
GWM389	0	0	6	33.5	32.8	1.23 (6)	5.50	5.07	1.08	–2.60	–3.10*	0.763	0.698
GWM408	0	1	7	27.1	27.9	–0.35 (8)	5.91	10.18	1.72*	1.29***	0.08	0.692	0.793
GWM415	0	0	1	19.0	19.0	–	0	0	–	–	–	0	0
GWM429	0	1	5	13.2	12.8	–0.00 (3)	3.12	2.74	1.14	3.89***	4.77****	0.529	0.416
GWM459	1	1	11	25.4	25.1	0.32 (10)	23.77	21.57	1.10	0.65	0.23	0.861	0.837
GWM537	0	0	6	37.5	37.4	1.51 (7)	33.04	8.12	4.07***	–1.38**	–2.31***	0.690	0.750
GWM540	0	1	8	18.4	19.0	0.41 (6)	4.74	5.15	1.09	1.50***	1.07*	0.730	0.782
GWM577	0	1	10	33.1	33.9	1.77c (8)	55.48	62.29	1.12	0.68	–0.04	0.772	0.853
GWM601	0	0	1	17.0	17.0	–	0	0	–	–	–	0	0
Overall	13	20	192									0.655	0.688

[†] *ARN* average repeat number, [1] *Shade* alleles observed only in the shady samples, *Sun* alleles observed only in the sunny samples
[2] Sign test was carried out for the pairs of ARN in the shade and sun niche across the 12 trees, *n* number of non-ties. The ARN difference was not tested for locus GWM136 due to missing data

Table 5.11. Summary of permutation test for niche specificity of microsatellite alleles (excluding rare alleles, observed in < 5 individuals) at microsatellite loci of *T. dicoccoides* in the sunny and shady microclimatic niches in Yehudiyya, Israel. (Li et al. 2001b)

Number of niche-specific alleles				Significance of niche specificity[b]
Significance level (χ^2 test)[a]	Shade	Sun	Total	
0.01	5	1	6	<0.0002
0.05	13	7	20	<0.0002
Total	18	8	26	

[a] Test for homogeneity of allele frequencies between the shady and sunny niches
[b] Significance was obtained in permutation test of 5,000 random shuffling sample sets

were also found under trees no. 1, 2, and 3, and another one under trees no. 11 and 12. The results of permutation tests (5,000 randomly shuffled sets) (Table 5.11) showed that the significance of the observed 26 *niche-specific* alleles was at least $p = 0.0002$. This result suggests with high confidence the *nonrandomness* of the revealed subpopulation divergence. Moreover, *niche-unique* alleles under neighboring trees reflect migration among close trees, but, without strong microclimatic selection, this *niche-uniqueness* of some alleles cannot be maintained.

Gene diversity, *He* (Nei 1973), based on allele frequencies, also evidenced obvious differences between the two climatically divergent subpopulations at some loci. The differences in *He* were as high as 0.10–0.19 at some loci, such as GWM99, GWM219, GWM251, GWM332a, GWM361, GWM408, and GWM429. These results indicated that allele distributions appeared to be *non-random* in the two subpopulations, only 2–4 m apart but associated with contrasting *sun-shade* climatic microniches. Figure 5.5 illustrates some typical examples of such allele distribution.

In order to test how much the distributions of repeat number (allele size) deviated from symmetry in the two microclimatic niches, the skewness was calculated for each locus. The allele size distributions significantly deviated from symmetry (skewness = 0) at 12 loci ($p <$ 0.05–0.00001). These results indicated that the allele size distributions at these loci were not symmetrical (skewness ≠ 0). In the *shady* and *sunny* subpopulations, allele distributions were significantly skewed to the right (skewness > 0) at seven and nine loci, and to the left (skewness < 0) at one and two loci, respectively ($p <$ 0.05–0.00001). This pattern indicated that the frequencies of alleles with relatively short repeats were significantly higher than those with long repeats (Fig. 5.6). This type of distribution does not meet a null hypothesis of pure random drift. The opposite direction of skewness was also observed at four loci between the *shady* and *sunny* subpopulations. See discussion in Li et al. (2001d).

At many loci, allele distributions seemed to be clustered at different allele sizes. At loci GWM415 and GWM601, alleles were limited in repeat numbers

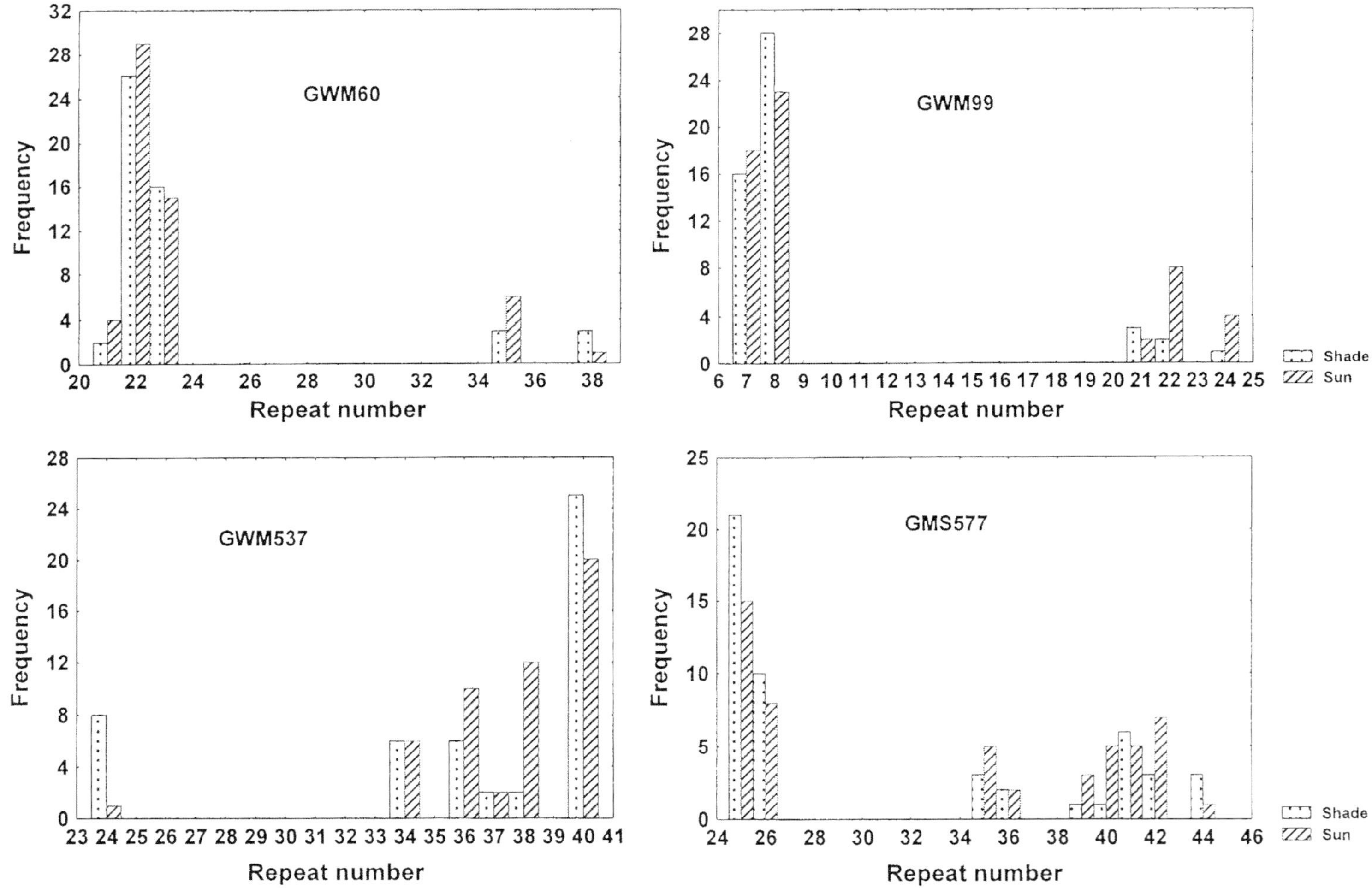

Fig. 5.6. SSR allele distributions with clusters at loci GWM60, GWM99, GWM537 and GWM577 in *shady* and *sunny* subpopulations of *T. dicoccoides* at the

19 and 17, respectively. Allele sizes at GWM95, GWM120, and GWM332a varied in the ranges of 15–20, 31–36, and 17–18 repeats, respectively, with different frequencies in the *shady* and *sunny* subpopulations. At GWM99 alleles were distributed in two clusters with a large gap of 12 repeats of (AC), one with small (7–8), the other with larger (21–24) repeat numbers. Similar patterns were observed at loci GWM60, GWM537, and GWM577 (Fig. 5.6). These results suggest that some constraints may exist on the repeat number at SSR loci. The last assumption is especially supported by the allele distribution at GWM537 (Fig. 5.6).

A sign test over each pair of *shady* and *sunny* niches around all 12 trees showed that the ARNs were significantly larger at nine loci in the *shade* niche than in the *sun* niche ($p < 0.05$–0.01). This suggests that microniche differences indeed existed across the 12 trees. Significantly larger variances in repeat number were found in the *sunny* than in the *shady* subpopulation at six loci (GWM18, GWM99, GWM294, GWM124, GWM332a, and GWM408), and the opposite pattern was observed at four other loci (GWM162, GWM219, GWM361, GWM537). These results also indicated significant differences in allele (repeat number) distributions between the two very close (a few meters apart) microclimatic subpopulations.

Since recombination is suppressed around the centromeres of chromosomes (Gill et al. 1996a,b), the distance of SSR loci from the centromeres (*D*) may affect the level of SSR diversity (*He*) and variance in repeat size. In this study, significant and positive correlation was observed between the *D* and variance in repeat size (Spearman's rank order correlation $r_s = 0.393$, $N = 54$, $p < 0.005$). The correlation between *D* and *He* was also positive and nearly significant ($r_s = 0.259$, $N = 54$, $p = 0.058$).

5.2.3.2
SSR Genetic Distance

The estimate of genetic distance (Nei 1972) was large, $D = 0.1057$, between the *shady* and *sunny* subpopulations (Li et al. 2001b). The results of a permutation test (1,000 randomly shuffled sets) showed that the significance of the observed genetic distance ($D = 0.1057$) was at least $p < 0.001$. This permutation test indicates that genetic divergence at the largely *noncoding* SSR loci is substantial between the neighboring *shady* and *sunny* subpopulations, only a few meters apart, as was also shown for largely *noncoding* RAPD loci and *coding* DNA for allozymes.

5.2.3.3
SSR Discriminant Analysis

Forward stepwise discriminant analysis was carried out in order to distinguish individuals from the two microclimatic niches according to multilocus analysis, including the 25 polymorphic loci (Li et al. 2001b). The result indicated that five out of the 25 SSR loci were sufficient to correctly classify 82%

of 105 individuals into their original microclimatic *sunny* and *shady* niches. Even with only two (GWM169, GWM332a) of the 25 loci, 70% of 105 individuals could be correctly assigned to their original microniches. These results clearly suggest that the observed genetic differentiation at SSR loci is substantial and microclimatic-*niche-associated*, i.e., ecologically determined by selection.

5.2.3.4
Linkage Disequilibrium (LD) Between SSR Loci

At the microclimatic *sun-shade* at Yehudiyya, 16,847, 13,530, and 14,541 possible pairwise associations were found in the total population, the *shady* subpopulation, and *sunny* subpopulation, respectively (Li et al. 2001b). Of these, 18.0, 21.6, and 14.2% were significant (χ^2 test: $p < 0.05$) linkage disequilibria (LDs) in the total population, the *shady* subpopulations, and the *sunny* subpopulation, respectively. Since rare alleles may cause biased estimation of LD (Lewontin 1997), further analyses were performed (Table 5.12) excluding the rare alleles (observed in ≤5 individuals in the *whole* population). In the *shady* niche, 86, 90, and 94% of 1,089, 594, and 267 LDs at $p < 0.05$, 0.01, and 0.001 significant levels, respectively, were *niche-specific*. Permutation tests, based on 5,000 randomly shuffled sets of the initial data, showed a high degree of significance of the LD *shade specificity* ($p < 0.002$–0.006) (Table 5.12). In the *sunny* subpopulation, after exclusion of rare alleles, 608, 296, and 131 LDs were also found to be significant at the levels of 0.05, 0.01, and 0.001, respectively. Among these significant linkage disequilibria, 74–88% were *sun-specific*. However, the permutation test did not show significance of this LD *sun specificity*. These results suggest that the possibility of observing *shade-specific* LDs

Table 5.12. Permutation test for niche specificity[a] of linkage disequilibria at microsatellite loci of *T. dicoccoides* in the sunny and shady niches in Yehudiyya, Israel. (Li et al. 2001b)

Niche	Significance level (χ^2 test)	Total LDs	Proportion of niche-specific LDs		No. cases $P_e \geq P_o$	Significance of niche specificity
			In samples (P_o)	In 5,000 permutations (P_e)		
Without rare alleles[b] ($\delta \leq 7$)						
Shade	0.001	267	0.94	0.52–0.96	29	<0.006
	0.01	594	0.90	0.49–0.93	19	<0.004
	0.05	1089	0.86	0.54–0.87	9	<0.002
Sun	0.001	131	0.88	0.63–0.98	3089	ns
	0.01	296	0.79	0.60–0.95	3752	ns
	0.05	608	0.74	0.57–0.90	3860	ns

ns Not significant
[a] Permutation test was carried out by 5,000 random shuffling samples
[b] Alleles observed in ≤5 individuals in the whole population were regarded as rare

Table 5.13. *F* ratios of analysis of variance for genetic effects on number of alleles (NA) per locus, average repeat number (ARN), and variance in repeat number of microsatellites of *T. dicoccoides* at Yehudiyya, Israel. (Li et al. 2001c)

Genetic factor[a]	*F* ratio			
	(df_1, df_2)	NA	ARN	Variance
Genome	1, 40	3.865@	4.987*	0.474
Chromosome	6, 40	1.886	3.848**	2.716*
Genome × chromosome	6, 40	5.896***	0.470	3.059*
Motif	6, 47	1. 473	2.877**	1.059
Locus	26, 27	37. 717****	114.132****	7.957****

*$p < 0.05$, **$p < 0.01$, ***$p < 0.0005$, ****$p < 0.00005$, @ $p < 0.10$

[a] Interaction of some genetic factors could not be estimated due to incomplete design

by chance was very low; the *shade-specific* stress may be responsible for the LD interniche specificity.

The log-linear model test showed that niche effect was very significant ($\chi^2_{df=1} = 34.37$, $p < 0.000005$), whereas genome effect was negligible ($\chi^2_{df=1} = 0.25$, $p > 0.50$). Linkage itself between loci within chromosomes also seemed not to affect percentages of significant LDs in both subpopulations, possibly reflecting the higher importance of selfing than linkage.

5.2.3.5
Genetic Effect on SSR Variation

Analysis of variance (ANOVA) was used to test the effects of genetic factors (Table 5.13). The results suggested that genome (A vs B) significantly ($p < 0.05$) affected the ARN, with genome B showing a larger ARN (30.2) than genome A (25.1) (Li et al. 2001c). Genome A showed a larger variance in repeat number ($\sigma^2 = 27.8$) than genome B ($\sigma^2 = 19.0$). The SSR ARN on chromosome 1 (including both 1A and 1B) were significantly ($F_{(6,40)} = 3.85$, $p < 0.01$) larger (ARN = 42.5) than those of other chromosomes (range: 22–31). Microsatellites on chromosome 7 showed the highest variance ($\sigma^2 = 53.5$) in repeat number ($F_{(6,40)} = 2.72$, $p < 0.05$) compared with those on other chromosomes. Compound SSRs, such as $(CA)_n(TA)_k$, $(CT)_n(GT)_k$, $(CT)_n(CA)_k$, had significantly ($F_{(6,47)} = 2.88$, $p < 0.01$) larger ARN (>33) than simple or perfect SSRs (ARN = 24–29), such as $(GA)_n$ and $(AT)_n$. The SSRs with $(CA)_n$ or $(GA)_n$ showed the smallest ARN (20). A locus effect was highly significant ($p < 0.00005$, Table 5.13) for NA, ARN and σ^2. Locus GWM294 showed the largest NA (13), whereas GWM415 and GWM601 each presented only one allele. Locus GWM136 had the largest ARN (70.0) and GWM99 the smallest (10.0). Loci GWM332a and GWM332b showed the smallest (0.27–0.55) and largest σ^2 (97.6–127.0), respectively.

5.2.3.6 Mutational Mechanisms for Producing SSR Variation

Replication slippage (Levinson and Gutman 1987; Wolff et al. 1991; Amos and Rubinstein 1996; Innan et al. 1997; Dermitzakis et al. 1998; Amos 1999) and unequal crossing over (Harding et al. 1992) were regarded as mechanisms for generating new alleles or diversity at SSR loci. If so, a longer repeat would tend to have larger diversity since the chance of replication errors is higher for a longer sequence (Levinson and Gutman 1987; Wolff et al. 1991). Likewise, microsatellite loci located farther from the centromeres should have larger variation, since recombination is suppressed around the centromeres of chromosomes (Gill et al. 1996a,b). In other words, the ARN and locus distance from the centromere (D) may affect the NA and σ^2 at SSR loci. Forward multiple regression was used to estimate contributions of ARN, D, and microniches to the A and σ^2 of different motifs (Table 5.14). The results indicated that ARN was the most important factor for $(GA)_n$ loci. Mutational mechanisms displayed indirectly by ARN, D and $D \times$ ARN could have significantly affected the σ^2 of $(GA)_n$ loci ($R^2 = 0.827$, $p < 0.00005$); and niche could also directly ($r = 0.466$, $p < 0.05$) or indirectly through the interaction niche $\times$ ARN ($r = -0.538$, $p < 0.05$) affect the σ^2 of $(GA)_n$ loci. Both the mutational mechanisms and ecological effects could have significantly determined the σ^2 of $(GA)_n$ loci ($R^2 = 0.889$, $p < 0.00005$). In $(CA)_n$ SSRs, $D \times$ ARN and microniche were highly responsible for the NA ($R^2 = 0.770$, $p < 0.01$). In $(CT)_n$ SSRs, mutational mechanisms (ARN, D and $D \times$ ARN) could have significantly regulated the σ^2 in repeat number ($R^2 = 0.997$, $p < 0.00005$); and niche $\times$ ARN also significantly explained the σ^2 in the repeat number ($r = -0.969$, $p < 0.0005$). In $(CA)_n(TA)_k$ SSRs, ARN $\times$ D and niche $\times$ ARN affected NA ($R^2 = 0.950$, $p <$ 0.0005), ARN $\times$ D was significant for σ^2 ($R^2 = 0.703$, $p < 0.05$). These results implied that the importance of mutational mechanisms and other factors for producing and maintaining population variation varied among different SSR motifs. The results also suggested that, apart from internal mechanisms, microniche ecological stress may directly or indirectly affect the origin and maintenance of SSR diversity. The dependence of SSR diversity may also reflect the effects of background selection or selective sweeping (see elaboration in Li et al. 2001d).

5.3 Edaphic, Topographical, and Temporal Factors Affecting Allozyme and DNA Differentiation of Wild Emmer Wheat at the Tabigha Microsite, Northern Israel

Is soil divergence at a microscale as selective as microclimatic selection? We chose a suitable microsite to seek an answer to this question, namely, at Tabigha (=Tabgha), northern Israel (Fig. 5.7), where we examined both protein and DNA divergence.

Table 5.14. Multiple regression of genetic factors, microniche and their interactions as independent variables with number of alleles per locus (NA) and variance (σ^2) in repeat number in the Yehudiyya population of *T. dicoccoides* as dependent variables. (Li et al. 2001c)

Motif	No. of cases	Step	Enter variable[a]	NA		Step	Enter variable	σ^2 in repeat number	
				r^b	R^{2c}			r^b	R^2
$(GA)_n$	22	1	ARN	0.427*	0.182*	1	ARN	0.862****	0.602****
		2				2	$D \times$ ARN	−0.738***	0.723****
		3				3	D	0.696**	0.827****
		4				4	Niche × ARN	−0.538*	0.858****
		5				5	Niche	0.466*	0.889****
$(CA)_n$	10	1	$D \times$ ARN	0.856**	0.621**	1	–[d]	–	–
		2	Niche	0.628*	0.770**				
$(CT)_n$	10	1	–	–	–	1	$D \times$ ARN	−0.989****	0.987****
		2				2	ARN	0.996****	0.989****
		3				3	D	0.994****	0.997****
		4				4	Niche × ARN	−0.969****	0.999****
$(CA)_n(TA)_k$	8	1	$D \times$ ARN	0.975***	0.924***	1	$D \times$ ARN	0.838*	0.703*
		2	Niche × ARN	0.651*	0.950***				

*$p < 0.05$, **$p < 0.01$, ***$p < 0.0005$, ****$p < 0.00005$. b – r Partial correlation coefficient at the last step with significance of t test; c – R^2 coefficient of multiple determination with significance of F test for multiple correlation

a – Variables: ARN (average repeat number), D (locus distance from the centromere in cM), niche-*sunny* or *shady*

d – No significant factors entered the multiple regression

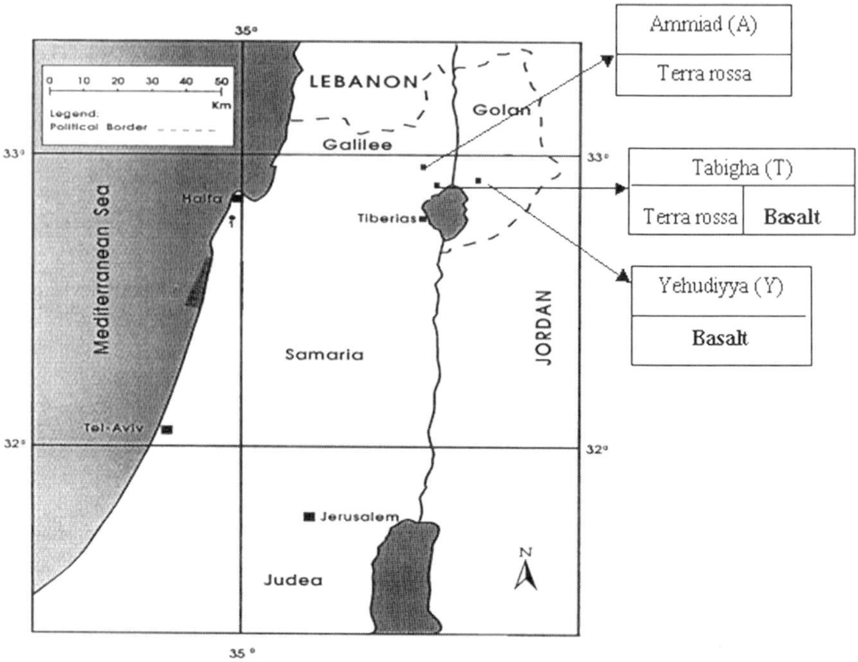

Fig. 5.7. The three microsites of wild emmer wheat located north of the Sea of Galilee that were studied for protein and DNA diversities across divergent microclimates, soils, and topographies. (Li et al. 2000b)

The studied microgeographic site at Tabigha, representing microscale edaphic contrasts, is north of the Sea of Galilee, at Mediterranean sea level (Nevo et al. 1988c). The red terra rossa sharply bordering the brown basalt soils has weathered from Middle Eocene hard limestone and Upper Pleistocene basalt flows, respectively (Atlas of Israel 1970). The brown basalt soil is subject to extremes of high and low water content due to its montmorillonitic clays. Hence, its water regime is more variable in terms of water content than that of the kaolinitic terra rossa soil, thereby representing a wider ecological niche than terra rossa. The terra rossa soil layer was also shallower and numerous bare rocks predominated over the terrain with pockets of soil varying in depth. The topography of the terra rossa is more hilly than that of the flat, deeper soil of the southwestern dipping basalt. The terra rossa soil niche clearly appeared drier, narrower, and more stressful than the basalt in terms of availability of water and moisture throughout the growing season (November to May). The differences were directly reflected in the different plant formations on the two soil types, which displayed long-term biotic differentiation (see Nevo et al. 1988c). On terra rossa, the vegetation, including wild emmer wheat, was sparser, lower in height, and ripened and dried earlier than that on the basalt. The physiological differences of wild barley on the contrasting soil types clearly express the dramatic differences of wild emmer growing on these two soil types (Ivandic et al. 2001).

Sampling for this study was conducted in 1984 and 1985 in two transects, 200 m apart (Nevo et al. 1988c). Each transect was 100 m long and equally subdivided into 50 m of red terra rossa and 50 m of brown basalt separated by a sharp geological boundary. Seeds were collected from 50 plants, about 1 m apart in each half of a transect.

5.3.1 Edaphic Allozyme Divergence

Allozymic variation in proteins encoded by 47 loci was analyzed electrophoretically in 1983–1984 and 1984–1985 in 356 individual plants of wild emmer wheat, *T. dicoccoides*, from a microsite at Tabigha, north of the Sea of Galilee, Israel (Nevo et al. 1988c). Each year the test involved two 100-m transects, each equally subdivided into basalt and terra rossa soil types. Comparisons were based on 16 common polymorphic loci. Significant genetic differentiation (Tables 5.15–5.17), genetic linkage disequilibria (Table 5.18), and genome organization (Table 5.19) according to soil type were found over very short distances (Figs. 5.8, 5.9). Our results suggest that allozyme polymorphisms in wild emmer wheat are partly adaptive. They differentiate, at both single and multilocus structures, primarily as a result of environmental heterogeneity and ecological stress factors. These include abiotic (edaphic and climatic, primarily aridity) and biotic (pathogens, parasites) factors.

5.3.1.1 Distribution of Polymorphism Among Loci

Of the 47 loci, 23 proved to be totally monomorphic *(Aat-2*B, *3*A, *3*B*; Adh-1*B, *2*A, 2B; *Cat-A,* B; *Est-4*A, *5*A, *Gdh*-A, B; *Mdh-l*B, *2; Nadh-l*A, *1*B, *2; Pepc; Pept-2; Pgi*-B, *Pgm*-B*; 6Pgd-l*A, *1*B), six nearly monomorphic *(Aat-1*A, 1B; *Est-4*B*; Ipor*-A, B; *Pept-1A),* and two polymorphic *(Gluc*-A, B) whose variants were distributed in a single patch, hence excluded from Table 5.15. The remaining 16 polymorphic loci (34%) appear in Table 5.15.

5.3.1.2 Distribution of Alleles

Several loci varied significantly in allele frequency between soil types, transects, and years (nine loci, out of 16, significant χ^2 tests for soils appear in Table 5.15). Figure 5.8 illustrates the spatial distribution of several polymorphic loci that differed in the two soil types: *Pgm*-A, *Pgi*-A, *Ipol, Hk, Pept-3,* and *Est-5*B. These loci all displayed a *nonrandom* soil distribution of genotypes across transects and years.

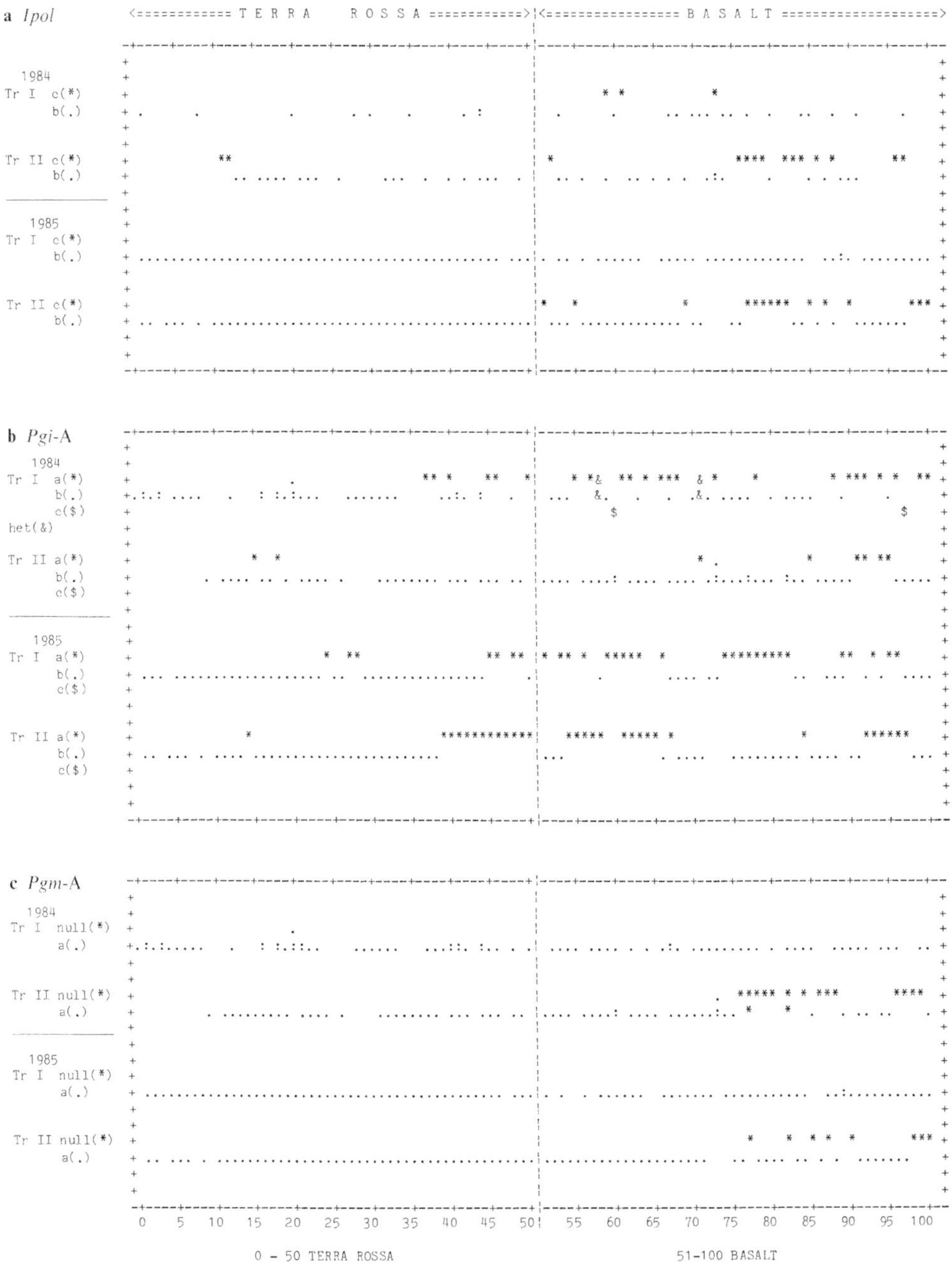

Fig. 5.8a–f. Allele distribution of selected allozyme loci along the 100-m, two-soil-type transects at Tabigha, Israel. **a** *Ipo*l, **b** *Pgi*-A, **c** *Pgm*-A, **d** *Hk*, **e** *Pep*t-3, **f** *Est*-5B. (Nevo et al. 1988c)

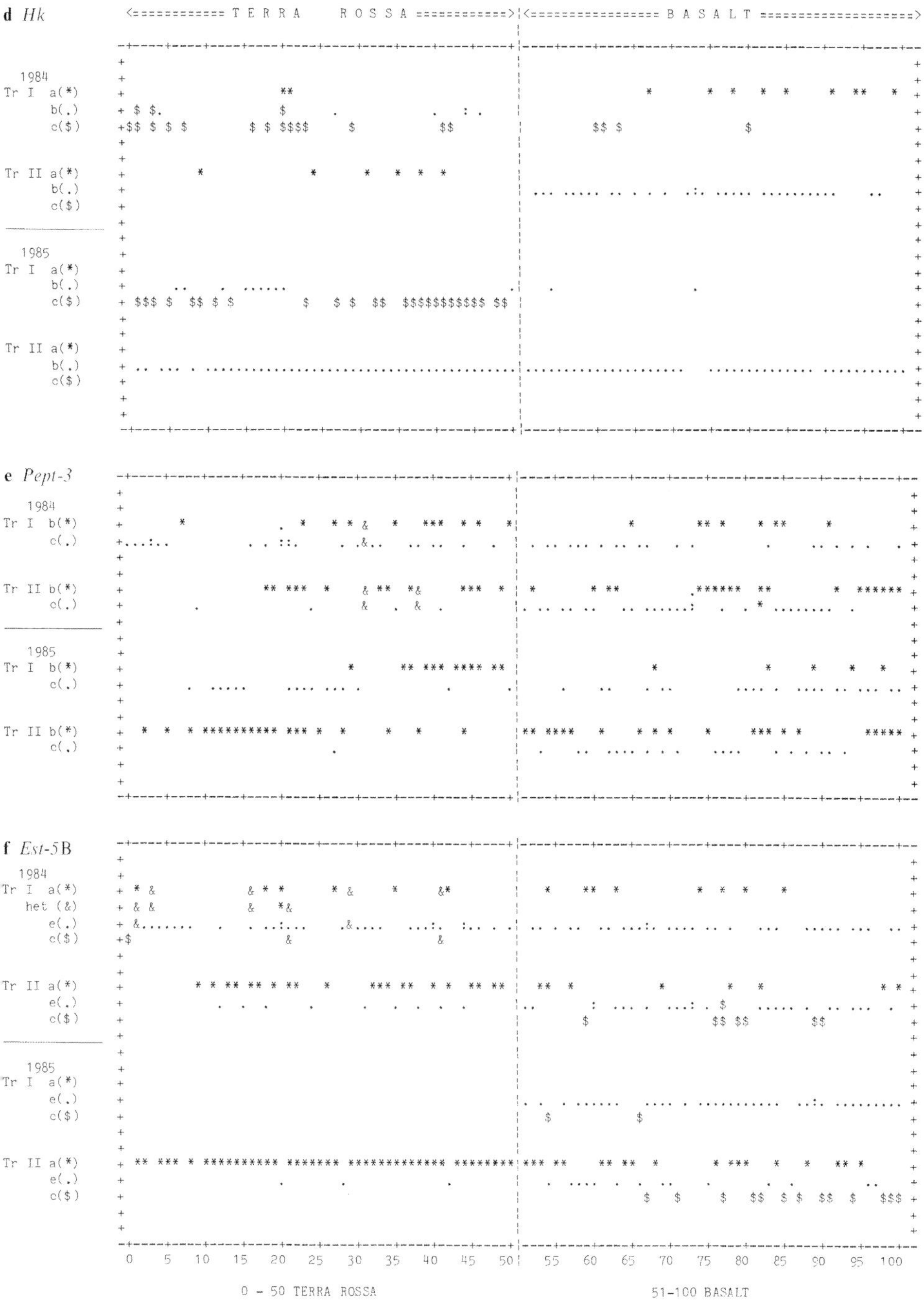

Fig. 5.8a–f. *Continued*

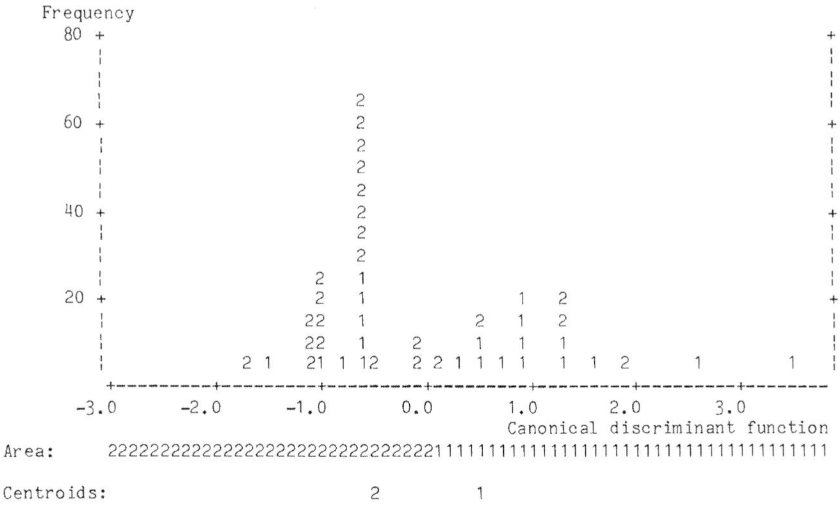

Fig. 5.9. Discriminant analysis of wild emmer wheat in Tabigha, Israel, according to soil type, based on seven polymorphic allozyme alleles. *1* Basalt, *2* terra rossa. (Nevo et al. 1988c)

5.3.1.3
Association with Soil Type

Several allozymes displayed consistent specific association (i.e., higher frequency) with soil type (Tables 5.15 here and 4 in Nevo et al. 1988c). Six alleles showed associations with basalt, and three alleles with terra rossa. Eight of these nine associations were statistically significant as indicated by a χ^2 test. Spearman rank correlations (r_s) were calculated between the presence of an allele and a dummy variable where basalt = 1, and terra rossa = 0. Likewise, estimates of r_s were also calculated with the position of the plant along the transect (first and third columns of statistical analyses in Table 5.15). The mean position (±SE) of each allele across the four transects is between the two r_s values in Table 5.15 (Nevo et al. 1988c). The analysis with position was sensitive not only to allele frequency for the soil types, but also to the distance from the contact zone between the two soil types.

All previous analyses showed five alleles (*Ipo*1^c, *Aat*-$2A^a$ *Est*-$5B^e$ *Pgi*-A^a, and *Pgm*-A^{null}) to be associated (i.e., more frequently) with basalt and two alleles (*Est*-$5B^a$ and *Hk*^{e}) with terra rossa. *Pept*-3^b also showed a consistent and significant association with terra rossa, but the analysis with position resulted in contradictory results (for the reason, see Fig. 5.8e). Therefore, the association of this allele with terra rossa needs confirmation.

Table 5.15. Allele frequencies at 16 allozyme polymorphic loci of eight subpopulations of *T. dicoccoides* (two soil types in two transects in 2 years in Tabigha, Israel. (Nevo et al. 1988c)

Year	($) + –	Complementary allele	1983/1984				1984/1985				Mean 356	Statistical analysis					
Transect			I		II		I		II			Spearman rank correlation with basalt[a]	Average position		N	Spearman rank correlation with position	χ^2 between soil types (1 df)
Soil type			TR	BA	TR	BA	TR	BA	TR	BA			$\bar{X}$	SE			
Sample size			47	42	31	50	49	44	47	46							
Locus	allele		Allele frequencies														
Aat-2A	a+++	(b)	0.087	0.205	0.0	0.120	0.0	0.0	0.149	0.239	0.103	0.129*	64	3.782	35	0.146**	4.989*
He			0.159	0.326	0.0	0.211	0.0	0.0	0.254	0.364	0.164						
N			46	39	31	50	49	41	47	46	349						
Acph-x	b–	(a)	0.568	0.111[b]	A	1.000	–	0.464	0.222	–	0.483	0.242**	54	3.087	69	0.279***	7.450**
He			0.491	0.198	A	0.0	–	0.497	0.346	–	0.306						
N			37	9		24		28	45		143						
Acph-3	a–	(b)	0.500	0.278	0.0	0.0	0.0	0.0	0.0	0.0	0.085	–0.112[c]	40	4.039	24	–0.111[c]	3.162[c]
	c++–+		0.0	0.185	0.032	0.036	0.526	0.286	0.0	0.022	0.136	–0.009 ns	50	4.412	40	–0.012 ns	0.0 ns
He			0.500	0.600	0.062	0.069	0.499	0.408	0.0	0.043	0.273						
N			35	27	31	28	38	42	47	46	294						
*Adh-1*A	a++	(b)	0.842	0.870	A	0.286[b]	0.051	0.571	0.0	0.0	0.262	0.133[c]	56	3.724	48	0.081 ns	2.653 ns
He			0.266	0.227	–	0.408	0.097	0.490	0.0	0.0	0.213						
N			19	23		21	39	7	31	46	183						
Est-1	b+–		0.250	0.818	1.000	0.938	–	–	–	–	0.896	0.050 ns	59	3.391	60	0.101 ns	0.0 ns
	c–		0.750	0.182	0.0	0.0	–	–	–	–	0.075	–0.143 ns	41	14.006	5	–0.165 ns	0.473 ns
(e = null)	e+		0.0	0.0	0.0	0.063	–	–	–	–	0.030	0.131 ns	71	11.000	2	0.073 ns	0.105 ns
He			0.375	0.298	0.0	0.117	–	–	–	–	0.197						
N			4	11	20	32					67						
*Est-2*A	a+	(b)	0.220	0.380	–	0.294[b]	–	–	–	–	0.283	0.146 ns	49	6.732	23	–0.002 ns	0.967 ns
He			0.343	0.471	–	0.415	–	–	–	–	0.410						
N			41	25		17					83						
*Est-2*B	a–		0.183	0.080	–	0.063[b]	–	–	–	–	0.128	–0.176 ns	37	11.780	10	–0.150 ns	1.113 ns
(a = null)	b+		0.0	0.120	–	0.0[b]	–	–	–	–	0.037	0.195[c]	78	11.724	3	0.203[c]	1.384 ns
	c–		0.549	0.520	–	0.500[b]	–	–	–	–	0.530	–0.036 ns	54	4.022	43	0.164 ns	0.014 ns
	d+		0.268	0.280	–	0.438[b]	–	–	–	–	0.305	0.080 ns	43	5.055	25	–0.153 ns	0.230 ns
He			0.593	0.630	–	0.555	–	–	–	–	0.593						
N			41	25		16					82						
*Est-5*B	a–––		0.202	0.195	0.700	0.190	–	0.0[b]	0.936	0.413	0.372	–0.338***	41	2.440	107	–0.330***	26.760***
(a = null)	c–++		0.043	0.0	0.0	0.190	–	0.049[b]	0.0	0.283	0.085	0.241***	80	4.387	22	0.282***	16.856***
	e+++		0.755	0.805	0.300	0.619	–	0.951[b]	0.064	0.304	0.543	0.184**	57	2.513	118	0.153*	8.011**
He			0.387	0.314	0.420	0.544	–	0.093	0.120	0.657	0.362						
N			47	41	30	42		41	47	46	294						
Hk	a+–		0.077	0.692	1.000	0.0	0.0	0.0	0.0	0.0	0.081	0.046 ns	58	7.628	17	0.100 ns	0.169 ns
	b–++		0.250	0.0	0.0	1.000	0.278	1.000	1.000	1.000	0.693	0.340***	53	2.327	145	0.293***	22.751***
	c––		0.673	0.308	0.0	0.0	0.722	0.0	0.0	0.0	0.226	–0.396***	27	2.889	47	–0.383***	31.088***

Table 5.15. *Continued*

Year	($)	Complementary	1983/1984				1984/1985				Mean	Statistical analysis					
	+ –	allele									356						
Transect			I		II		I		II			Spearman rank correlation with basalt[a]	Average position		N	Spearman rank correlation with position	χ^2 between soil types (1 df)
Soil type			TR	BA	TR	BA	TR	BA	TR	BA							
Sample size			47	42	31	50	49	44	47	46			X̄	SE			
Locus	allele					Allele frequencies											
He			0.479	0.426	0.0	0.0	0.401	0.0	0.0	0.0	0.163						
N			26	13	6	34	36	2	47	46	210						
Ipol	c+++	(b)	0.0	0.158	0.091	0.387	0.0	0.0	0.0	0.326	0.121	0.309***	75	3.735	32	0.304***	23.464***
He			0.0	0.266	0.165	0.475	0.0	0.0	0.0	0.440	0.168						
N			9	19	22	31	49	42	47	46	265						
*Mdh-1*A	b++––	(a)	0.0	0.189	0.233	0.800	0.277	0.0	0.021	0.0	0.202	0.175***	58	3.346	68	0.116*	9.405**
He			0.0	0.307	0.358	0.320	0.400	0.0	0.042	0.0	0.178						
N			38	37	30	50	47	41	47	46	336						
*Pept-1*B	b+++–	(a)	0.170	0.500	0.0	0.261	0.238	0.459	0.489	0.261	0.324	0.071 ns	60	2.606	94	0.178**	1.272 ns
He			0.283	0.500	0.0	0.386	0.363	0.497	0.500	0.386	0.364						
N			44	33	6	46	42	37	45	46	299						
Pept-3	b––––	(c)	0.311	0.286	0.737	0.408	0.429	0.185	0.955	0.512	0.448	–0.181**	54	2.424	111	–0.073 ns	7.263**
He			0.428	0.408	0.388	0.483	0.490	0.302	0.087	0.500	0.386						
N			37	28	19	49	28	27	22	41	251						
Pgi-A	a++++	(b)	0.136	0.463	0.065	0.120	0.146	0.571	0.277	0.409	0.274	0.243***	65	2.251	94	0.286***	19.632***
	c+		0.0	0.049	0.0	0.0	0.0	0.0	0.0	0.0	0.006	0.075 ns	79	18.500	2	0.072 ns	
He			0.236	0.545	0.121	0.211	0.249	0.490	0.400	0.483	0.342						
N			44	41	31	50	48	42	47	44	347						
Pgm-A null	++	(a)	0.0	0.0	0.0	0.340	0.0	0.0	0.0	0.174	0.072	0.281***	87	1.699	24	0.353***	23.400***
He			0.0	0.0	0.0	0.449	0.0	0.0	0.0	0.287	0.092						
N			45	38	31	50	49	42	47	46	348						
6Pgd-2	a–––+	(c)	0.261	0.225	0.267	0.220	0.190	0.125	0.021	0.043	0.164	–0.036 ns	48	3.380	55	–0.071 ns	0.097 ns
He			0.386	0.349	0.391	0.343	0.308	0.219	0.042	0.083	0.265						
N			46	40	30	50	42	40	47	46	341						

TR Terra rossa,*BA* basalt. $ consistency of predominant +, – Comparison of the frequency on BA with that on TR in the same year and transect; + higher frequency on BA, – higher frequency on TR; – not available (missing data in section of allele frequencies)

$*p < 0.05$, $**p < 0.01$, $***p < 0.001$, $c\ p < 0.10$, $ns\ p > 0.10$

Monomorphic loci: *Aat-2*B, 3A, 3B, *Adh-1*B, 2A, 2B, *Cat*-A, B, *Est-4*A, *Est-5*A, *Gdh*-A, B, *Mdh-1*B, *2*, *Nadh-1*A, *1*B, *2*, *Pepc*, *Pept-2*, *Pgi*-B, *Pgm*-B, *6Pgd-1*A, *1*B

Low polymorphic loci: *Aat-1*A, *1*B, *Est-4*B, *Gluc*-A, B, *Ipor*-A, B, *Pept-1*A

Missing data: In most cases sample size ranged from N = 19–50; however, due to technical reasons, some enzymes had low N. This does not include monomorphic loci, except *Pept-2*, whereas *Est-4*A, B did not show upon gels of transect I-1984

[a] BA was designated as 1 and TR as 0 for the correlation analysis; degrees of freedom are N of that locus minus 2

[b] Excluded from analyses based on subpopulation frequencies to balance missing data

Table 5.16. Summary of genetic variation based on 47 allozyme loci of *T. dicoccoides* in eight subpopulations according to soil, transect, and year at Tabigha. (Nevo et al. 1988c)

Subpopulation		Sample size	Mean no. of alleles	Mean proportion of loci				Gene diversity
Transect	Soil type	(*N*)	per locus (*A*)	Polymorphic per population (*P*)		Heterozygous per individual (*H*)		(*He*)
				(1%)	(5%)	Mean	S.E.	
1983/1984								
I	Basalt	42	1.457	0.370	0.370	0.005	0.002	0.130
I	Terra rossa	47	1.447	0.362	0.298	0.008	0.003	0.112
II	Basalt	50	1.279	0.256	0.233	0.001	0.001	0.084
II	Terra rossa	31	1.171	0.171	0.146	0.003	0.003	0.046
1984/1985								
I	Basalt	44	1.175	0.175	0.175	0.0	0.0	0.073
I	Terra rossa	49	1.308	0.308	0.308	0.0	0.0	0.100
II	Basalt	46	1.233	0.209	0.163	0.0	0.0	0.075
II	Terra rossa	47	1.182	0.182	0.114	0.0	0.0	0.041
Total		356	1.281	0.254	0.226	0.002	0.001	0.083

5.3.1.4 Genetic Summary

The genetic summary of the eight subpopulations appears in Table 5.16. The estimates of gene diversity, *He* (Nei 1975), based on all 47 loci were, on average, higher in the basalt than in the terra rossa (0.091, across transects and years, and 0.075, respectively). Our results suggest that genetic diversity and differentiation at both single and multilocus structures in Tabigha are subject to selection. This is indicated by: (1) single loci (Figs. 5.8, 5.9, see next section), and multilocus soil differentiations; (2) high *He* in basalt and transect I in accordance with the prediction of the niche-width variation hypothesis; (3) soil-specific two-locus linkage disequilibria (Table 5.18); and (4) soil-specific multilocus genome organization (Table 5.19).

5.3.1.5 Discriminant Analysis

We conducted stepwise discriminant analysis, maximizing the overall multivariate *F* ratio of the two soil microniches, based on a multilocus analysis involving seven polymorphic loci and nine alleles (*Pgm*-A^{null}, *Pgi*-A^{a},^{b}; *Mdh*-1A^{b}, *Pept*-1B^{b}; *Aat*-2A^{a}, *Acph*-3^{a},^{c}; *6Pgd*-2^{a}). These seven loci were selected out of the 16 polymorphic loci tested in order to minimize the reduction in

Table 5.17. Partition of genetic diversity of *Triticum dicoccoides* between two soil types, years and transects at Tabigha, based on 12 polymorphic allozyme loci: *Gst* analysis. (Nevo et al. 1988c)

Locus	Alleles	Sample	*Ht*	*Hs*	*Dst*								*Gst*	*Dm*	*Rst*
					Total	(%)	Transect	(%)	Years	(%)	Soil	(%)			
*Aat-2*A	2	349	0.1850	0.1697	0.0153	100	0.0024	16	0.0001	1	0.0031	20	0.0826	0.0175	0.1030
Acph-3	3	294	0.3676	0.2572	0.1104	100	0.0656	59	0.0159	14	0.0022	2	0.3003	0.1261	0.4904
Gluc-A	2	295	0.0904	0.0640	0.0264	100	0.0042	16	0.0030	11	0.0042	16	0.2920	0.0302	0.4713
Gluc-B	2	295	0.0965	0.0706	0.0259	100	0.0049	19	0.0024	9	0.0048	19	0.2683	0.0296	0.4191
Hk	3	210	0.4622	0.1544	0.3078	100	0.2092	68	0.0317	10	0.0521	17	0.6660	0.3518	2.2784
Ipol	2	265	0.2123	0.1646	0.0478	100	0.0149	31	0.0070	15	0.0203	42	0.2249	0.0546	0.3316
*Mdh-1*A	2	336	0.3228	0.1752	0.1477	100	0.0120	8	0.0365	25	0.0099	7	0.4575	0.1688	0.9637
*Pept-1*B	2	299	0.4383	0.4031	0.0352	100	0.0000	0	0.0031	9	0.0022	6	0.0804	0.0403	0.0999
Pept-3	2	251	0.4946	0.4087	0.0859	100	0.0380	44	0.0048	6	0.0157	18	0.1738	0.0982	0.2403
Pgi-A	3	347	0.4059	0.3447	0.0612	100	0.0049	8	0.0095	16	0.0241	39	0.1508	0.0700	0.2030
Pgm-A	2	348	0.1334	0.1025	0.0309	100	0.0103	33	0.0018	6	0.0101	33	0.2317	0.0353	0.3446
6pgdh-2	2	341	0.2745	0.2583	0.0162	100	0.0028	17	0.0112	69	0.0002	2	0.0592	0.0186	0.0719
Mean	2.3	307	0.2903	0.2144	0.0759	100	0.0308	27	0.0106	16	0.0124	18	0.2615	0.0868	0.4049

Ht Total gene diversity, *Hs* average gene diversity within populations, *Dst* average gene diversity between populations, *Dm* average of interpopulational diversity only, *Gst* gene diversity between populations relative to *Ht*, *Rst* interpopulational diversity relative to *Hs*

Table 5.18a. Representative linkage disequilibria [(LD) and (LD′)] between paired loci of *T. dicoccoides* in Tabigha reported on the entire sample and separately for each soil type. (Nevo et al. 1988c)

Locus and allele	Locus and allele	Overall		Basalt		Terra rossa	
		LD	(LD′) p	LD	(LD′) p	LD	(LD′) p
Consistent disequilibria (overall)							
*Adh-1*A a	*Acph-3* a	0.082	(0.83)***	0.066	(0.91)***	0.099	(0.80)***
*Adh-1*A a	*Est-5*B e	0.111	(0.65)***	0.079	(0.54)**	0.171	(0.81)***
Hk b	*Acph-3* b	0.092	(0.81)***	0.083	(0.88)***	0.087	(0.75)***
Hk a	*Est-5*B e	0.048	(0.76)***	0.048	(0.81)**	0.049	(0.72)**
6Pgd-2 a	*Adh-1*A a	0.044	(0.37)**	0.038	(0.37)@	0.053	(0.38)*
*Adh-1*A a	*Pept-3* c	0.076	(0.57)***	0.086	(0.77)**	0.062	(0.38)@
Pept-3 b	*Est-2*A b	0.055	(1.0; c)*	0.054	(1.0; c) n.s.	0.036	(1.0; c) n.s.
Pept-3 b	*Est-2*B c	0.061	(0.62)*	0.043	(0.59) n.s.	0.069	(0.61) n.s.
Mainly on basalt							
*Adh-1*A a	*Pept-1*B b	0.023	(0.14) n.s.	0.060	(0.34)*	–0.016	(0.20) n.s.
Pept-3 c	*Pept-1*B b	0.056	(0.50)***	0.075	(0.69)***	0.020	(0.18) n.s.
Hk c	*Acph-3* a	0.034	(0.68)***	0.039	(1.0; c)***	0.023	(0.44) n.s.
Ipol b	*Pgm*-A a	0.062	(0.86)***	0.104	(0.85)***	–	
Ipol b	*Pgi*-A a	0.022	(0.66)*	0.060	(0.73)**	0.003	(1.0; c) n.s.
Ipol c	*Est-5*B c	0.042	(0.47)***	0.055	(0.41)**	–	
Ipol c	*Pept-3* b	0.017	(0.21) n.s.	0.058	(0.36)**	–	
*Mdh-1*A a	*Acph*-x a	0.087	(0.92)***	0.152	(1.0; c)***	–0.005	(1.0; c) n.s.
*Pept-1*B b	*Est-2*A b	0.054	(0.69)*	0.079	(0.60)*	0.040	(1.0; c) n.s.
Pgi-A b	*Mdh-1*A b	0.031	(0.53)**	0.061	(0.57)***	0.016	(0.78) n.s.
Mainly on terra rossa							
Pgi-A b	*Acph*-x a	0.039	(0.24)@	–0.052	(0.24) n.s.	0.088	(0.64)***
*Adh-1*A a	*Acph*-x a	0.037	(0.16) n.s.	–0.099	(0.69) n.s.	0.095	(0.52)*
Pept-3 b	*Est-5*B a	0.070	(0.40)***	0.014	(0.10) n.s.	0.130	(0.65)***
Hk c	*Est-5*B e	0.023	(0.29) n.s.	–0.008	(0.41) n.s.	0.066	(0.45)*
Hk b	*Pept-1*B b	0.054	(0.64)***	–0.005	(0.07) n.s.	0.107	(0.79)***
Hk c	*Pept-1*B a	0.036	(0.59)*	–0.016	(1.0; c) n.s.	0.084	(0.75)***
Hk b	*Pept-3* b	0.027	(0.17) n.s.	–0.004	(0.07) n.s.	0.092	(0.50)**
*Mdh-1*A a	*Gluc*-B a	0.031	(0.75)***	–		0.064	(0.76)***
Acph-3 a	*Est-5*B e	0.024	(0.46)***	0.003	(0.14) n.s.	0.064	(0.61)***
Acph-3 b	*Est-5*B a	0.046	(0.60)***	0.020	(0.44) n.s.	0.071	(0.67)***
6Pgd-2 a	*Est-2*A a	0.047	(0.24)@	0.028	(0.13) n.s.	0.063	(0.39)@
Acph-x a	*Est-5*B a	0.050	(0.28)*	–0.009	(0.21) n.s.	0.051	(0.23)@

*$p < 0.05$, **$p < 0.01$, ***$p < 0.001$, @ $p < 0.10$, n.s. $p > 0.10$ by χ^2 with Yates correction, or by Fisher's exact test; (c) complete association: one gamete type is missing (Clegg et al. 1976)

sample size for this analysis due to missing data. The results based on the seven chosen alleles are given in Fig. 5.9.

The analysis succeeded in significantly differentiating plants from the two soil types by a pairwise comparison ($F = 7.894$, df = 7/218, $p < 0.0001$) and by a canonical discriminant function (Fig. 5.9). Essentially the same results were

Table 5.18b. Estimates of allozyme multilocus genetic organization of *Triticum dicoccoides* in two soil types at Tabigha. (Nevo et al. 1988c)

Soil type	Polymorphic loci	Expected variance of *K*	Upper limit of 95%	Estimate of variance of *K*	X(2)
Entire sample	18	0.791	<1.31	2.54*	2.21
Overall					
Terra rossa	17	3.193	6.28	6.91*	1.16
Basalt	16	3.490	7.17	6.41	0.84
Transect I					
Terra rossa	15	3.228	7.63	6.17	0.91
Basalt	15	3.149	7.35	6.87	1.18
Transect II					
Terra rossa	10	1.752	2.71	3.70*	1.11
Basalt	14	2.858	4.35	4.57*	0.60

*$p < 0.05$; *K* number of heterozygous loci in two randomly chosen gametes; *X(2)* measure of multilocus structure

obtained by analysis of each year separately based on four to five alleles as with the three alleles originally chosen, i.e., *Pgm-A*, *Mdh-1A*, and *Pept-1b*. The correct classification of plants into their soil microniche was 73% for the overall analysis (167 out of 229 plants analyzed); 63% on basalt, and 84% on terra rossa. Similarly, the overall correct classification was 79% and 73% for 1983–1984 and 1984–1985, respectively. These figures are greater than those expected by chance, on the assumption of a random sample of neutral alleles.

A second discriminant analysis conducted on the allele frequencies in the eight subpopulations (2 soils, 2 transects, and 2 years) succeeded in significantly differentiating the two soil types using three (Pgm-A^{null}, *6Pgd*-2^{a}, and *Pgi*-A^{b}) out of 11 alleles that went into the analysis ($F = 134.27$, df = 3/4, $p = 0.0002$).

5.3.1.6
Analysis of Genetic Diversity, *He*, Within and Between Populations

Gene diversity (*Ht*) of a subdivided population can be analyzed as explained previously in the allozyme and RAPD analyses into its components, i.e., gene diversities within and between subpopulations. The results provide measures of the average (*Dst*), absolute (*Dm*), relative (*Gst*) and inter- vs intrapopulational (*Rst*) degree of gene differentiation among subpopulations, where *Hs* is the mean gene diversity within a population; *Ht is* mean gene diversity in the total population; and $Ht = Hs + Dst$ (Nei 1973). *Dm* can be used for comparing the degrees of gene differentiation in different organisms. Considering each of the four "half transects" as a subpopulation, Table 5.17 gives the *Dst* analysis of gene diversity of the 12 polymorphic loci that were scored in both transects and both years. On average, 73.9% of the total genetic diversity in the eight subpopulations exists within, and 26.1% between, subpopulations. These results indicate a strong genetic differentiation between subpopulations which is higher than the degree of relative differentiation found in the microgeographic analysis of allozymes (Nevo et al. 1981) and hordein (Nevo et al. 1983), conducted in wild barley at the same site in the early 1980s. The contrasting soils contributed 18% to the differentiation, but in certain loci *(Ipol, Pgi-A, Pgm-A)* the soil contribution was considerably greater. The ranking of the contribution of the three main factors to genetic differentiation between subpopulations was transect 27%, soil 18%, and year 16%. The unaccounted percentages are due to various levels of interaction. Thus, allozyme differentiation in wild emmer wheat between soil types and other factors is substantial, as was also displayed in the previous discriminant analysis (Nevo et al. 1998c).

5.3.1.7 Genetic Differentiation at the Two-Locus Level

Linkage disequilibria (LD) and relative LD (LD′) for the entire sample and for each soil type separately are given in Table 5.18a. LDs can be classified into general (occurring in both soil types) and specific (occurring either on basalt or on terra rossa). Examples of general Ds were: *Adh-l*A^{a}-*Acph-3*^{e}; *Hk*^{a}-*Est*-5B^{e}. Specific LDs for basalt were: *lpol*^{b}-*Pgm*-A^{a} (*LD* = 0.104***); *Mdh-1*^{a}-*Acph-x*^{a} (*LD* = 0.152***); *Pept-3*^{c}-*Pept-1*B^{b} (*LD* = 0.075***); *Hk*^{c}-*Acph-3*^{a} (*LD* = 0.039***). Specific *LD*s for terra rossa were: *Hk*^{b}-*Pept*-1B^{b} (*LD* = 0.107***); *Mdh-l*A^{a}-*Gluc*-B^{a} (*LD* = 0.064***); *Pept-3*^{b}-*Est*-5B^{a} (*LD* = 0.130***); Pgi-A^{b}-Acph-x^{a} (*LD* = 0.088***) (significance levels, *$p < 0.05$; **$p < 0.01$; ***$p < 0.001$, calculated by χ^2 with Yates correction and excluding all heterozygotes). LD′ displayed a pattern similar to LD except at very low values of LD. Notably, in some cases, such as *Pgi*-A^{b}-*Acph-x*^{a}, the sign of LD was opposite in the two soil types, a pattern not revealed by LD′(Nevo et al. 1988c).

5.3.1.8 Multilocus Organization – Terra Rossa versus Basalt

Multilocus organization indices (Brown et al. 1980) were calculated for the entire data set of wild emmer at Tabigha, as well as for each transect separately (Table 5.18b). Notably, samples of the 2 years represent the same local seed pools. The observed variance in the entire data set was 221% above the expected. Multilocus organization appeared to be higher in emmer growing on terra rossa than on basalt. The standardized intensity index of multilocus structure X(2) was 1.16 for wild emmer on terra rossa, and 0.84 for that on basalt. In other words, the observed variance in *K* on terra rossa was 116%, significantly higher than expected, and on basalt only 84%, not significantly higher than expected. The difference between the observed and expected variance of σ^2K was significant in transect II in both years (1983/4–1984/5), but this difference was nonsignificantly opposite in transect I (Table 5.18b). We conclude that, on average, multilocus organization intensity is higher in wild emmer growing on terra rossa than on basalt (Nevo et al. 1988c).

5.3.1.9
Natural Selection of Allozymic Diversity

Our data seem inconsistent with the hypothesis that allozymic diversity is neutral and that its fate in nature is primarily determined by mutation input and random fixation, as suggested by the neutral theory of molecular evolution (Kimura 1983). The allozymic patterns discovered at Tabigha are structured in accordance with the ecological heterogeneity of the test site, presumably, predominantly by the aridity index. The genetic differentiation discovered is unlikely to be the result of random forces. The population size at Tabigha involves thousands of plants. The largest stands of wild emmer in Israel occur in the neighboring areas of the Golan Heights and eastern Upper Galilee Mountains, and Tabigha, located within this region. Neither migration nor neighborhood effects in the framework of a neutral model can explain the pattern revealed. Migration is indeed low (Golenberg 1986a), but it is continuous over many generations, and neighborhood effects vary adaptively between both soil types. Thus, the genetic differentiation encountered here on a very low scale suggests that natural selection may be an important differentiating factor of wild emmer at Tabigha, as it is elsewhere.

The higher levels of genetic diversity, *He*, and polymorphism, *P*, in 1983–1984, on basalt, and in transect I, may be regarded as an adaptive strategy for increasing fitness in a year, soil type, and/or topography subject to a larger spatiotemporal microclimatic heterogeneity, in accordance with the niche-width variation hypothesis (Van Valen 1965). This result supports regional (Nevo 1983c) and global (Nevo et al. 1984a) patterns in which allozymic diversity is, in general, higher with increasing spatiotemporal ecological heterogeneity and unpredictability (Nevo et al. 1988c).

Our empirical results are consistent with other microsite experimental studies (e.g., Hamrick and Allard 1972; Hamrick and Holden 1979). They also corroborate the theoretical expectation that the existence of a protected polymorphism is more likely in a more heterogeneous environment (e.g., Karlin 1981, 1982). We hypothesize that the patterns observed are affected by diversifying and balancing selection on allozyme polymorphisms. Transplant experiments could test this hypothesis by studying the fitness of allozyme genotypes from alternative microniches (Jain and Bradshaw 1966), as well as substantiate their biochemical specificities and physiological functions (Nevo 1983b). Likewise, allele frequencies of gene pools in the soil should be compared and contrasted with those of the adults over several years, in order to assess the temporal operation of selection (Karlin and Levikson 1974). Transplant experiments were conducted by Golenberg (1986a) on the Yehudiyya and nearby Qazrin alternative multilocus allozyme genotypes, but they did not reveal differential fitness, possibly due to the short testing period and technical disturbances. In wild barley, *Hordeum spontaneum,* at "Evolution Canyon", Mount Carmel, we did demonstrate interslope differential fitness by transplant experiments (Lavie et al. 1993).

Our results not only have theoretical, but, as already emphasized, also practical application. If allozymic diversity indeed varies dynamically with the environment, then not only its *macro-* but also its *micro*geographic structure is tractable by ecological factors. Such clues may help in optimizing the exploitation, conservation, and utilization of wild emmer wheat in breeding (Brown and Clegg 1983; Nevo 1983a, 1989a, 1995, 2001b).

5.3.2 Edaphic SSR DNA Divergence in Wild Emmer Wheat at the Tabigha Microsite

How does largely noncoding SSR DNA diversity react to the soil contrast? To answer this question, we analyzed (Li et al. 2000a) genetic diversity in DNA dinucleotide microsatellites of the same two edaphic subpopulations of *T. dicoccoides* from the two contrasting soil types, terra rossa and basalt, at the Tabigha microsite in northern Israel (Fig. 5.10a,b); this site had been previously analyzed for allozymes (Nevo et al. 1988b). In this second study, we demonstrated microgeographic molecular-genetic divergence caused by ecological stress and edaphic selection, now at the *noncoding* DNA level (Li et al. 2000a).

Twenty-eight microsatellite markers (Table 5.9) were used to analyze genetic divergence in tandem with dinucleotide repeated DNA regions between the two edaphic subpopulations of *T. dicoccoides* dwelling on the contrasting terra rossa and basalt soils. The terra rossa soil niche was drier, more stressful, and *narrower* in niche breadth than the temporally more variable basalt throughout the growing season (November to May). In this study (Li et al. 2000a), 155 individuals from the two transects (76 from the terra rossa and 79 from the basalt) were analyzed for their microsatellite (SSR) diversity.

Significant microsatellite divergence in allele distribution, repeat length, genetic diversity, and LDs was found between emmer wheat from the two soil types over two short transects of 100 m each. Soil-*specific* and -*unique* alleles and LDs were observed in terra rossa and basalt subpopulations. Permutation tests (Table 5.19c) showed that the effects of random genetic drift were very low for the significant genetic diversity at the microsatellite loci between the two subpopulations, suggesting that an adaptive molecular pattern derived from edaphic selection may influence variation of microsatellites, i.e., in primarily noncoding genomic regions. The details are discussed in the following sections.

5.3.2.1
Distribution of Allelic Diversity at Polymorphic SSR Loci

The 28 microsatellite primer pairs produced DNA amplifications except for one marker, GWM637 (Li et al. 2000a). Of the 27 markers, each of 26 amplified a single fragment in each individual, regarded as 26 loci; one marker, GWM332, amplified two fragments in many individuals, and we regarded the two fragments as two alleles at two loci (GWM332a and 332b). In all, 28 loci were detected in this study; only GWM601 was monomorphic for emmer wheat in both soils. With the exception of the locus GWM332b due to "null allele(s)," the total alleles amplified over 27 loci were 170 in the terra rossa (TR) subpopulation and 177 in the basalt (BA) subpopulation; 142 alleles were shared by the two subpopulations. The allele distributions at microsatellite loci were not equal in the two edaphic subpopulations (Table 5.19a). For example, Fig. 5.10b illustrates different allele distributions at two microsatellite loci between the closest sampling points in the TR and BA soils.

There were distinct alleles and higher allelic diversity in the BA section than in the TR section. In particular, GWM332a and GWM332b were completely linked on the long arm of chromosome 7A in the population of *T. dicoccoides* × *T. durum* (Peng et al. 2000c). At GWM332a, similar alleles were detected in many individuals of wild emmer in both subpopulations. At GWM332b, however, *soil-unique* alleles (including a "null" allele) were observed (Fig. 5.10b).

This locus was not included in other genetic analyses because it is hard to estimate the repeat number of a null allele. It is likely that locus GWM332b was a duplication of the common locus GWM332a. According to the allele distribution in the TR and BA subpopulations (Table 5.19a), we could expect some *soil-specific* mutation to occur in the tandem repeat region and also in the flanking area of GWM332b, particularly in the BA subpopulation.

5.3.2.2
SSR Permutation Test

Noncoding sequences are considered by many researchers as neutral, nonfunctional, or "junk" DNA. This hypothesis would predict that allele frequencies should be equal between the TR and BA subpopulations. We tested this

Fig. 5.10. a Allele distribution at the GWM095, GWM120, GWM162, GWM169, and GWM218 microsatellite loci on the spatially close parts of the terra rossa (TR) and basalt (BA) soils for *T. dicoccoides*. Allele sizes are numbered by base pairs including tandem-repeated and flanking regions. The *first lane on the left* is the *T. aestivum* cultivar Chinese Spring. (Li et al. 2000a) **b** Wild emmer microsatellite DNA allele distribution detected by the GWM332 primer pairs on the TR and BA soils. Allele sizes are marked by base pairs including tandem-repeated and flanking regions. The first lane on the left is the *T. aestivum* cultivar Chinese Spring. (Li et al. 2000a). **c** Plot showing the relative performance (water stress toleránce) of 52 wild barley, *Hordeum spontaneum*, genotypes from Tabigha. Pc1 scores (yield-related traits) are plotted against the genotype number equivalent to the position on the 100 m transect. (Ivandic et al. 2001)

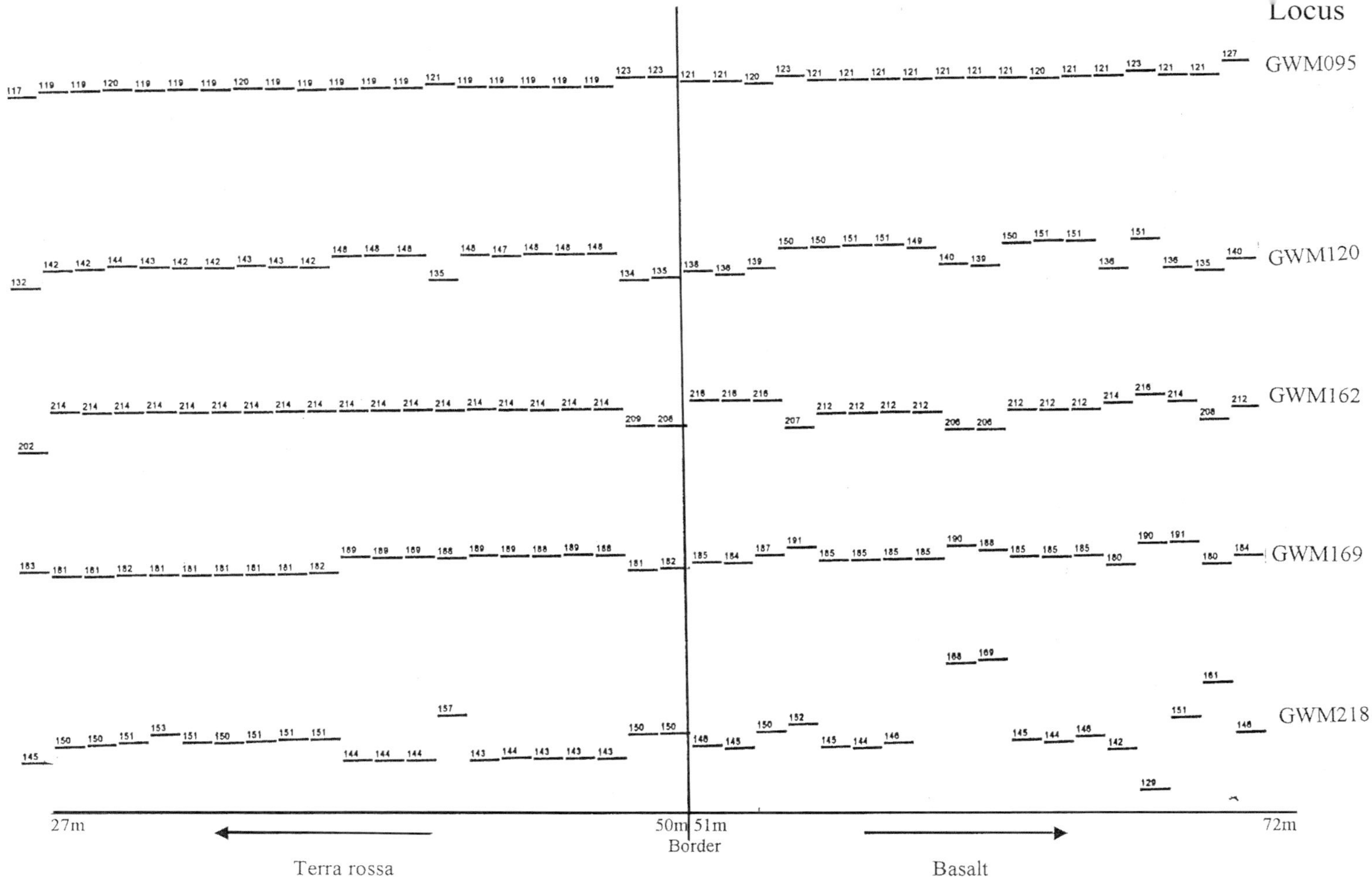

Locus
GWM095
GWM120
GWM162
GWM169
GWM218
27m
50m
51m
72m
Border
Terra rossa
Basalt

a

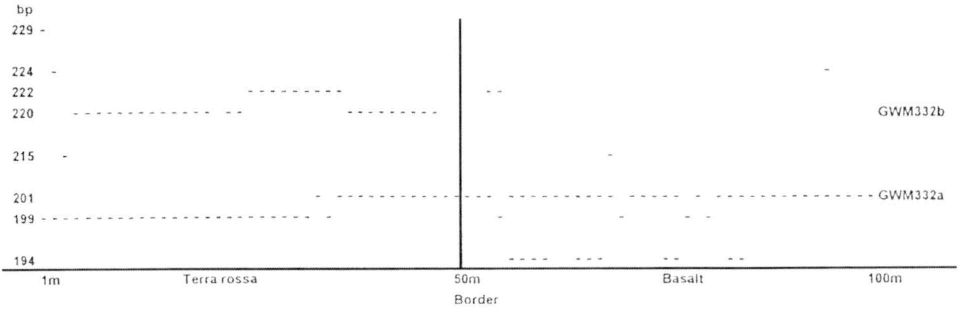

Fig. 5.10b. *Continued*

null hypothesis by the χ^2 test in an attempt to compare and contrast the difference of each allele distribution between the two edaphic subpopulations (Li et al. 2000a). After exclusion of rare alleles (observed in ≤5 individuals in the whole population), 39 and 30 TR-*specific*, and 42 and 25 BA-*specific* alleles were found at $p < 0.05$ and 0.01 significance levels, respectively. Among these alleles, 15 alleles were TR-*unique* and 14 were BA-*unique*.

Out of 5,000 random permutations of sampled individuals, there was no instance of more, or the same number of, soil-*specific* alleles than observed in the real data of both subpopulations. The results indicated that the possibility of obtaining the observed soil-*specific* alleles by chance was < 0.0002 (Table 5.19b).

5.3.2.3
SSR Genetic Diversity and Distance Between the Terra Rossa and Basalt Soils

Although the difference of overall gene diversities (*He*, Nei 1973) was not large between the BA (0.683) and the TR (0.661) subpopulations, gene diversities were obviously higher in the BA than in the TR subpopulation at 16 microsatellite loci (Table 5.19a). In particular, gene diversities were 0.188, 0.209, 0.169, significantly higher in the BA than in the TR subpopulations at loci GWM095, GWM340, and GWM577, respectively. However, an opposite pattern of *He* estimates between the BA and TR subpopulations was observed at nine other loci. At locus GWM415, genic diversity, *He*, was significantly higher on the TR soil than on the BA soil.

The genetic distance between the two subpopulations in the TR and BA soils was estimated using the allele-sharing distance index (D_{AS} = 1–proportion of shared alleles; range: 0–1; Chakraborty and Jin 1993) for each microsatellite locus (Table 5.19a). The largest D_{AS} was 0.726 at GWM537, followed by nine other loci ($D_{AS} > 0.510$); in 12 other loci, D_{AS} ranged from 0.232 to 0.487. At the remaining five polymorphic loci, the genetic distances between the TR and BA subpopulations ranged from 0.138 to 0.187. The overall D_{AS}

Table 5.19a. Number of alleles, genetic diversity (*He*; Nei 1973) and allele-sharing distance (D_{AS}) at microsatellite loci of *T. dicoccoides* between terra rossa (TR) and basalt (BA) soil types at the Tabigha microsite, upper Galilee, Israel. (Li et al. 2000a)

Locus	Number of alleles[a]				ARN		K-W test H(1, *N*)	Variance		Hartley's *F* (df_1, $df_2 > 70$)	*He*		D_{AS}
	Total	TR	BA	Shared	TR	BA		TR	BA		TR	BA	
GWM018	7	1	2	4	32.9	33.2	16.078***	28.55	2.53	11.3****	0.731	0.789	0.519
GWM060	9	2	2	5	21.2	22.7	1.522	16.36	37.86	2.3***	0.713	0.762	0.349
GWM095	6	0	1	5	15.5	17.6	32.766****	3.88	3.13	1.2	0.587	0.775	0.650
GWM099	3	0	1	2	7.7	8.4	4.265*	0.22	8.21	37.3****	0.441	0.369	0.138
GWM120	10	0	3	7	32.7	33.9	6.612**	4.26	7.83	1.8*	0.681	0.722	0.298
GWM124	12	3	1	8	51.6	53.7	7.825**	20.30	27.78	1.9*	0.827	0.865	0.470
GWM136	11	3	1	7	62.0	58.2	0.813	61.45	76.76	0.3	0.179	0.675	0.449
GWM162	6	0	0	6	23.6	22.4	21.252****	3.08	2.42	1.1	0.700	0.769	0.523
GWM169	8	1	0	7	23.1	22.0	0.381	6.03	2.19	2.7***	0.777	0.760	0.142
GWM186	8	3	1	4	23.9	22.7	1.139	36.36	9.24	3.9***	0.714	0.780	0.400
GWM218	12	2	1	9	21.6	23.2	0.191	56.34	26.75	2.1**	0.813	0.874	0.487
GWM219	6	1	2	3	22.1	26.1	12.244***	5.68	36.22	6.4****	0.644	0.751	0.336
GWM251	8	2	3	3	23.4	18.6	33.664****	41.46	34.91	1.2	0.670	0.587	0.533
GWM294	9	2	0	7	20.7	19.3	9.383**	15.48	8.41	1.8*	0.800	0.819	0.584
GWM332	5	0	1	4	20.4	20.3	0.091	1.45	2.74	1.9*	0.653	0.692	0.232
GWM340	12	0	5	7	27.0	29.1	9.491**	32.39	38.47	1.2	0.659	0.868	0.559
GWM361	6	1	1	4	22.0	21.6	11.343***	1.56	0.66	2.4***	0.643	0.562	0.465
GWM368	7	1	0	6	23.0	22.9	0.412	9.88	12.33	1.2	0.715	0.780	0.264
GWM389	8	1	0	7	32.6	31.7	0.694	11.73	3.37	3.5***	0.794	0.760	0.252
GWM408	6	0	1	5	28.2	25.0	25.306****	9.39	9.05	0.1	0.719	0.647	0.557
GWM415	2	0	0	2	19.6	19.1	10.565**	2.16	0.20	10.6****	0.266	0.025	0.145
GWM429	6	1	2	3	15.6	15.1	4.735*	19.43	26.04	1.3	0.606	0.609	0.187
GWM459	9	0	1	8	25.3	26.2	2.701	16.95	9.39	1.8*	0.834	0.845	0.299
GWM537	16	4	6	6	33.9	36.3	0.470	11.28	50.22	4.5****	0.804	0.885	0.726
GWM540	4	0	0	4	18.1	17.4	10.711***	2.13	0.81	2.6***	0.645	0.632	0.177
GWM577	8	0	0	8	28.7	34.4	20.093****	31.53	66.18	2.1**	0.665	0.834	0.537
GWM601	1	0	0	1							0.000	0.000	0.000
Total	205	28	35	142									
Mean	7.1	1.0	1.3	5.3							0.661	0.683	0.421

$^*p < 0.05$, $^{**}p < 0.01$, $^{***}p < 0.001$; $^{****}p < 0.0001$
[a] TR = terra rossa-*unique*, BA = basalt-*unique*

Table 5.19b. Niche specificity of linkage disequilibria at microsatellite loci of *T. dicoccoides* in the terra rossa and basalt subpopulations at the Tabigha microsite, Israel. (Li et al. 2000a)

Subpopulation	Significance of χ^2 test for allele frequencies[a]	Number of actual observed niche-specific alleles (n_s)	Cases of $n_p \geq n_o$ in 5,000 permutations	Significance of allele niche specificity
Terra rossa	0.05	39	0	0.0002
	0.01	30	0	0.0002
Basalt	0.05	42	0	0.0002
	0.01	25	0	0.0002

An allele was excluded if the observation of an allele was ≤5; n_p the number of niche-specific alleles observed in the permutations

[a] χ^2 test for homogeneity of allele frequencies between the two subpopulations

between the TR and BA subpopulations was also high (0.421). If the distance (200 m) between transect I and II was taken into account, the genetic distances between TR and BA subpopulations were very similar in transect I (D_{AS} = 0.480) and transect II (D_{AS} = 0.501). This result served as a double check, as the genetic distance between the TR and BA subpopulations combined the two transects.

5.3.2.4
Microsatellite Diversity of Repeat Numbers Between the Terra Rossa and Basalt Subpopulations

The ARNs of microsatellite core sequences were significantly (Kruskal-Wallis test, $p < 0.05$–0.00005) larger at eight loci in the BA than in the TR subpopulations (Li et al. 2000a). By contrast, at nine other loci ARNs were significantly ($p < 0.05$–0.00005) smaller in the BA than in the TR subpopulation (Table 5.19a). The difference of repeat numbers between the two subpopulations did not seem to relate to motif types. For example, among five loci with motif $(CA)_n$ or $(AC)_n$, ARNs at two loci (GWM095 and GWM099) were significantly ($p < 0.00005$ and < 0.05, respectively) larger in the BA than in the TR subpopulation; at loci GWM162 and GWM251, however, the ARNs were significantly ($p < 0.00005$) smaller in the BA than in the TR subpopulation. Among nine loci with motif $(GA)_n$, the ARNs at three loci (GWM294, GWM361, and GWM415) were significantly ($p < 0.01$–0.0005) larger in the TR than in the BA subpopulation; but at loci GWM219 and GWM340, the ARNs were significantly ($p < 0.01$–0.0005) larger in the BA than in the TR subpopulation. Notably, most loci with compound or imperfect motifs, such as $(CT)_n(GT)_k$, $(CA)_n(TA)_k$, $(CA)_nAA(CA)_k$, etc., showed different repeat numbers in the TR and BA subpopulations.

Variances (σ^2) in repeat number were significantly (Hartley's F test, $p < 0.05$–0.00005) larger at ten loci in the TR than in the BA subpopulations;

Table 5.19c. Summary of permutation test for the observed microsatellite LD specificity, excluding the rare alleles, based on 5,000 random shufflings of samples from the terra rossa and basalt subpopulation at Tabigha, Israel. (Li et al. 2000a)

Subpopulation	Significance level (χ^2 test)	Total LDs	Proportion of niche-specific LDs		No. cases $P_o \geq P_e$	Significance of niche specificity or uniqueness[a]
			In real data (P_o)	In 5,000 permutations (P_e)		
With rare alleles						
Terra rossa	0.001	902	0.904	0.475–0.886	0	0.0002
	0.01	1612	0.849	0.500–0.798	0	0.0002
	0.05	2665	0.794	0.488–0.795	0	0.0002
	Total	5179	0.830			
Basalt	0.001	316	0.747	0.158–0.760	1	0.0002
	0.01	750	0.676	0.226–0.655	0	0.0002
	0.05	1516	0.648	0.248–0.601	0	0.0002
	Total	2582	0.660			
Without rare alleles ($n_o \leq 5$)[b]						
Terra rossa	<0.001	881	0.904	0.519–0.912	0	0.0002
	<0.01	1523	0.848	0.549–0.834	0	0.0002
	<0.05	2490	0.797	0.528–0.772	0	0.0002
	Total	4894	0.832			
Basalt	<0.001	521	0.661	0.239–0.834	28	0.005
	<0.01	597	0.613	0.310–0.729	65	0.013
	<0.05	1228	0.589	0.336–0.679	80	0.016
	Total	2348	0.650			

Random test performed by 5,000 permutation experiments

[a] Significance based on 5,000 permutations

[b] Alleles observed in ≤5 individuals in the whole population were considered rare alleles

by contrast, at eight other loci the opposite was true. The results suggest that distributions of repeat numbers (alleles) were *not equal* between the two subpopulations.

5.3.2.5 SSR Discriminant Analysis

Stepwise discriminant analysis was conducted to attempt to classify individuals into the TR and BA subpopulation by microsatellite variation (Li et al. 2000a). At the third step, three selected loci (GWM095, GWM408, and GWM120) could correctly classify 75% of the 129 individuals into their original subpopulations (Li et al. 2000a). In this case, the difference among the subpopulation centroids was also significant ($F_{(3,125)} = 31.08$, $p < 0.00005$). At the fifth step, five loci (GWM095, GWM408, GWM120, GWM124, and GWM429) chosen from the 21 polymorphic loci were sufficient to correctly classify most individuals into their original soil niches. The overall multivariate F ratio for the test of differences among the subpopulation centroids was significant ($F_{(5, 123)} = 17.92$, $p < 0.00005$). According to the five loci, 83% of 129 sampled individuals were correctly classified in their original subpopulations.

5.3.2.6 Linkage Disequilibrium (LD) Between SSR Loci

Nonrandom association of alleles between loci is said to be in linkage disequilibrium (LD), which may arise from selfing, selection, etc. (Hartl and Clark 1997). The LD values were computed by the methods of Weir (1979). With all alleles, 5,179 and 2,582 linkage disequilibria (LDs, $p < 0.05$–0.001) were observed between loci in the TR and BA subpopulations, respectively (Li et al. 2000a). The higher the significance of the LDs, the larger the proportion of *niche-specific* LDs in both subpopulations. In the TR subpopulation, the proportions of *niche-specific* LDs were larger (0.796–0.904) than those in the BA (0.648–0.747) at $p < 0.05$, 0.01, and 0.001 (Table 5.19c).

Permutation test. Out of 5,000 permutations of random shuffling of sampled individuals, only one case showed higher LD *niche specificity* in the BA subpopulation at $p < 0.001$ level (Table 5.19c). In no permutation case was there a *niche specificity* that was higher than in the real data. The results of the permutation test indicated that the significance of LD *niche specificity* was very high ($p < 0.0002$) in both subpopulations. The proportions of *niche-specific* LDs in the TR subpopulation did not change at $p < 0.05$, 0.01, and 0.001 levels due to exclusion of rare alleles (observed in ≤5 individuals) in the whole population, and the LD *niche specificity* was still highly significant ($p < 0.0002$) based on 5,000 permutations. In the BA subpopulation, although the proportions of *niche-specific* LDs decreased after the rare alleles were excluded, the *niche specificity* for LDs was also significant ($p < 0.016, 0.013, 0.005$ at the 0.05, 0.01, and 0.001 levels, respectively) based on 5,000 random permutations. These results suggested that obtaining LD *niche specificity* by stochastic effects was highly improbable (Table 5.19c).

The log-linear model test indicated that genome and chromosomes did not affect the LDs, but that soil type strongly affected LD significance ($\chi^2 = 336.52$, df = 1, $p < 0.00005$). Sample size could not be the cause of a significant difference between the two subpopulations in the percentages of the significant LDs, since the sample sizes were similar between the TR (76) and BA (79) soils.

5.3.2.7
Differential Physiological Response in Wild Barley Across the Edaphically Subdivided Transect at Tabigha

At the Tabigha microsite, edaphically subdivided into BA (50 m) and TR (50 m), we also demonstrated significant intersoil genotypic divergence of allozymes in wild barley (Nevo et al. 1981), hordeins (Nevo et al. 1983), and RAPDs (Owuor et al. 1999). Can the intersoil genotypic divergence of wild cereals also be demonstrated phenotypically? We tested this intriguing question in wild barley, *Hordeum spontaneum*, at the edaphically subdivided Tabigha microsite (Ivandic et al. 2000). From TR, 24 genotypes were sampled, and 28 from BA. Due to different soil water-holding capacities, plants growing on TR naturally experience more intense drought than plants growing on BA. In a glasshouse experiment, water was withheld from plants for two periods (10 and 14 days) after flag leaf emergence. A total of 15 agronomic, morphological, developmental, and fertility related traits were examined by analysis of variance (ANOVA). Ten of these traits were significantly affected by the treatment. A high degree of phenotypic variation was found in the population with significant genotype × treatment and soil type × treatment interactions. Principal component analysis (PCA) was performed by means of combined control and stress treatment data sets. The first three principal components (pc) explained 88.8% of the variation existing in the population, with pc1 (47.9%) comprising yield-related and morphological traits, pc2 (22.9%) developmental characteristics, and pc3 (18.0%) fertility-related traits. The relative performance of individual genotypes was determined by comparing pc scores. The yield traits of TR genotypes were significantly less affected by the imposed water stress than BA genotypes. Moreover, TR genotypes showed accelerated development under water-deficit conditions (Fig. 5.10c). The data indicate that specific genotypes demonstrating differential responses may be useful for comparative physiological studies, and that TR genotypes exhibiting yield stability may have value for breeding barley that is better adapted to drought. In other words, on the drier TR, the wild barley has evolved on a microscale differential fitness components that enable it to excel in the narrower and drier TR ecological niche. We expect the same intersoil phenotypic divergence in wild emmer wheat.

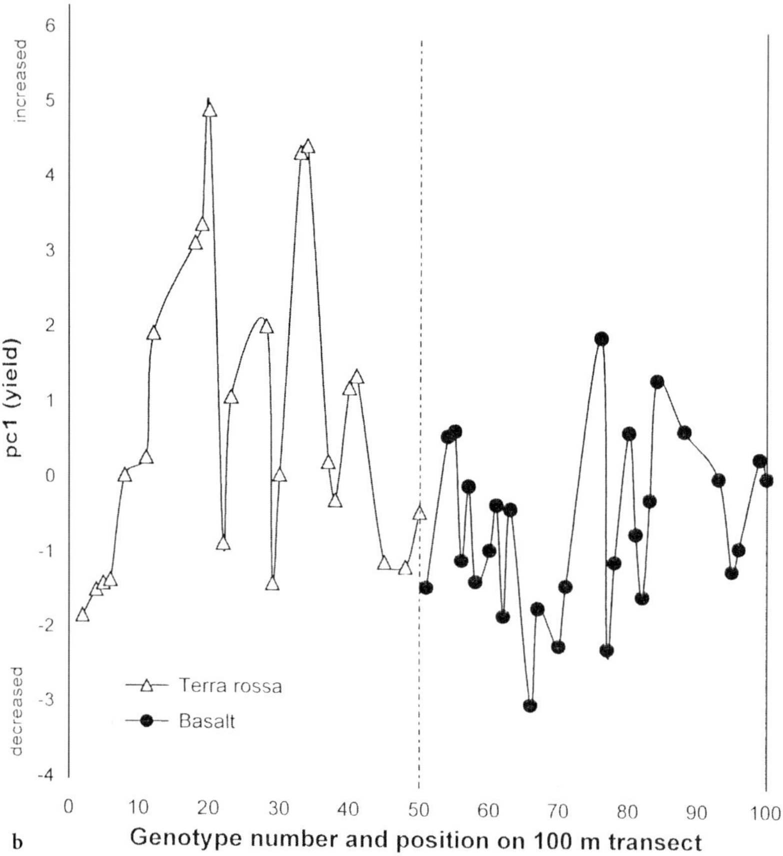

Fig. 5.10c. *Continued*

5.4 Population Dynamics of *Triticum dicoccoides* in a Natural Habitat in Eastern Galilee: The Ammiad Microsite Study

The third allozyme microsite study of wild emmer is described later, in the Ammiad multidisciplinary study (Nevo et al. 1991c) (Fig. 5.11). The Ammiad study described below represents a national effort to unravel the ecological, genetic, and morphological aspects over time of *T. dicoccoides* at a microsite differentiated lithologically and topographically.

The spatiotemporal genotypic and phenotypic dynamics of *T. dicoccoides* have been studied in the Ammiad microsite and are described in-depth in a special volume of the Israel Journal of Botany, edited by A. Horovitz and M. Feldman (1991). A site for intensive and long-term population-genetic, ecological, morphological, pathological, and other studies of wild emmer was set up at Ammiad, eastern Galilee, Israel (Anikster and Noy-Meir 1991). The

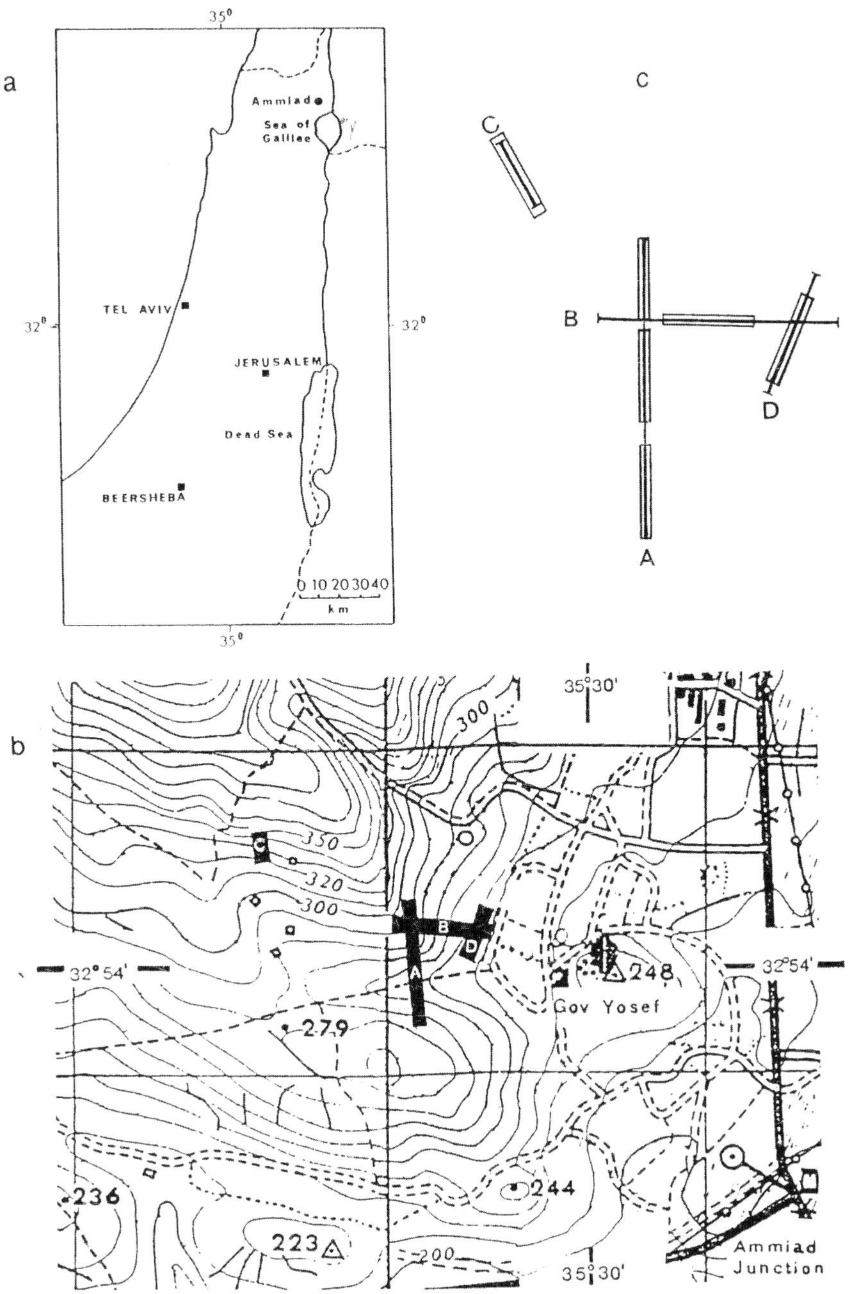

Fig. 5.11a–c. The Ammiad study site. **a** Map of Israel showing the location of the Ammiad area. b Map of the Ammiad area (1:23,800) showing transects *A*, *B*, *C*, and *D*. **c** Diagram of the transects showing parts that were fenced-in in 1985 to exclude cattle. (From Anikster and Noy-Mair 1991)

objective of the study was to gain basic scientific information on a variable and representative population of this annual selfing wild cereal, the progenitor of *durum* and bread wheats, in the center of its local distribution. The information gathered was to aid the rationalization of in situ conservation and sampling practices. An area of grassland on terra rossa soil overlying Middle Eocene hard limestone was chosen because of its heterogeneous topography, rock microrelief and plant species composition, which promised a diversity of habitats, including variable lithological and topographic aspects, based on an in-depth ecological study. The survey was conducted on the following traits: population density, floristic and ecological differentiation (Noy-Meir et al. 1991a, b), phenotypic variation (Anikster et al. 1991), fungal disease resistance (Dinoor et al. 1991), allozymic diversity (Nevo et al. 1991c; Li et al. 2000b), and glutenin high-molecular-weight (HMW) variation (Felsenburg et al. 1991).

The site was divided into four major habitats and eleven subhabitats (Fig. 5.11). Environmental variation was also caused by wide differences in annual total and distribution of the Mediterranean winter rainfall. In the years 1984 to 1989, plants were monitored at closely spaced, permanent sampling points along four topographically diverse transects. Single-spike seed collections were made at each point annually, and progenies from these collections were distributed for parallel ex situ studies of different traits of the phenotype and genotype. Below, we briefly survey these studies, ecologically, phenotypically, phytopathologically, and genetically.

5.4.1
General Characteristics of the Ammiad Microsite

The 250 sampling sites at which the population of wild emmer wheat at Ammiad was sampled since 1984 for multidisciplinary studies were subjected to a floristic and ecological analysis (Noy-Meir et al. 1991a). The local habitat at each of the sampling points, to a radius of 50 cm, was characterized by measurement of topographic position, microrelief created by rock outcrops, associated vegetation, soil depth, soil moisture, and soil nitrogen. An initial topographical classification of habitats was revised on the basis of the first three components of a nodal ordination analysis of the floristic composition of associated vegetation. Four major habitat units were distinguished: Valley (including also lower slopes), Ridge (various slopes and aspects), North (a north-facing slope) and Karst (with deeply dissected rock relief) (Fig. 5.11, b–c: A–D). At a finer level of resolution, eleven subunits were recognized. The units thus defined showed strong and highly significant differentiation in the other ecological factors measured, in particular cover, proximity, and height of rocks, and surface soil moisture after early rains. Soil moisture was significantly higher at the sampling points close to or between rocks, in particular tall smooth rocks. These results suggest that the dominant ecological factor causing heterogeneity within the site and the vegetation and, presumably, the wild wheat populations, is the rock microrelief, operating through soil moisture and possibly also through local modification of grazing pressure and competition. Soil nitrogen shows a different pattern and may be a secondary factor.

5.4.1.1 Population Dynamics

The dynamics of local populations of *T. dicoccoides* were observed over four growing seasons (1985–1988) in circles with a span of about 1 m at 229 permanent wheat sampling points in Ammiad (Noy-Meir et al. 1991b). Germination after light (25–45 mm) early rain in autumn was more frequent near rock outcrops, where surface soil moisture was subsequently higher. Following the heavier winter rains, more widespread germination occurred. Differences between habitats in the density of fertile wild wheat plants in spring were not consistent, but varied over years. Mean population densities in habitats varied in the range of 1–15 plants per m^2. A marked decline in population density over 2 years, followed by a recovery the next year, can be related to seasonal rainfall patterns interacting with local soil moisture regimes. Population densities in areas heavily grazed by cattle gradually fell below those in areas in similar habitats that were protected from grazing. The annual rates of extinction in the 1-m neighborhoods were 11–30% and the rates of recolonization were 41–57%. It was proposed that rainfall, rockiness, and grazing are the primary factors that interact to affect demographic processes in wild wheat populations; hence they are probably the major selection forces also affecting genetic structure. These factors should be given prime attention in the design and management of areas for *in situ* conservation of wild wheat (Maxted et al. 1997; Nevo 1998a).

5.4.1.2 Phenotypic Patterns

Morphological, phenological, and yield traits were studied in two or more consecutive years in accessions of wild emmer, derived from permanent sampling points at Ammiad (Anikster et al. 1991). In addition to being examined annually in container nurseries, each accession was also studied in different open-field plantings. Qualitative traits defining spikelet color, leaf sheath pigmentation and pubescence, leaf shape and growth habit varied *nonrandomly* among descendants of different habitats. In all quantitative traits, the differences scored between plantings were greater than those that could be ascribed to provenance specificity. Variation in the productivity traits, number of tillers, dry weight of the shoot, and number of spikes, in particular, had a very large environmental component, and the environment also modified the correlation among these traits. However, in characters quantifying the width of the flag leaf, the time of appearance of the first awn and spike size, significant differences among subpopulations from different habitats were maintained in most plantings. Different traits tended to occur in association with each other. This was especially marked in a phenotype common in the Lower-Karst habitat.

5.4.1.3
Disease Resistances

Diverse fungal disease agents were present at Ammiad on wild emmer during each growing season in the 6-year study (Dinoor et al. 1991). Although the pathogens were virulent, and selfed progenies of the host plants were susceptible when inoculated ex situ, virtually no epidemic occurred in the natural population. Three pathogens not recorded previously on wild emmer in its natural habitat were discovered: a species of *Ascochyta* causing a leaf spot was found for the first time on any wheat in Israel; *Gaeumannomyces graminis* var. *tritici* J. Walker, the take-all pathogen that devastates cultivated wheat worldwide; and *Septoria nodorum* (Berk.) Berk, the glume blotch pathogen. In experimental seedling inoculations with pathogenic races under controlled conditions, high resistance to a race of *Puccinia striiformis* Westend was found in one Ammiad accession and moderate resistance in another. Of 225 accessions tested for seedling response, 22 were resistant to the *Erysiphe graminis* DC isolate "En Hanaziv", to which locally grown cultivated wheats are susceptible. The same lines were all susceptible to a local isolate. Of 192 seedling accessions inoculated with a local *Ascochyta* sp., 2% of the isolates were immune and 10% were resistant to the disease.

Moderate seedling resistance to a cultivated wheat isolate of the take-all fungus *Gaeumannomyces graminis* var. *tritici* J. Walker was found in two of 178 accessions. Juvenile and adult plants of a set of accessions inoculated with an isolate of *Septoria nodorum*, taken from bread wheat, showed considerable variation in disease severity. Local races of *Puccinia recondita* Dietel et Holw were compatible with *Thalictrum speciosissimum* C.A. Meyer, the alternate host of the fungus, although this host species does not occur in Israel or in contiguous countries. The detection of resistance to a non-local isolate of powdery mildew in accessions susceptible to a local isolate, confirms previous findings that natural host populations may contain hidden, valuable resistance to alien pathogens. The wild wheat population at Ammiad harbors many virulent pathogens, but, even though most of the plants are susceptible, the population is virtually free of disease. This situation demonstrates the major importance of environmental conditions in the dynamic equilibrium between hosts and pathogens, and the efficiency of defense strategies of wild populations against their pathogens and parasites.

5.4.1.4
High-Molecular-Weight Glutenin Polymorphisms: Spatiotemporal Ecological Factors

Variation in the electrophoretic pattern of the HMW glutenin subunits was studied at Ammiad during a 5-year period 1983–1984 to 1987–1988 (Felsenburg et al. 1991). These storage proteins were analyzed through seeds collected annually from individual plants at 249 defined sampling sites distributed in the aforementioned 11 subhabitats. Altogether 1,108 accessions were analyzed. The population was found to be highly polymorphic: the HMW glutenin loci of genome A, Glu-A1-1 and Glu-A1-2, had four and two alleles, respectively, and those of genome B, Glu-B1- and Glu-B1-2, had five and seven alleles, respectively. The A-genome alleles appeared in four combinations, and the B-genome alleles appeared in 12 combinations. There were 18 intergenomic combinations (A and B genotypes), some of which were very rare, while others were abundant and distributed along transects in clusters. The spatial distribution of these genotypes was nonrandom, each of the 11 subhabitats was characterized by different genotype frequencies. Yearly changes in genotypes, mostly occurring in the last 2 years of the study, had little effect on the total frequencies of the various genotypes. A high affinity was found between specific HMW glutenin genotypes and certain habitats. This affinity may have resulted from a random fixation of specific genotypes in different habitats (founder effect) or from natural selection, thus indicating either linkage between HMW glutenin alleles and adaptive genes (hitchhiking effect) or fitness of some of these allele combinations to specific microenvironments.

We now return from the general description of the Ammiad microsite study to the genetic studies of *T. dicoccoides* at this site, at the *coding* (allozyme) and largely *noncoding* (RAPD and SSR DNA) genomes.

5.4.1.5
Allozyme Diversity: Spatiotemporal Ecological Factors

Allozymic diversity in proteins encoded by 43 loci was analyzed electrophoretically from 1983–1984, during four consecutive growing seasons, in 812 individual plants of wild emmer wheat, *T. dicoccoides*, from the Ammiad microsite (Nevo et al. 1991c). Twenty-seven (= 63%) of the loci were polymorphic and 87 alleles were found at Ammiad. Significant genetic differentiation according to vegetationally and topographically defined habitats and subhabitats (described earlier in Noy-Meir et al. 1991a) was found primarily in space over very short distances, and secondarily over time (Fig. 5.12; Tables 5.20–5.22). The mean genetic indices (allele diversity *A*, polymorphism, *P-1%*, *P-5%*, and gene diversity *He*) in the four habitats over the 4 years of the study were: 1.478, 0.396, 0.300, and 0.110. The highest gene diversity occurred in the Karst formation, where soil moisture was most variable, i.e., in the habitat dis-

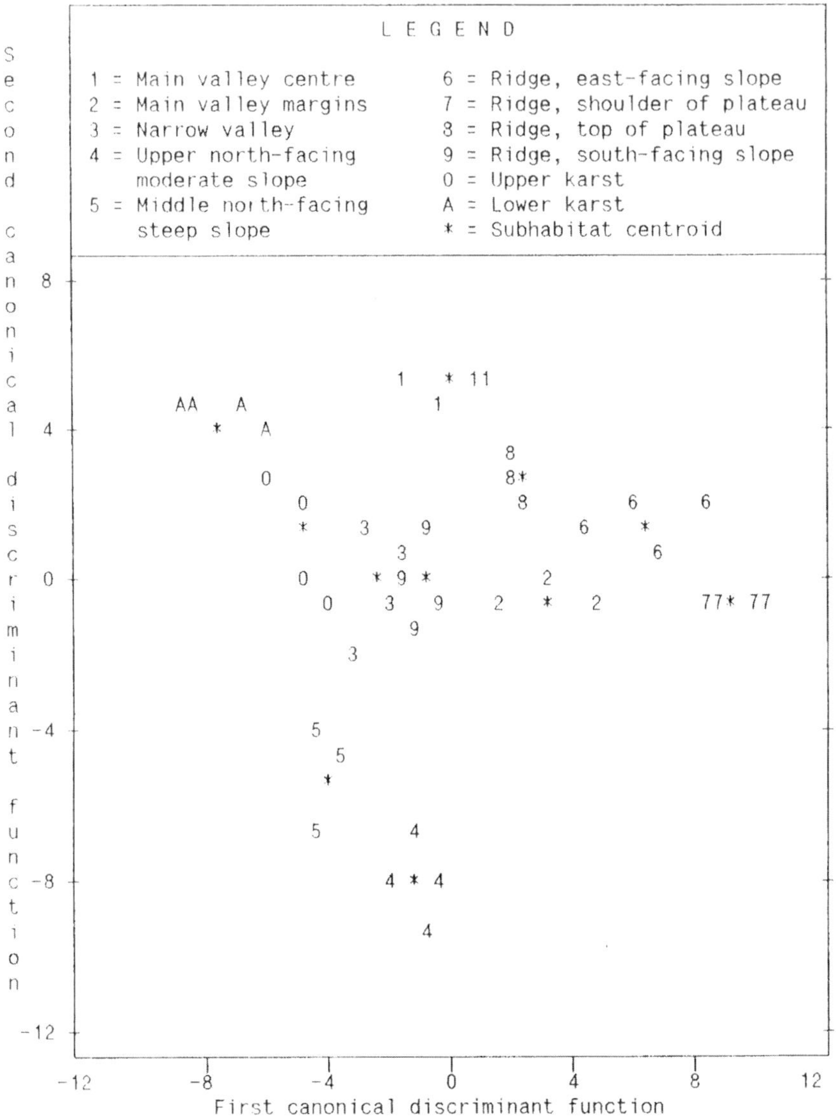

Fig. 5.12. Discriminant analysis by allele frequencies of 11 subpopulations in topographically and vegetationally defined subhabitats in four growing seasons. (Details in Table VIII in Nevo et al. 1991c)

playing the broadest ecological niche. Our results (Table 5.23) suggest that allozyme polymorphisms in wild emmer wheat are partially adaptive. Genetic differentiation appears to be primarily affected by environmentally variable factors related to topography and temporal climatic changes, probably

Table 5.20. Genetic indices based on 43 loci in four subpopulations of wild emmer wheat at Ammiad in vegetationally and topographically defined habitats during four consecutive growing seasons (numbers of accessions are given in parentheses). (Li 2000)

Year	Valley	North	Ridge	Karst	Mean	Total
A						
1983–1984	1.465 (85)	1.465 (29)	1.535 (88)	1.372 (30)	1.459 (232)	1.674 (232)
1984–1985	1.581 (73)	1.410 (24)	1.488 (62)	1.395 (24)	1.469 (183)	1.744 (183)
1985–1986	1.512 (62)	1.368 (23)	1.581 (75)	1.372 (19)	1.458 (179)	1.721 (179)
1986–1987	1.643 (74)	1.476 (28)	1.476 (90)	1.500 (26)	1.524 (218)	1.786 (218)
Mean	1.550 (294)	1.430 (104)	1.520 (315)	1.410 (99)	1.478 (812)	1.731 (812)
Total	1.767 (294)	1.721 (104)	1.744 (315)	1.674 (99)	1.727 (812)	2.023 (812)
P-1%						
1983–1984	0.349 (85)	0.372 (29)	0.442 (88)	0.302 (30)	0.366 (232)	0.419 (232)
1984–1985	0.488 (73)	0.359 (24)	0.419 (62)	0.326 (24)	0.398 (183)	0.535 (183)
1985–1986	0.442 (62)	0.263 (23)	0.535 (75)	0.326 (19)	0.392 (179)	0.512 (179)
1986–1987	0.476 (74)	0.429 (28)	0.405 (90)	0.405 (26)	0.429 (218)	0.524 (218)
Mean	0.439 (294)	0.356 (104)	0.450 (315)	0.340 (99)	0.396 (812)	0.497 (812)
Total	0.535 (294)	0.535 (104)	0.488 (315)	0.465 (99)	0.506 (812)	0.512 (812)
P-5%						
1983–1984	0.233 (85)	0.209 (29)	0.326 (88)	0.279 (30)	0.262 (232)	0.372 (232)
1984–1985	0.256 (73)	0.308 (24)	0.349 (62)	0.279 (24)	0.298 (183)	0.395 (183)
1985–1986	0.279 (62)	0.211 (23)	0.372 (75)	0.326 (19)	0.297 (179)	0.395 (179)
1986–1987	0.381 (74)	0.310 (28)	0.357 (90)	0.333 (26)	0.345 (218)	0.429 (218)
Mean	0.287 (294)	0.260 (104)	0.351 (315)	0.304 (99)	0.300 (812)	0.398 (812)
Total	0.395 (294)	0.256 (104)	0.395 (315)	0.372 (99)	0.355 (812)	0.442 (812)
He						
1983–1984	0.069 (85)	0.099 (29)	0.093 (88)	0.121 (30)	0.096 (232)	0.102 (232)
1984–1985	0.111 (73)	0.110 (24)	0.122 (62)	0.142 (24)	0.121 (183)	0.145 (183)
1985–1986	0.096 (62)	0.075 (23)	0.110 (75)	0.131 (19)	0.103 (179)	0.121 (179)
1986–1987	0.129 (74)	0.099 (28)	0.114 (90)	0.148 (26)	0.123 (218)	0.148 (218)
Mean	0.101 (294)	0.096 (104)	0.110 (315)	0.136 (99)	0.110 (812)	0.129 (812)
Total	0.111 (294)	0.110 (104)	0.115 (315)	0.164 (99)	0.125 (812)	0.135 (812)

A Mean number of alleles per locus; *P-1%* polymoprhism, 1% criterion; *P-5%* polymorphism, 5% criterion

through drought, i.e., aridity stress or soil-water availability (see a review of drought resistance in cereals in Baker 1989).

5.4.1.6
Spatiotemporal Allozyme Divergence Caused by Aridity Stress at the Ammiad Microsite: Extended Study over the 6 Years

Spatiotemporal diversity at 35 allozyme loci was assayed in 1,207 individuals of wild emmer wheat (*Triticum dicoccoides*) over 6 years from a microgeographic microsite, Ammiad, northern Israel (Li et al. 2001a). This analysis used, in addition to the local ecological parameters, two additional sample sets (1988 and 1993) and the allozymic data (1984–1987) mentioned above and analyzed by Nevo et al. (1991c). As described above, the Ammiad

Table 5.21. Coefficients of genetic distance, *D*, among subpopulations of *T. dicoccoides* in four vegetationally and topographically defined habitats in each of four consecutive growing seasons at the Ammiad microsite, Israel. (Nevo et al. 1991c)

Subpopulation	1	2	3	4	5	6	7	8	9	10	11	12	13	14	15	16
1 1983–1984 Valley		0.042	0.006	0.033	0.014	0.061	0.026	0.068	0.027	0.023	0.018	0.053	0.018	0.039	0.016	0.094
2 1983–1984 North			0.033	0.065	0.047	0.027	0.061	0.075	0.063	0.008	0.048	0.076	0.050	0.013	0.047	0.109
3 1983–1984 Ridge				0.042	0.017	0.051	0.022	0.073	0.031	0.022	0.014	0.060	0.021	0.033	0.012	0.106
4 1983–1984 Karst					0.047	0.068	0.056	0.051	0.063	0.035	0.053	0.029	0.048	0.064	0.049	0.051
5 1984–1985 Valley						0.044	0.015	0.074	0.016	0.028	0.012	0.061	0.006	0.040	0.010	0.101
6 1984–1985 North							0.061	0.090	0.075	0.028	0.058	0.090	0.062	0.012	0.062	0.134
7 1984–1985 Ridge								0.091	0.024	0.050	0.012	0.073	0.015	0.053	0.011	0.121
8 1984–1985 Karst									0.083	0.056	0.079	0.035	0.061	0.091	0.069	0.044
9 1985–1986 Valley										0.054	0.011	0.063	0.015	0.056	0.019	0.118
10 1985–1986 North											0.038	0.053	0.037	0.019	0.038	0.086
11 1985–1986 Ridge												0.061	0.010	0.041	0.005	0.111
12 1985–1986 Karst													0.053	0.080	0.054	0.030
13 1986–1987 Valley														0.050	0.006	0.089
14 1986–1987 North															0.045	0.121
15 1986–1987 Ridge																0.095
16 1986–1987 Karst																

D: Mean 0.049, range 0.005–0.0134; SE: range 0.002–0.011

Table 5.22. Partition of gene diversity, based on 43 loci, between and within subpopulations of *T. dicoccoides* in four vegetationally and topographically defined habitats at Ammiad in each of four consecutive growing seasons. (Nevo et al. 1991c)

Locus	Alleles	Sample	*Ht*	*Hs*	*Dst*	*Gst*
Aat-1A	1	770	0.0	0.0	0.0	–
Aat-1B	1	771	0.0	0.0	0.0	–
Aat-2A	2	808	0.0074	0.0073	0.0001	0.0162
Aat-2B	1	808	0.0	0.0	0.0	–
Aat-3A	2	807	0.0148	0.0145	0.0003	0.0181
Aat-3B	1	807	0.0	0.0	0.0	–
Acph-3	3	727	0.3190	0.2672	0.0518	0.1624
Adh-1A	3	588	0.2362	0.1756	0.0606	0.2566
Adh-1B	3	589	0.0368	0.0250	0.0117	0.3188
Adh-2A	2	740	0.0081	0.0079	0.0002	0.0188
Adh-2B	1	740	0.0	0.0	0.0	–
Cat-A	1	207	0.0	0.0	0.0	–
Cat-B	1	207	0.0	0.0	0.0	–
Est-4A	1	605	0.0	0.0	0.0	–
Est-4B	2	605	0.2625	0.2318	0.0307	0.1169
Est-5A	3	789	0.4247	0.3654	0.0594	0.1397
Est-5B	5	785	0.3314	0.2822	0.0492	0.1484
Gdh-A	1	794	0.0	0.0	0.0	–
Gdh-B	1	794	0.0	0.0	0.0	–
Gluc-A	2	759	0.3058	0.2563	0.0495	0.1619
Gluc-B	2	758	0.3315	0.2770	0.0545	0.1644
Hk-A	3	447	0.3154	0.2589	0.0566	0.1793
Hk-B	4	345	0.2146	0.1612	0.0535	0.2492
Ipol	2	178	0.1060	0.0872	0.0189	0.1781
Ipor-A	2	777	0.0102	0.0097	0.0005	0.0507
Ipor-B	2	776	0.1001	0.0816	0.0185	0.1847
Mdh-1A	2	787	0.2312	0.1680	0.0632	0.2734
Mdh-1B	1	787	0.0	0.0	0.0	–
Mdh-2	2	795	0.0150	0.0146	0.0004	0.0280
Nadh-1A	3	634	0.4693	0.3512	0.1182	0.2518
Nadh-1B	3	635	0.5132	0.4139	0.0993	0.1934
Nadh-2A	2	729	0.1134	0.0687	0.0448	0.3948
Nadh-2B	2	729	0.1348	0.0929	0.0419	0.3108
Pept-1A	1	754	0.0	0.0	0.0	–
Pept-1B	2	754	0.0715	0.0697	0.0018	0.0257
Pept-3	2	568	0.0712	0.0641	0.0071	0.1000
Pgi-A	5	795	0.5284	0.4705	0.0579	0.1096
Pgi-B	3	799	0.3911	0.2519	0.1392	0.3558
Pgm-A	1	810	0.0	0.0	0.0	–
Pgm-B	1	810	0.0	0.0	0.0	–
6Pgd-1A	1	809	0.0	0.0	0.0	–
6Pgd-1B	1	809	0.0	0.0	0.0	–
6Pgd-2	3	762	0.2611	0.1938	0.0673	0.2578
Mean			0.1355	0.1086	0.0269	0.1986
Mean between four habitats (across seasons)			0.1355	0.1189	0.0166	0.1224
Mean between growing seasons (across habitats)			0.1355	0.1305	0.0050	0.0366

Ht Total gene diversity, *Hs* average gene diversity within populations, *Dst* average gene diversity between populations, *Gst* gene diversity between populations, relative to *Ht*

Table 5.23. Coefficients of multiple determination (R^2) of allozyme genetic diversity (*He*, *P*, and selected allelic frequencies as the dependent variables) and 22 micro-ecological variables (as independent variables) in 11 vegetationally and topographically defined subhabitats of *T. dicoccoides* in three consecutive growing seasons at Ammiad. (Nevo et al. 1991c)

	Stepwise model; micro-ecological variables		
Genetic indices			
He	*MNDB* 0.343***	*MNDB DBS* 0.547***	*MNDB DBS ATW* 0.585***
P-5%	*MNDB* 0.313***	*MNDB DBN* 0.467***	*MNDB DBN Access* 0.558***
Locus; allele			
Aat-2A a	*Access* 0.112@	*Access* F_c 0.390***	*Access* F_c F_a 0.566***
Adh-1A a	F_c 0.533***	F_c *HTW* 0.581***	
Est-4B c	*DBS* 0.141*	*DBS ATS* 0.248*	*DBS ATS* F_a 0.301*
Est-5A a	F_b 0.366***	F_b *HTS* 0.550***	F_b *HTS ATE* 0.602***
Est-5A e	F_b 0.366***	F_b *HTE* 0.550***	F_b *HTE Access* 0.602***
Est-5B a	*HTE* 0.121*	*HTE DBW* 0.283**	*HTE DBW MNDB* 0.336**
Est-5B d	*DBS* 0.155*	*DBS HTS* 0.510***	*DBS HTS* F_b 0.563***
Est-5B f	F_c 0.341***		
Gluc-A a	F_c 0.526***	F_c *AVH* 0.751***	F_c *AVH* F_a 0.784***
Gluc-B a	F_c 0.529***	F_c *DBN* 0.730***	F_c *DBN MXH* 0.761***
Hk-A b	*AVA* 0.321**	*AVA* F_a 0.402**	*AVA* F_a *ATE* 0.448**
Hk-B b	F_c 0.525***		
Mdh-1A a	F_c 0.709***	F_c *ATE* 0.768***	
Nadh-1B c	*HTW* 0.195**	*HTW Access* 0.232*	*HTW Access MXA* 0.391**
Nadh-2A a	*HTE* 0.436***	*HTE ATW* 0.566***	*HTE ATW HTS* 0.591***
Nadh-2B a	*HTE* 0.248**	*HTE* F_c 0.396***	*HTE* F_c *ATE* 0.448***
Pept-1B a	*ATS* 0.108@	*ATS DBS* 0.321**	*ATS DBS HTE* 0.424***
Pept-3 b	*Access* 0.070	*Access ATS* 0.176@	*Access ATS ATE* 0.304*
Pgi-A a	F_c 0.473***	F_c *DBS* 0.528***	F_c *DBS HTN* 0.552***
Pgi-A b	*HTS* 0.222**	*HTS DBE* 0.287**	*HTS DBE* F_b 0.474***
Pgi-A d	*MXH* 0.485***	*MXH* F_b 0.549***	*MXH* F_b *DBW* 0.696***
Pgi-B b	*HTS* 0.546***	*HTS DBS* 0.727***	*HTS DBS HTE* 0.781***
6Pgd-2 a	F_c 0.636***	F_c *ATE* 0.719***	F_c *ATE MXA* 0.740***

@ $p < 0.10$; * $p < 0.05$, ** $p < 0.01$, *** $p < 0.001$

Legend of ecological variables:

average rock angle, *AVA*, average rock height, *AVH*, and harmonic average distance of a wheat sampling point to the base of the nearest rock, *HADB* (Noy-Meir et al. 1991). These variables could be further designated as follows: average angle between sampling points of rocks to the north, east, south and west as *ATN*, *ATE*, *ATS* and *ATW*; height of rocks in the four compass directions as *HTN*, *HTE*, *HTS* and *HTW*; and distance to rock bases in the four directions as *DBN*, *DBE*, *DBS* and *DBW*. The maximum angle was denoted *MXA*, maximum rock height as *MXH*, maximum distance as *MXDB* and the corresponding minimum values as *MNA*, *MNH* and *MNDB*. Following Noy-Meir et al. (1991) a five-point scale was used to denote the accessibility, *Access*, of the sampling site to grazing animals. Affinities of the vegetation to three vegetation components, designated by Noy-Meir et al. (1991) by the generic names of their leading species as *Aegilops*, *Brachypodium* and *Cephalaria*, were estimated as functions F_a, F_b and F_c, respectively microsite includes four major habitats (North-Facing Slope, Valley, Ridge, and Karst) which show topographic and ecological heterogeneity (Fig. 5.11).

Significant temporal and spatial variations of allele frequencies and levels of genetic diversity were detected in the four major subpopulations. Significant associations were observed among allele frequencies and gene diversities at different loci, indicating that many allele frequencies change *over time* and *space* in the same or opposite directions. Multiple regression analysis (Tables 5.23, 5.28) showed that variation in soil-water content and rainfall distribution in the growing season significantly affected ten allele frequencies, genetic diversity (*He* and *P*-5%), and gene diversity at four loci. As detailed below, random genetic drift and hitchhiking models may not explain such locus-specific spatiotemporal divergence, strong allelic or locus correlation, and the functional importance of allozymes. Natural ecological selection, presumably through water stress, might be an important force adaptively directing spatiotemporal allozyme diversity and divergence in wild emmer wheat at the Ammiad microsite.

Most importantly, we presented direct estimates for correlation between spatiotemporal allozyme diversity and soil moisture, including rain distribution during the growing season of wild emmer, coordinating changes between alleles/locus, and for the differential effect of protein functional groups. The detailed results are described in the following sections.

5.4.1.7
Allozymic Correlations with Rainfall and Soil Moisture over the 6 Years

Rainfall and soil-moisture content during the period 1983–1993 appear in Table 5.24. We examined electrophoretically 35 allozymic loci in wild emmer collected from 1984 to 1993 (Li et al. 2001a). Out of the 35 allozymic loci, allele frequencies varied significantly (χ^2, $p < 0.05$–0.00005) over time at 7, 9, 11, and 11 allozymic loci in the North, Valley, Ridge, and Karst subpopulations, respectively (Table 5.25). Alleles $Est\text{-}5B_{null}$, $Nadh\text{-}2B_a$, $Pgi\text{-}A_b$, and $Pgi\text{-}B_a$, showed significant temporal variations in their frequencies in all four subpopulations. In four other alleles, $Adh\text{-}1B_a$, $Gluc\text{-}A_b$, $Gluc\text{-}B_b$, and $Mdh\text{-}1A_b$, significant temporal changes in frequencies were only found in the Karst subpopulation (Li et al. 2001a).

Among the four subpopulations from the North, Valley, Ridge, and Karst habitats, the mean allele frequencies over the 6 years significantly varied at 12 loci (Kruskal-Wallis test, $p < 0.05$–0.001). Likewise, significant spatial variation was found for between-year variance in allele frequencies at eight loci. The largest differences were found not only between the farthest Karst and the other three habitats (e.g., $Gluc\text{-}A_b$, $Gluc\text{-}B_b$, $Mdh\text{-}1A_b$), but also among the neighboring North, Valley, and Ridge subpopulations (e.g., at loci $Pgi\text{-}B_a$, $Nadh\text{-}2A_a$). The results suggested that allele distributions at allozymic loci were *nonrandom* on the microgeographic scale as demonstrated earlier (1984–1987) using 43 allozymic loci by Nevo et al. (1991c).

Table 5.24. Rainfall (mm) distribution, surface (0–5 cm) soil-moisture content (%) after early rains in the growing season of wild emmer, and samples sizes in different years at Ammiad microsite. (1984–1988 data from Noy-Meir et al. 1991a)

Year	Rainfall (mm)				Soil-moisture content (%)				Sample size				
	Annual	Oct.– Nov.	Dec.– Feb.	Mar.– Apr.	North	Valley	Ridge	Karst	North	Valley	Ridge	Karst	Tota
1983–1984	505	97	203	205					30	89	86	28	23[illegible]
1984–1985	623	124	416	83					24	75	61	22	18[illegible]
1985–1986	415	43	321	25	30.8b	29.0c	32.7a	31.9ab	23	62	74	18	17[illegible]
1986–1987	792	288	365	139	23.6b	19.9c	21.3bc	30.3a	28	81	89	24	22[illegible]
1987–1988	598	33	468	97	25.7b	23.6c	27.8b	29.9a	30	68	99	21	21[illegible]
1992–1993	649	116	476	49					23	61	72	19	17[illegible]
Total									158	436	481	132	120[illegible]

Significance based on Duncan's multiple range test among microhabitats ($p < 0.05$) for each of the 3-year measurement separately. Values with a common letter are not significantly different. The analysis of variance for soil moisture-conten among microhabitats was highly significant ($p < 0.00001$, see Noy Meir et al. 1991a)

5.4.1.8 Genetic Diversity of Allozymic Loci: Summary over the 6 Years

The level of genetic diversity, including number of alleles per locus (*A*) averaged over the 35 loci, proportion of polymorphic loci (*P*), and gene diversity (*He*), fluctuated over the 6 years in the four subpopulations (Fig. 5.13a–c) (Li et al. 2001a). They showed different amounts of temporal variation in *A*, *P*, and *He*. The *A* (Fig. 5.13a) and *P* (Fig. 5.13b) variation over the 6 years in the Karst subpopulation was relatively lower (SD = 0.076, 0.058 for *A* and *P*, respectively) than in the other three subpopulations (SD = 0.085–0.093, 0.069–0.080 for *A* and *P*, respectively). However, variation in gene diversity, *He*, over sampling years was larger in the Karst (SD = 0.035) than in the other three subpopulations (SD = 0.009–0.025). In particular, the Karst *He* value became much lower than those of the other subpopulations in the 1993 samples (Fig. 5.13c); such a pattern was inconsistent with those found in 1984–1988 by Nevo et al. (1991c) and by Li et al. (2001a; Fig. 5.13c). Thus, while some yearly changes were parallel in some habitats, others differed considerably.

5.4.1.9 Partition of Allozymic Diversity over the 6 Years

Significant gene differentiation (*Gst*) among sampling years was observed at 18 allozymic loci (χ^2 test, $p < 0.05$–0.00005, Table 5.26) (Li et al. 2001a). In particular, at locus *Ipor*-B, significant gene differentiation ($Gst = 0.151$, $p < 0.00005$) was found only among sampling years, not among microgeographic

Table 5.25. Temporal and spatial variations in allele frequencies at polymorphic allozymic loci over 6 years in four natural subpopulations of *T. dicoccoides* at the Ammiad microsite (from Li et al. 2001a). The allele frequencies in 1984–1987 were extracted from Nevo et al. (1991c)

$Locus_{allele}$[a]	Temporal variation								Spatial variation[b]:	
	North ($N = 158$)		Valley ($N = 436$)		Ridge ($N = 481$)		Karst ($N = 132$)		Of mean H(3, 24)	Of variance Bartlett's test (df = 3)
	Mean	SD	Mean	SD	Mean	SD	Mean	SD		
*Aat-3*A_a	0.97	0.034	0.97	0.029	0.99	0.006	1.00	0.000	7.36	0.20
*Adh-1*A_a	0.93	0.117	0.93	0.092****	0.92	0.105****	0.60	0.332***	3.45	4.96
*Adh-1*B_a	0.95	0.033	1.00	0.000	1.00	0.000	0.93	0.167**	11.42**	–
*Adh-2*A_a	1.00	0.000	0.99	0.011	0.99	0.006	1.00	0.000	2.09	–
*Est-5*A_a	0.90	0.045	0.71	0.102***	0.52	0.089***	0.98	0.024	20.80****	0.88
*Est-5*B_{null}	0.58	0.209****[c]	0.86	0.110***	0.87	0.094****	0.56	0.408****	9.01*	9.03*
Gluc-A_b	0.95	0.061	0.77	0.025	0.94	0.039	0.34	0.217**	19.80****	17.74**
Gluc-B_b	0.95	0.054	0.75	0.033	0.93	0.042	0.31	0.187*	19.46****	14.00**
Ipor-B_a	1.00	0.000	0.95	0.124***	0.96	0.086***	0.96	0.067	2.50	–
*Mdh-1*A_b	0.97	0.032	0.92	0.062	0.92	0.022	0.36	0.198**	16.46****	10.62*
*Nadh-1*A_a	0.81	0.260****	0.48	0.305****	0.56	0.230****	0.80	0.137	6.27	1.65
*Nadh-1*B_a	0.50	0.234***	0.65	0.241****	0.57	0.130****	0.86	0.114	9.17*	2.57
*Nadh-2*A_a	0.59	0.200****	0.98	0.028	0.99	0.017	0.98	0.023	14.82**	4.36
*Nadh-2*B_a	0.61	0.210***	0.95	0.090****	0.89	0.189****	0.82	0.402***	9.11*	7.85*
*Pept-1*B_a	0.97	0.047	0.96	0.030	0.98	0.020	1.00	0.000	7.40	–
Pept-3$_a$	0.97	0.054	0.96	0.029	0.82	0.208***	0.72	0.327****	3.58	15.60**
Pgi-A_b	0.40	0.208****	0.57	0.283***	0.55	0.277****	0.16	0.120****	9.46*	1.92
Pgi-B_a	0.65	0.336****	0.14	0.085****	0.09	0.048****	0.56	0.313****	10.13*	10.78*
6Pgd-2$_a$	0.98	0.022	0.95	0.037	0.88	0.076**	0.44	0.321****	10.62*	12.65**

[a] The following loci were monomorphic in all samples collected in different years: *Aat-1*A, *Aat-1*B, *Aat-2*B, *Aat-3*B, *Adh-2*B, *Mdh-1*B, *Pept-1*A, The loci *Pgm*-A, *Pgm*-B, *6Pgd-1*A, *6Pgd-1*B were monomorphic in all except the 1988 collection

[b] Difference of means among the four subpopulations was tested using nonparametric Kruskal-Wallis test

[c] The χ^2 test for homogeneity of allele frequency: * $P < 0.05$, ** $P < 0.01$, *** $P < 0.001$, **** $P < 0.0001$

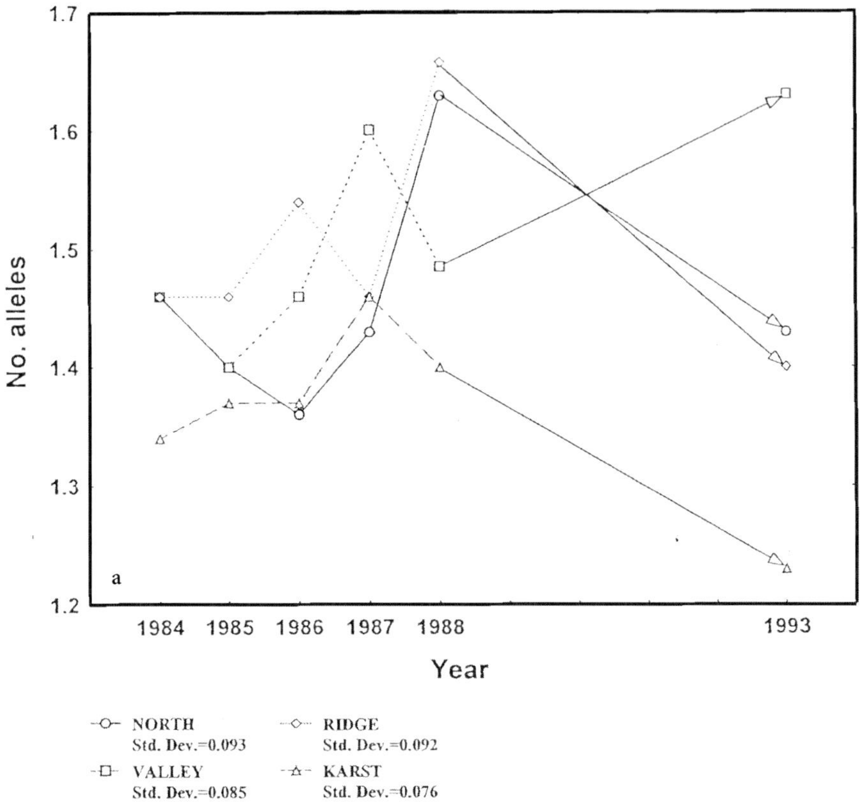

Fig. 5.13a–c. Temporal changes of number of alleles per allozymic locus (**a**) proportion of polymorphic loci (**b**) and gene diversity (**c**) in the North, Valley, Ridge, and Karst subpopulations of *T. dicoccoides* at the Ammiad microsite. (Li et al. 2001a)

habitats; and in 1984, 1988, and 1993, locus *Ipor*-B was monomorphic with the same allele *Ipor*-B_a. Over all the 35 loci, the variation due to differences in allele frequency among sampling years within habitats could account for 17.3% of the total gene diversity, whereas that between habitats accounted for 15.2% (Table 5.26).

Over the 6 years, the Karst subpopulation showed relatively lower *A* (1.8) and *P* (0.54) values than the North ($A = 2.0$, $P = 0.66$), Valley ($A = 2.0$, $P = 0.66$) and Ridge subpopulations ($A = 2.0$, $P = 0.77$). But *He* over all loci and 6 years was relatively higher in the Karst ($He = 0.180$) than in the other three subpopulations ($He = 0.136$–0.138). Significant gene differentiations (*Gst*) at 11–14 loci were found among the four subpopulations in each of the sampling years. The *Gst* values among the four subpopulations fluctuated through the sampling years. At four loci, *Est-5A*, *Est-5B*, *Gluc-A*, and *Gluc-B*, gene differentiations among the four subpopulations were highly significant (χ^2 test, $p < 0.05$–0.0001) in all six sampling years. Over all the polymorphic loci, a sig-

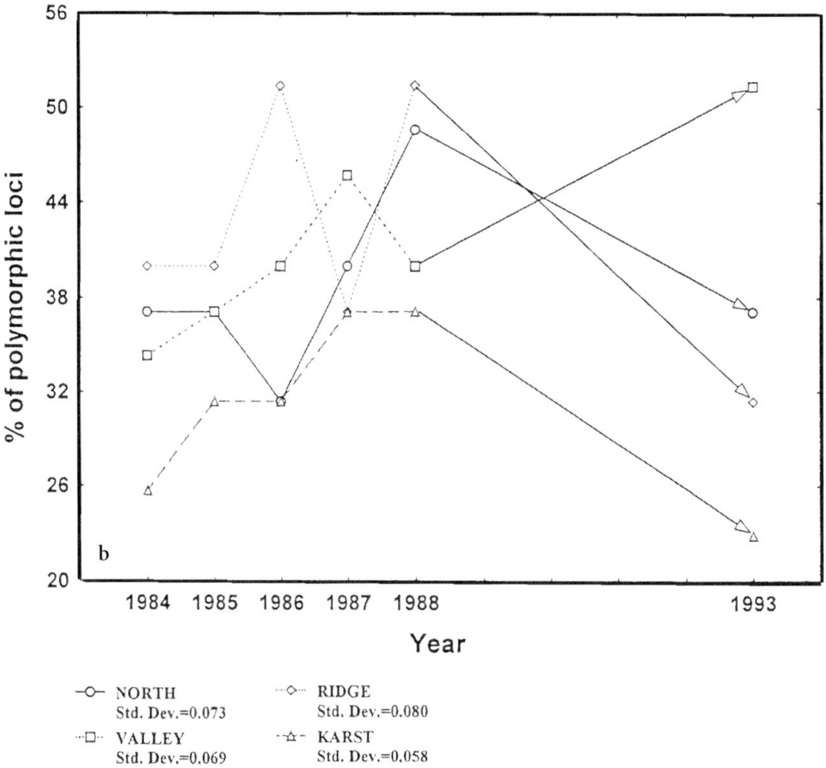

Fig. 5.13a–c. *Continued*

nificant correlation was found between the amount of early rains (Oct.–Nov., Table 5.24) and *Gst* value (Pearson's $r = 0.82$, $N = 6$, $p < 0.05$). This correlation may suggest that early seed germination due to early rains, and microgeographic difference in soil moisture after early rains (Table 5.24) influence genetic divergence among habitats. On the average of the six sampling years, the variation among the four habitats could account for 15.2% of the total gene diversity (Table 5.26). This is a lower estimate than in the study of Nevo et al. (1991c) examining 43 enzymatic loci but only over a 4-year period, where the *Gst* between habitats was 19.8% (Table 5.22).

5.4.1.10
Correlation of Allele Frequencies and *He* Among Allozymic Loci over the 6 Years

For the data in the four subpopulations taken over 6 years (1984–1993), correlations of allele frequencies and gene diversities at different loci were calculated, including only the most common alleles at polymorphic loci (Li et al. 2001a). A large number of correlation coefficients between allele frequencies

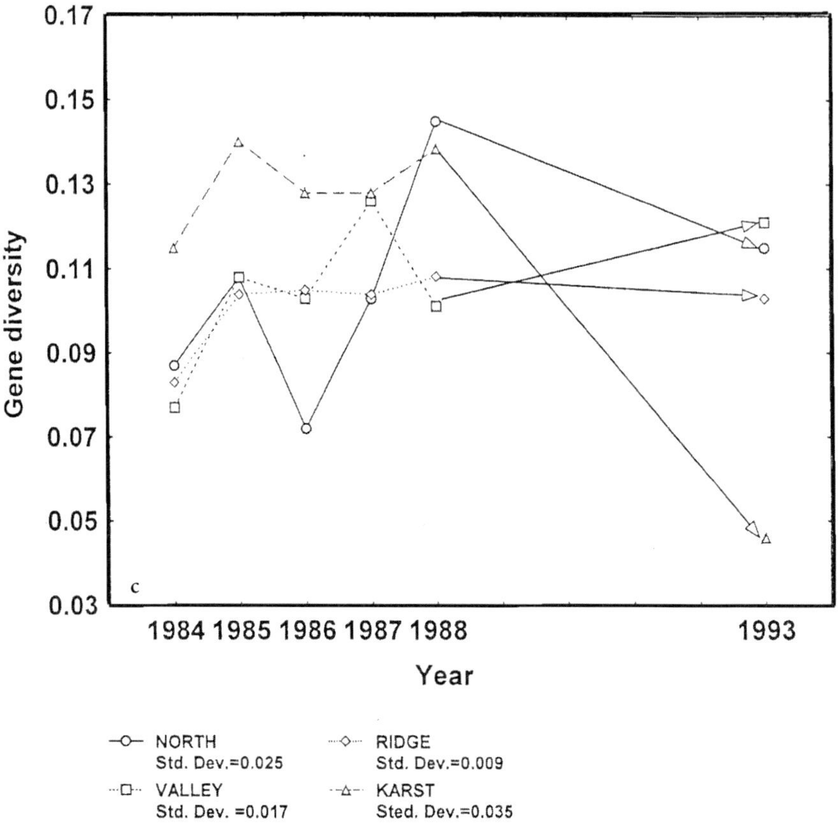

Fig. 5.13a–c. *Continued*

(26.8% of the 153 cases examined) at different loci were significant ($p <$ 0.05–0.001). Very close correlations were observed between some loci that belonged to the same allozymic groups, e.g., alleles *Gluc-A*$_b$ and *Gluc-B*$_b$ (r = 0.99, $p < 0.001$). Likewise, very strong associations were found between allozymic groups, such as between *Mdh-1A*$_b$ and the *Gluc* group ($r = 0.91$, $p < 0.001$). In general, however, allele-frequency changes within the same allozymic groups showed a higher likelihood ($\chi^2 = 4.65$, df = 1, $p < 0.05$) of being significantly ($p < 0.05$) associated with each other (55% of 11 cases examined) than those among different allozymic groups (33% of 142 cases examined).

For gene diversity (*He*), 15.7% of the 153 cases examined were significantly correlated among a variety of loci. Obviously, locus pairs *within* the same allozymic group (57% of 11 cases examined) showed more significant correlations ($p < 0.01$) than those *between* allozymic groups (4% of the 142 cases examined). These two proportions are significantly different ($\chi^2 = 19.89$, df = 1, $p < 0.00005$). For instance, very strong correlations were observed

Table 5.26. Partition of allozyme genetic differentiation, *Gst*, among the North, Valley, Ridge and Karst microhabitats in samples of *T. dicoccoides* collected at Ammiad, Israel, from 1984 to 1993 sampling years. The 1984–1987 data were from Nevo et al. (1991c). (Li et al. 2001a)

Locus	Gst among the four subpopulations						Overall H_T	Relative gene diversity (Gst)[a]	
	1984	1985	1986	1987	1988	1993		Between years	Between habitats
*Aat-3*A	0.026	0.010	0.010	0.016	0.045	0.041	0.026	0.008	0.011**
*Adh-1*A	–	0.320****	Missing	0.269***	0.089**	–	0.298	0.096****[b]	0.141****
*Adh-1*B	0.033	–	–	0.291***	0.028	0.070	0.066	0.026****	0.051****
*Adh-2*A	–	0.020	0.010	–	–	–	0.004	0.006	0.002
*Est-5*A	0.170****	0.242****	0.171****	0.198****	0.256****	0.150*	0.363	0.057****	0.167****
*Est-5*B	0.055**	0.373****	0.260****	0.089***	0.097***	0.466****	0.456	0.054****	0.095****
Gluc-A	0.190****	0.122**	0.279****	0.431****	0.304****	0.810****	0.373	0.003	0.299****
Gluc-B	0.260****	0.121**	0.268****	0.418****	0.314****	0.810****	0.388	0.003	0.315****
Ipor-B	–	0.024	0.087*	0.019	–	–	0.054	0.151****	0.010
*Mdh-1*A	0.132***	0.392****	0.404****	0.446****	0.310	0.712****	0.330	0.011**	0.368****
Mdh-2	0.026	0.020	–	0.028	0.015	0.024	0.019	0.005	0.009*
*Nadh-1*A	–	0.119*	0.282***	0.316****	0.161****	0.126*	0.466	0.174****	0.107****
*Nadh-1*B	0.246***	0.081*	0.225****	0.190****	0.068	0.221***	0.498	0.071****	0.080****
*Nadh-2*A	0.244****	0.465****	0.111	0.418****	0.142****	0.433****	0.202	0.033****	0.231****
*Nadh-2*B	0.101**	0.465****	0.111	0.418***	0.142****	0.322****	0.296	0.195****	0.114****
*Pept-1*B	0.025	0.027	0.015	0.038	–	0.012	0.041	0.014**	0.011*
Pept-3	0.168**	0.395**	0.015	0.013	0.222****	0.521****	0.209	0.086****	0.078****
Pgi-A	0.046	0.236****	0.123***	0.199***	–	0.267****	0.664	0.312****	0.057****
Pgi-B	0.444****	0.316****	0.340****	0.423****	0.032	0.438****	0.658	0.462***	0.158****
Pgm-A	–	–	–	–	0.368****	–	0.060	0.086****	0.043****
Pgm-B	–	–	–	–	0.168****	–	0.127	0.257***	0.019****
*6Pgd-1*A	–	–	–	–	0.161****	–	0.199	0.703***	0.021****
6pgd-2	0.164****	0.397****	0.442****	0.431****	–	0.564****	0.300	0.039***	0.296****
Mean	0.182	0.264	0.249	0.306	0.196	0.428	0.175	0.173	0.152

[a] Calculated over all 35 allozymic loci
[b] Significance level of the relative gene differentiation was obtained from a χ^2 for allele frequency heterogeneity at each hierarchical level: $*p < 0.05$, $**p < 0.01$, $***p < 0.001$, $****p < 0.0001$

between *He* for *Gluc-A*$_b$ and *Gluc-B*$_b$ ($r = 0.99$, df $= 22$, $p < 0.001$), and for *Nadh-2A*$_a$ and *Nadh-2B*$_a$ ($r = 0.84$, df $= 22$, $p < 0.001$)

No obvious pattern was found when we attempted to extend the foregoing group-related test to the level of protein groups, such as transferases, oxidoreductases, hydrolases, and isomerases. This suggests a high specificity of the factors that caused the similarity of within-allozymic groups.

5.4.1.11
Variable Selection of Protein Groups over the 6 Years

Relative environmental and protein effects were further analyzed by ANOVA for frequency of the most common allele and *He* at polymorphic loci (Li et al. 2001a). The allele-frequency data were transformed by means of ArcSin($\sqrt{x}$) before the ANOVA analysis. Loci coding for proteins with similar functions are defined as an allozymic group, and proteins with similar metabolic function belong to a metabolic group (showing a similar metabolic mechanism, such as transferase, oxidoreductase, hydrolase, and isomerase groups). Table 5.27 summarizes this analysis, which showed that habitat, allozymic group, metabolic group, and interaction of habitat × allozymic group and year × allozymic group significantly affected allele frequencies and *He* over all 21 polymorphic loci ($p < 0.05$–0.00005, Table 5.27). The yearly effect seemed to be low over all 21 polymorphic loci. This finding may imply that possible allozymic-specific effects of temporal environmental fluctuation were not taken into account.

Accordingly, further analysis was carried out for each allozymic group. The results showed that sampling years significantly affected allele frequency and/or *He* at *Gluc, Pgi, Pgm*, and *Nadh* loci ($p < 0.05$–0.00005), and habitat effect and the interaction of year × habitat also significantly affected allele frequency and/or *He* at *Gluc* and *Pgm* loci ($p < 0.00005$, Table 5.27). At the loci *Gluc-A* and *Gluc-B*, the most different year was 1993 for both allele frequency and gene diversity in all four habitats. At the *Pgi* and *Pgm* loci, the most contrasting allele frequencies and *He* values were found in 1988 in all four habitats. At loci *Nadh-1A* and *Nadh-1B*, the extreme allele frequency and *He* value were shown in 1993 in the North, and in 1984 in the Ridge and Karst subpopulations. In the Valley subpopulation, the extreme allele frequency and *He* value were found in 1988 at the locus *Nadh-1A* and in 1986 at locus *Nadh-1B*. Thus, the temporal dynamics differed among habitats and loci, suggesting selection directly on these loci.

5.4.1.12
Allozymic Diversity and Water Availability in Critical Growth Periods

To determine the existence of any significant associations between genetic variables and environmental water supply, stepwise multiple regression was conducted using amounts of annual early and late rains and soil-moisture content of 1986–1988 (data for other years are not available) as independent

Table 5.27. Results of ANOVA test for temporal and spatial effects on allozyme frequencies of the most common alleles (AF) and gene diversities (*He*) at polymorphic loci in the North, Valley, Ridge, and Karst subpopulations of *T. dicoccoides* at Ammiad, Israel, for 6 years (1984–1988 and 1993). (Li et al. 2001a)

Variables[a]	*F* ratio											
	Over 21 polymorphic loci			df1, df2	*Gluc*		*Pgi*		*Pgm*		*Nadh*	
	df1, df2	AF	*He*		AF	*He*	AF	*He*	AF	*He*	df1, df2	AF
1. Year	5, 238	1.15	0.61	5, 24	13.41****	158.35****	4.79**	2.50@	18.77****	19.01****	5, 72	2.95*
2. Habitat	3, 238	9.85****	2.18	3, 24	2244.86****	816.01****	1.13	2.47@	1.94	0.67	3, 72	3.25*
3. Allozymic group[b]	10, 238	34.76****	18.32****									
4. Metabolic group[a]	3, 406	60.88****	21.60****									
1 × 2	15, 238	0.37	0.93	15, 24	110.52****	48.56****	0.07	1.19	6.13****	0.67	15, 72	0.33
1 × 3	50, 406	1.52*	1.91**									
2 × 3	30, 238	3.92****	3.16****									
1 × 2 × 3	150, 238	0.38	0.39									
1 × 4	15, 406	1.69	1.51									

$*p < 0.05$, $**p < 0.01$, $****p < 0.00005$, $@p < 0.10$

[a] Variables include genome (A vs B), but genome effect and its interaction with four other factors are not presented due to nonsignificant effect for AF and *He*

[b] Allozymic group: proteins with the same enzymatic name, such as esterase group, peptidase group

[c] Metabolic group: enzymes with similar metabolic mechanisms, such as oxidoreductase, hydrolase, isomerase groups

variables, and allele frequencies, number of alleles and *He* as dependent variables (Li et al. 2001a). The results showed that ten allele frequencies were significantly ($p < 0.05$–0.01) correlated with rain distribution in the growing season and soil-water content (Table 5.28). Early rain and soil moisture during 1986–1988 in the four habitats appeared to be the most important ecological factors for four allele distributions (*Adh-1*B_a, *Nadh-1*A_a, *Pgi-*B_a, *6Pgd-*2_a), and could predict 35.5–42.1% of the variation of allele frequencies. Early and late rains in the growing season seemed to be important factors for three other alleles (*Ipor-*B_a, *Pgi-*A_b, *Pept-*3_a) (Table 5.28). Soil moisture and late rain could account for 31.7%–43.4% of the variation of the frequencies of alleles *Adh-1*A_a, *Mdh-1*A_b. Early rain alone could account for 31.1% of the variation of the frequency of allele *Pept-1*B_b.

The number of alleles at five loci was significantly correlated with yearly rain distribution and soil moisture in the four habitats (Table 5.28). Early rain and soil moisture significantly affected *A* at loci *Adh-1A* ($R^2 = 0.407$, $p < 0.05$) and *Pept-3* ($R^2 = 0.507$, $p < 0.05$). Early and late rains could significantly predict 65.8% and 63.3% of the variation of *A* at loci *Pgi-A* and *6Pgd-2*, respectively. Late rain alone also significantly affected the number of alleles at locus *Nadh-1B* ($R^2 = 0.381$, $p < 0.01$).

Annual rainfall and its early and late distribution in the growing season, and soil moisture also affected gene diversities at six loci (*Adh-1B*, *Est-5A*, *Ipor-B*, *Pept-1B*, *Pept-3*, *Pgi-B*) (Table 5.28). One or two of these ecological factors could predict 38.0%–58.6% of the variation in gene diversity (*He*) at the six loci. For example, in 1986–1987, the early and late rains were extremely high; the gene diversity was lowest at locus *Ipor-B* and highest at locus *Pept-1B*. In all, allozyme diversity appears to be intimately influenced by water availability of early and late rains in critical growth periods at germination and spiking, respectively.

5.4.2 SSR Divergence in Wild Emmer Wheat at Ammiad

Genetic diversity was analyzed in 28 microsatellite loci (see list in Fig. 5.9) in the 1993 collection of *T. dicoccoides* at the Ammiad microsite, north of the Sea of Galilee, Israel (Li et al. 2000c). This microsite was subdivided into four major habitats, North, Valley, Ridge and Karst, and further subdivided into nine subhabitats, as previously described in our allozyme studies (Nevo et al. 1991c and section 5.4.1 above). The units thus defined showed strong and highly significant differentiation in ecological factors, in particular with respect to cover, proximity and height of rocks, and surface soil moisture after early rains (Anikster and Noy-Meir 1991a). The results showed that allele distributions at microsatellite loci were *nonrandom* and associated with habitats, as was shown by coded allozymes. Significant genetic differentiation and variation in repeat number were found among subpopulations in the four major habitats and nine subhabitats. Habitat-*specific* and *unique* alleles and

Table 5.28. Significant adjusted coefficients of stepwise multiple regression (R^2) of allozyme allele frequencies, number of alleles, and gene diversity (*He*) (as the dependent variable) and water conditions (as independent variables) in the four habitats of *T. dicoccoides* in three consecutive growing years (1986–1988) at Ammiad microsite, Israel. (Li et al. 2001a)

Locus-allele	First Variable	R^2	Two Variables	R^2
Allele frequency				
*Adh-1*A_a	Soil moisture[a]	0.007	Soil moisture, late rain	0.317*
*Adh-1*B_a	Early rain	0.124	Early rain, soil moisture	0.421**
*Mdh-1*A_b	Soil moisture	0.067	Soil moisture, late rain	0.434*
*Nadh-1*A_a	Soil moisture	0.043	Soil moisture, early rain	0.355*
Ipor-B_a	Late rain[b]	0.460**	Late rain, early rain	0.515*
*Pept-1*B_b	Early rain[c]	0.311*		
Pept-3$_a$	Early rain	0.040	Early and late rain	0.392*
Pgi-A_b	Early rain	0.120	Early and late rain	0.578**
Pgi-B_a	Early rain	0.025	Early rain, soil moisture	0.411*
6Pgd-2$_a$	Soil moisture	0.027	Soil moisture, early rain	0.407*
Number of alleles				
*Adh-1*A	Early rain	0.083	Early rain, soil moisture	0.407*
*Nadh-1*B	Late rain	0.381**		
Pept-3	Soil moisture	0.233*	Soil moisture, early rain	0.507*
6Pgd-2	Early rain	0.517**	Late rain, early rain	0.658**
Pgi-A	Early rain	0.116	Late rain, early rain	0.633**
Gene diversity, *He*				
*Adh-1*B	Early rain	0.165	Early rain, soil moisture	0.415*
*Est-5*A	Soil moisture	0.231[d]	Soil moisture, total rain	0.381*
Ipor-B	Late rain	0.477*	Late rain, annual rain	0.543*
*Pept-1*B	Early rain	0.308*		
Pept-3	Early rain	0.057	Early and late rains	0.524*
Pgi-B	Early rain	0.018	Early and late rains	0.586**

* $p < 0.05$, ** $p < 0.01$; [d] $p < 0.10$
[a] *Surface (0–5 cm) soil-moisture content (%) after early rains (data from Noy-Meir et al. 1991a)
[b] Sum of rainfall during March and April
[c] Sum of rainfall during Oct. and Nov.

linkage disequilibria were observed in the Karst subpopulation. The subpopulations dwelling in drier habitats and subhabitats showed higher genetic diversities at microsatellite loci. These results suggest that natural selection, presumably through aridity stress, acts upon microsatellite divergence on *noncoding* sequences just as it does on *coding* sequences, thereby contributing to differences in fitness. Noncoding sequencess seem to be involved, at least partly, in gene regulation.

5.4.2.1 Distribution of Alleles at Polymorphic SSR Loci

Of the 28 SSR primer pairs, 24 amplified DNA fragments in all samples (Li et al. 2000c). Each of 23 markers produced only one band per genotype and was subsequently regarded as one locus. GWM332 amplified two fragments in all samples, which were considered to be two loci, defined as GWM332a and GWM332b. Four markers (GWM368, GWM408, GWM577, and GWM637) were excluded from the data analysis because of missing data or no amplification products. In total, 25 loci were included in the following genetic analyses. Of these, two loci (GWM601 and GWM361) were monomorphic in all habitats and 23 loci were polymorphic. The allele frequencies for the 23 polymorphic loci appear in the Appendix of Li et al. (2000c). The proportions of polymorphic loci were $P = 0.76, 0.92, 0.92$, and 0.84 in the North, Valley, Ridge, and Karst habitats, respectively (Table 5.29).

One to 12 alleles were amplified at the 25 loci, and the number of alleles differed at all 23 polymorphic loci of the four habitats. The total number of alleles over all loci were 62, 87, 118, and 63 in the North, Valley, Ridge, and Karst subpopulations, respectively (Table 5.29). The average number of alleles per locus was 2.5, 3.5, 4.7, and 2.5 in North, Valley, Ridge, and Karst, respectively (Table 5.29). The habitats significantly affected the number of SSR alleles [(Kruskal-Wallis test), $H_{(3,96)} = 19.555$, $p < 0.001$]. Excluding rare alleles (observed in ≤5 individuals), 19 alleles were found to be Karst-*specific* (i.e., predominating in the Karst habitat, χ^2 test, $p < 0.01$–0.00005). Among them, six alleles were Karst-*unique*: $GWM018_{36}$, $GWM060_{17}$, $GWM095_{22}$, $GWM136_{58}$, $GWM162_{21}$, and $GWM218_{26}$. These alleles were found at the six loci with the motifs $(CA)_n$ and $(CT)_n$. The Karst-*unique* alleles were almost fixed in the Karst subpopulation (allele frequency > 0.82, see Li et al. 2000c). In the Ridge subpopulation, two alleles ($GWM060_{22}$, $GWM136_{67}$) were Ridge-*unique*. The permutation test showed that, out of 5,000 random shuffling samples, no cases were found with more habitat-*specific* alleles than actually observed. The probability of obtaining the observed habitat-*specific* and -*unique* alleles by chance was < 0.0002. Particularly, in most (> 70%) of the individuals tested from the Karst habitat, alleles were distributed in clusters at all polymorphic loci except GWM136, GWM332b, GWM340, and GWM537 (Li et al. 2000c). These results indicated that allele distributions were *nonrandom* and associated with habitats across the small area of the Ammiad microsite.

5.4.2.2
SSR Variation in Repeat Number

The Kruskal-Wallis test indicated that the ARN was significantly different across the four habitats at 16 loci (Table 5.30). At six loci (GWM099, GWM120, GWM162, GWM219, GWM415, and GWM540), the ARNs were significantly smaller in the Karst than in the other subpopulations. However, at the other ten loci, the ARNs in the Karst subpopulation were mostly significantly larger than in the others (Li et al. 2000c).

The variances in repeat number were also significantly (Levene's test, $p <$ 0.05–0.00005) different among the four subpopulations at most loci (Table

Table 5.29. Microsatellite DNA markers used in the regional and local studies of wild emmer wheat in this book: repeat types and chromosomal locations in the *T. aestivum* cultivar, and number of alleles amplified at each locus in the subpopulations from the four major habitats at the Ammiad microsite, Israel. (Li et al. 2000c)

Marker	Repeat type	Chromosomal location	Number of alleles				
			North	Valley	Ridge	Karst	Oveall
GWM018	$(CA)_nGA(TA)_k$	1BS	1	2	5	2	6
GWM060	$(CA)_n$	7AS	1	3	4	2	5
GWM095	(AC)	2AS	2	2	3	4	6
GWM099	$(AC)_n$	1AL	1	3	2	1	4
GWM120	$(CT)_n(CA)_k$	2BL	2	4	4	2	4
GWM124	$(CT)_n(GT)_k$imp	1BL	2	5	7	4	7
GWM136	$(CT)_n$	1AS	2	4	7	3	9
GWM162	$(CA)_nAA(CA)_k$	3AL	3	3	4	3	5
GWM169	$(GA)_n$	6AL	4	6	10	3	12
GWM186	$(GA)_n$	5AL	4	6	7	2	9
GWM218	$(CT)_n$	3AS	3	4	5	2	6
GWM219	$(GA)_n$imp	6BL	5	2	5	2	6
GWM251	$(CA)_n$	4BL	2	2	4	1	4
GWM294	$(GA)_nTA(GA)_k$	2AL	1	2	6	3	6
GWM332a	$(GA)_n$	7AL	3	3	3	2	3
GWM332b	$(GA)_n$	7AL	4	5	6	4	9
GWM340	$(GA)_n$	3BL	2	2	3	3	4
GWM361	$(GA)_n$imp	6BS	1	1	1	1	1
GWM389	$(CT)_n(GT)_k$	3BS	3	5	7	3	8
GWM415	$(GA)_n$imp	5AS	3	2	2	2	4
GWM429	$(CT)_n$	2BS	2	2	3	2	3
GWM459	$(GA)_n$	6AS	3	4	5	3	5
GWM537	$(CA)_n(TA)_k$	7BS	3	8	8	5	8
GWM540	$(CT)_n(CC)(CT)_k$	5BS	4	6	6	3	8
GWM601	$(CT)_n$	4AS	1	1	1	1	1
Total			62	87	118	63	143
Mean allele no. per locus			2.5	3.5	4.7	2.5	5.7
Habitat-unique alleles			5	2	22	9	–
Proportion of polymorphic loci			0.76	0.92	0.92	0.84	0.92
Sample size			7	14	37	17	75

n and *k* Number of repeats, *imp* imperfect (GWM368, 408, 577, and 637 were excluded from the list due to no amplifications in most or all samples)

5.30). The order of the variances among the four subpopulations varied at different loci. In the Karst subpopulation, those loci carrying the largest or smallest ARNs also showed the highest variance, e.g., GWM018, GWM095, GWM218, GWM332b, and GWM415. These results suggest that the distributions of repeat number (alleles) varied among the different subpopulations, i.e, they were differentiated according to habitat.

Table 5.30. Average and variance of repeat number, gene diversity (*He*) and genetic differentiation (*Gst*) at polymorphic microsatellite loci in *T. dicoccoides* of the four habitats at the Ammiad microsite, Israel. (Li et al. 2000c)

Loci	Average repeat number[a]				K-W[b] test	Variance				Levene's test	Gene diversity (*He*)				*Gst*
	North	Valley	Ridge	Karst		North	Valley	Ridge	Karst		North	Valley	Ridge	Karst	
GWM018	21.0	20.8	22.1	33.4	33.92****	0.00	0.18	11.02	34.74	8.28***	0.000	0.335	0.606	0.291	0.433
GWM060	9.0	6.5	14.5	15.2	10.85*	0.00	34.73	9.73	24.82	28.59****	0.000	0.561	0.737	0.208	0.485
GWM095	13.1	14.3	14.2	21.1	52.69****	2.14	0.53	0.58	5.49	4.91**	0.408	0.245	0.156	0.311	0.485
GWM099	17.0	18.9	19.1	7.0	42.43****	0.00	156.59	21.34	0.00	6.93***	0.000	0.520	0.142	0.000	0.665
GWM120	33.3	32.4	32.6	32.1	17.19****	0.24	0.71	0.52	0.24	5.01**	0.408	0.541	0.614	0.111	0.324
GWM124	21.3	33.1	42.1	49.4	22.02****	128.57	283.98	216.68	70.76	9.08****	0.245	0.674	0.814	0.464	0.236
GWM136	50.7	41.7	45.7	56.8	9.65*	736.33	315.90	264.95	12.74	7.80***	0.444	0.612	0.809	0.480	0.249
GWM162	25.4	25.4	26.0	22.2	8.70*	14.95	13.80	11.83	7.94	0.97	0.571	0.571	0.647	0.305	0.257
GWM169	31.0	31.9	32.2	27.8	2.76	44.00	35.76	42.27	7.40	15.44****	0.694	0.776	0.863	0.215	0.227
GWM186	17.0	18.2	16.3	16.6	0.29	96.33	99.57	165.77	55.74	4.43**	0.612	0.745	0.776	0.291	0.238
GWM218	7.7	7.4	7.6	22.1	21.87****	2.91	4.42	20.74	74.74	6.27**	0.571	0.663	0.733	0.291	0.276
GWM219	26.0	32.9	30.1	21.4	36.04****	40.14	0.07	20.19	13.34	10.69****	0.776	0.133	0.676	0.206	0.375
GWM251	24.1	25.1	24.8	26.0	1.16	24.14	4.75	9.29	0.00	6.67***	0.245	0.245	0.468	0.000	0.086
GWM294	16.0	13.9	15.4	17.3	16.19***	0.00	8.90	18.14	7.97	5.00**	0.000	0.459	0.662	0.388	0.405
GWM332a	17.6	18.1	17.9	18.1	1.25	0.95	1.06	0.97	0.06	55.16****	0.653	0.602	0.647	0.111	0.230
GWM332b	25.7	29.2	26.9	45.4	28.57****	7.91	47.41	16.40	64.74	1.87	0.735	0.735	0.722	0.533	0.182
GWM340	30.9	34.2	33.6	32.9	6.14	9.14	4.31	4.08	23.26	6.35***	0.245	0.497	0.495	0.623	0.227
GWM389	27.6	28.4	30.9	31.6	6.50	42.95	34.86	18.71	0.38	7.03***	0.653	0.674	0.642	0.491	0.116
GWM415	23.3	22.8	22.5	19.7	31.38****	0.14	0.18	1.92	2.47	3.49*	0.449	0.337	0.234	0.291	0.382
GWM429	16.4	15.0	11.1	16.9	50.18****	0.29	6.46	10.69	0.11	24.20****	0.490	0.245	0.440	0.208	0.419
GWM459	26.6	27.6	27.2	28.6	20.97****	0.62	1.50	1.34	0.62	2.08	0.449	0.694	0.736	0.398	0.181
GWM537	34.9	33.8	30.4	36.5	4.21	48.48	53.72	77.33	41.13	6.22**	0.571	0.806	0.769	0.711	0.063
GWM540	22.0	22.2	21.4	16.4	35.52****	6.67	5.87	4.06	1.63	2.16	0.694	0.776	0.619	0.215	0.302
Mean											0.397	0.498	0.560	0.286	0.291

$^{*}p < 0.05$, $^{**}p < 0.01$, $^{***}p < 0.001$, $^{****}p < 0.0001$

[a] The average repeat numbers were 17 and 19 at the monomorphic GWM601 and 361, respectively

[b] Kruskal-Wallis one-way analysis of variance

5.4.2.3
SSR Gene Diversity Among the Four Habitats

At each SSR locus, gene diversity was different among the four habitats (Table 5.30). The highest gene diversity across all loci was observed in the Ridge subpopulation (0.560), followed by Valley (0.498), North (0.397), and Karst (0.286). A Wilcoxon signed-rank test showed that gene diversities were significantly different in four of six pairs of comparisons for the four habitats, except those between North and Karst, and between Valley and Ridge (Li et al. 2000c).

5.4.2.4
SSR Genetic Differentiation Among the Four Habitats

The total gene diversity of a subdivided population (*Ht*) was analyzed according to mean gene diversity in a population (*Hs*), the average ($Dst = Ht{-}Hs$), and the relative (*Gst*) degree of gene differentiation among subpopulations (Nei 1973). At all SSR loci except GWM251 and 537, the *Gst* ranged from 0.12 to 0.67, with an average of 0.291 (Table 5.30), suggesting that 12–67% of the total gene diversity of SSRs existed *between* these very close microscale habitats at the SSR loci (Li et al. 2000c). The average *Gst* (0.291) indicated that 29% of the total gene diversity existed *among* the four habitats. This pattern reveals on a microscale the "archipelago" macroscale genetic pattern of wild emmer described across the Near East Fertile Crescent (Nevo and Beiles 1989). This microscale divergence in SSRs suggests that *selection overrides* the ongoing *migration* between these close microhabitats.

5.4.2.5
Genetic Distances Among the Four Subpopulations

Genetic distances among the four subpopulations were estimated by the allele-sharing distance index (D_{AS}, Chakraborty and Jin 1993). Karst is the most remote subpopulation compared with the three other habitats ($D_{AS} \cong 0.7$). The genetic distances among the other habitats were almost half those of the Karst habitat. The smallest D_{AS} (0.288) was observed between Valley and Ridge, which are closer ecologically (Li et al. 2000c).

5.4.2.6
SSR Genetic Diversity in Subpopulations from Eight Subhabitats

Within the four habitats, the number of alleles (*A*), the proportion of polymorphic loci (*P*), and the gene diversities (*He*) differed in subpopulations from subhabitats (Table 5.31). Within the Karst habitat, higher genetic diversity (*A*, P, and *He*) was found in the Upper Karst (K_1) subpopulation than in the Lower Karst (K_2), although the sample size in the K_1 ($N = 6$) was smaller than in the K_2 subpopulation ($N = 11$). Within the Valley habitat, genetic diversity in the Narrow Valley (Vn) was higher than in the Valley center (Vc) subpopulation, only partly explained by the larger sample size. Within the Ridge habitat, higher genetic diversities were observed on the south-facing slope (Rs) and the top of the plateau (Rp_2) than on the east-facing slope (Re) and the shoulder of the plateau (Rp_1). The North habitat was not subdivided here because of the small sample size. In the nine subhabitats, the gene diversities (*He*) were ordered as follows: Rp_2 (0.474) > Rs (0.453) > Re (0.433) > North (0.397) > Vn (0.396) > Rp_1 (0.384) > K_1 (0.350) > Vc (0.314) > K_2 (0.199). This was a substantial divergence in a microscale. The Wilcoxon signed-rank test showed that within the Ridge subpopulation the gene diversity in Rp_1 was significantly ($p < 0.05$) lower than in Rp_2, Re, and Rs; and within Karst the *He* was significantly ($p < 0.01$) higher in the K_1 than in the K_2 subpopulation (Table 5.31). This divergence again suggests that selection overrides migration and that the microscale SSR pattern is adaptively heavily structured, presumably by natural selection and in accordance with the microsite ecological divergence (Li et al. 2000c).

5.4.2.7 SSR Spatial Autocorrelation

Spatial autocorrelation analysis tests whether the observed values (here allele frequencies) at one locality are significantly dependent on those at neighboring localities. This analysis can summarize the pattern of geographic variation exhibited by the response surface of any given variable. The autocorrelation coefficients (e.g., Moran's *I*) and their significance levels are a function of the distance between pairs of localities (Sokal and Oden 1978a, b). We computed the Moran's I autocorrelation coefficient of relatively common alleles across the entire microgeographic range in our study, including the nine subpopulations. The space was partitioned into three distance classes, so that each class contained equal numbers of locality pairs. The

Table 5.31. Genetic diversity of microsatellites in subpopulations of *T. dicoccoides* at subhabitats[a] at the Ammiad microsite, Israel. (Li et al. 2000c)

Genetic diversity	Karst			Valley			Ridge						North
	K_1 ($N = 6$)[b]	K_2 ($N = 11$)	Shared alleles	Vc ($N = 5$)	Vn ($N = 9$)	Shared alleles	Re ($N = 9$)	Rs ($N = 9$)	Shared alleles	R_{p1} ($N = 6$)	R_{p2} ($N = 13$)	Shared alleles	($N =$
No. alleles (*A*)													
Total	51	48	37	51	66	29	64	66	32	59	76	42	62
Mean	2.0	1.9	1.5	2.0	2.6	1.2	2.0	2.6	1.3	2.4	3.0	1.7	2.5
P	0.80	0.76		0.60	0.76		0.76	0.84		0.84	0.88		0.76
He	0.350	0.199		0.314	0.396		0.433	0.453		0.384	0.474		0.39

For abbreviations, see text. *N* sample size, *P* proportion of polymorphic loci, *He* gene diversity (Nei 1973)
[a] The North habitat was not subdivided because of the small sample size

results are presented in Table 5.32 (Li et al. 2000c). All the average autocorrelation coefficients over all tested alleles in all three distance classes were low and nonsignificant. In each distance class, positive or negative values of low, medium, and sometimes high estimates were found for different alleles, suggesting that there is no similar pattern across loci. Positive and negative autocorrelations were intermixed across loci. No positive and significant autocorrelations were found. Although all negative autocorrelations in the lowest distance class were not significant, their abundance suggests the existence of a pattern opposite to the expected highest positive autocorrelations at low distance. This indirectly suggests that microlocal selection in each of the four microhabitats and 11 topographical units overrides any actual migration between these units.

5.4.2.8
Multiple Regression Analysis of Ecological Factors and SSR Diversity

A test of the best predictors of *A*, *P*, and *He* over all loci in the nine subpopulations was conducted by stepwise multiple regression using these genetic indices as dependent variables and ecological factors as independent variables (Li et al. 2000c). The following ecological factors were included in the analysis: distance to rock base (HAVDB), soil-moisture content (M) averaged for 3-year measurements, average height to rock top (AVH), and angle to rock top (AVA). The best two variable predictors of *A*, *P*, and *He* were HAVDB and M, explaining significantly ($p < 0.01$) 0.86, 0.87, and 0.80 of their variances, respectively (Table 5.33). The best three predictors including HAVDB, M, and AVH could significantly ($p < 0.01$) explain 0.89 and 0.92 of the SSR variances of *P* and *A* (Table 5.33). These results suggested that a substantial amount of genetic diversity at the Ammiad microsite is associated with some of the microecological variables, i.e., that selection may operate on the largely *noncoding* and presumably partly regulatory SSRs.

5.4.2.9
SSR Discriminant Analysis

Stepwise discriminant analyses were conducted to distinguish between the four subpopulations and classify individuals into the four habitats. The program computed three canonical functions based on eight (GWM095, GWM124, GWM162, GWM186, GWM219, GWM389, and GWM429) out of 23 polymorphic loci. The overall multivariate *F* ratio for differences among the subpopulation centroids was highly significant ($F_{(24, 185)} = 17.08$, $p < 0.00005$). These functions were sufficient to correctly classify 85% of 70 individuals into their original four habitats. Figure 5.14 shows the plot of these 70 individuals in the space of the three canonical discriminant functions (Li et al. 2000c).

With respect to only five loci (GWM095, GWM124, GWM186, GWM219, and GWM429), 81% of the 70 individuals could be correctly classified into

Table 5.32. Spatial autocorrelation analysis of microsatellite allele frequencies at 50 alleles of nine subpopulations of *T. dicoccoides* at the Ammiad microsite, Israel. Moran's coefficients for each allele in three distance classes are given. The expected values of *I* were –0.143 for classes 1 and 3, and – 0.125 for class 2. (Li et al. 2000c)

Locus	Allele[a]	Distance (m)		
		0–200	200–500	500–800
GWM060	2	–0.16	–0.11	–0.06
	9	0.17	–0.32	–0.23
GWM120	32	–0.11	0.00	–0.36
	33	–0.23	–0.14	–0.34
	34	–0.18	–0.15	–0.07
GWM124	17	–0.03	–0.03	–0.27
	49	–0.13	–0.55**	0.00
	50	–0.09	0.17	–0.39
GWM162	30	–0.01	–0.19	–0.19
GWM169	24	–0.30	–0.32	–0.10
	25	0.03	–0.33	–0.16
	30	–0.10	–0.25	–0.05
	36	–0.28	0.03	–0.13
	38	0.01	–0.15	–0.21
	39	–0.13	–0.09	–0.15
GWM186	1	–0.30	–0.12	0.00
	19	–0.08	–0.26	–0.06
	32	–0.08	–0.06	–0.21
GWM218	3	–0.26	–0.02	–0.07
	4	–0.26	–0.09	–0.07
	8	–0.27	–0.16	–0.21
GWM219	32	–0.30	–0.42	0.00
GWM251	13	–0.14	–0.46*	–0.05
	20	–0.02	–0.51*	–0.26
	26	–0.15	–0.26	–0.17
GWM294	10	–0.02	–0.28	–0.06
GWM332a	17	0.02	0.15	–0.54*
GWM332b	23	–0.21	–0.12	–0.03
	26	–0.04	–0.17	–0.15
	28	–0.14	–0.03	–0.19
GWM340	32	0.16	–0.50*	–0.09
GWM389	18	0.07	–0.28	–0.16
	30	–0.22	–0.12	–0.05
	31	–0.18	–0.25	–0.13
GWM429	9	–0.06	0.11	–0.36
	16	–0.16	0.15	–0.30
GWM459	25	–0.16	–0.04	–0.16
	26	0.04	–0.40	–0.12
	27	–0.02	–0.22	–0.12
	28	0.05	–0.32	–0.14
GWM537	20	–0.04	–0.49*	–0.05
	25	–0.30	0.02	–0.08
	34	–0.23	–0.12	–0.11
	37	–0.34	–0.08	–0.04
	38	–0.12	0.04	–0.29
	39	–0.30	0.00	–0.08
	41	–0.41	0.02	0.00
GWM540	23	0.01	–0.26	–0.11
	24	–0.08	–0.08	–0.18
	25	–0.16	–0.12	–0.07
Mean		–0.122	–0.165	–0.151
SE		0.019	0.026	0.016

*$p < 0.05$, **$p < 0.01$

[a] This analysis was carried out only for the relatively common alleles in the subpopulation

Table 5.33. Coefficient of multiple determination (R^2) of microsatellite genetic diversity and microecological variables (as independent variables) in the nine vegetationally and topographically defined subhabitats of *T. dicoccoides* at Ammiad, Israel. (Li et al. 2000c)

Step	Genetic diversity					
	P		Number of alleles, *A*		*He*	
	Variable[a]	R^2 ($N = 9$)	Variable	R^2 ($N = 9$)	Variable	R^2 ($N = 9$)
1	HAVDB	0.711**	HAVDB	0.203	M	0.196
2	HAVDB, M	0.862**	HAVDB, M	0.873**	M, HAVDB	0.796**
3	HAVDB, M, AVH	0.887*	HAVDB, M, AVH	0.920**	–	–

P Proportion of polymorphic loci, *He* gene diversity (Nei 1973), $^{**}p < 0.01$

[a] Four ecological variables are included in this multiple regression analysis: *HAVDB* distance to rock base (harmonic average, cm), *M* soil-moisture-content-averaged (%) 3-year measurements, *AVH* average height to rock top, *AVA* angle to rock top. All data of these variables were extracted from Noy-Meir et al. (1991b)

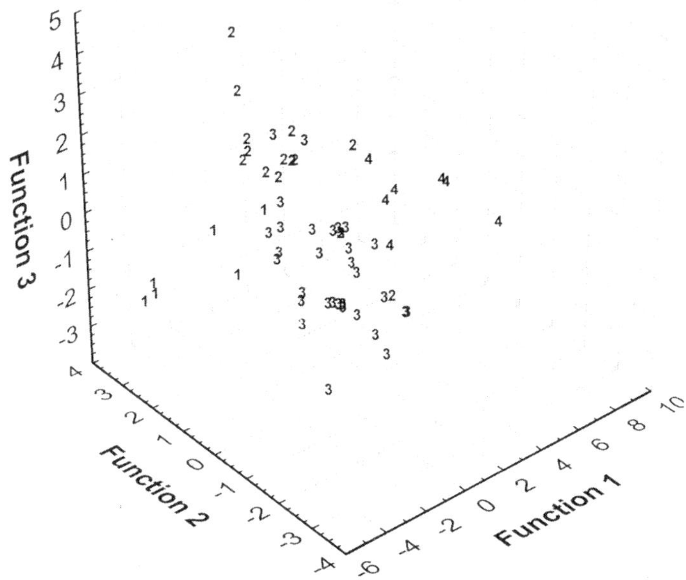

Fig. 5.14. Scatter-plot of 70 plants of *T. dicoccoides* from the four major habitats at the Ammiad microsite in the spaces of canonical discriminant functions 1, 2, and 3 based on eight polymorphic microsatellite loci. 1 *North*, 2 Valley, 3 Ridge, 4 Karst. (Li et al. 2000c)

Table 5.34a. Pairs of significant ($p < 0.05$) linkage disequilibria (LD)[a] of microsatellite alleles in the four major subpopulations of *T. dicoccoides* at the Ammiad microsite, Israel. (Li et al. 2000c)

Habitat	North	Valley	Ridge	Karst	Overall
Total pairs	1413	3297	6259	1626	9532
Significant LDs	114	405	819	586	1720
% of the total[a]	8.0**	12.3****	13.1****	36.0****	18.4****
Within genome A					
Total pairs	324	838	1635	428	3088
Significant LDs	20	123	232	195	641
% of the total	6.2	14.7****	21.0****	46.5****	20.8****
Within genome B:					
Total pairs	273	625	1199	293	1496
Significant LDs	21	85	238	73	255
% of the total	7.7	13.6****	19.8****	24.9****	17.0****
Between A and B:					
Total pairs	680	1605	3097	791	4968
Significant LDs	71	185	535	286	824
% of the total	10.4***	11.5****	17.3****	36.2****	16.6***
Within chromosome					
Total pairs	43	103	206	60	216
Significant LDs	4	17	24	22	38
% of the total	9.3	16.5*	11.7*	36.7***	17.8****

[a] χ^2 test for difference of percentage of significant LDs from the 5% expected by chance. *$p < 0.05$, **$p < 0.01$, ***$p < 0.0005$, ****$p < 0.00005$

their original habitats, the difference among the subpopulation centroids being highly significant ($F_{(15,184)} = 23.46$, $p < 0.00005$). Moreover, even for three loci (GWM095, GWM219, and GWM429), 75% of the 70 individuals could be correctly assigned to their original habitats; the difference among the subpopulation centroids was also highly significant ($F_{(15,175)} = 33.88$, $p < 0.00005$). These results suggested that the genetic differentiation was *substantial*, unaffected by migration among the four major habitats and supporting *microscale selection* in each (Li et al. 2000c).

5.4.2.10 Linkage Disequilibrium (LD) Between SSR Loci

In the present study, 8.0, 12.3, 13.1, and 36.0% of the total 1413, 3297, 6259, and 1626 two-allele pairs between loci in North, Valley, Ridge, and Karst, respectively, displayed significant ($p < 0.05$) LDs (Table 5.34a). In the four subpopulations, the percentages of significant LDs between loci over all genomes, within genome A, within genome B, between genomes A and B, and within

Table 5.34b. Habitat-*unique* microsatellite alleles linkage disequilibrium in the Ammiad population of *T. dicoccoides*, Upper Galilee, Israel. (Li et al. 2000c)

$Locus_{allele}$-$locus_{allele}$	Linkage disequilibrium (D)			
	North	Ridge	Valley	Karst
$GWM294_{10}$-$GWM415_{23}$	No[a]	0.040	−0.138**	0.097***
$GWM294_{10}$-$GWM186_{21}$	No	No	0.092*	−0.097***
$GWM294_{10}$-$GWM332b_{43}$	No	−0.025	−0.128*	0.104***
$GWM218_{4}$-$GWM332b_{66}$	No	No	0.061**	−0.114*
$GWM415_{23}$-$GWM124_{48}$	No	−0.051	−0.056*	0.048*
$GWM186_{1}$-$GWM332b_{66}$	No	No	0.061**	−0.114**
$GWM186_{21}$-$GWM124_{48}$	No	No	0.061**	−0.048*
$GWM169_{10}$-$GWM540_{14}$	No	−0.004	0.061**	−0.052**
$GWM169_{30}$-$GWM219_{33}$	No	0.073**	−0.061**	No
$GWM060_{9}$-$GWM540_{19}$	Fixed	0.063**	−0.153*	No
$GWM095_{15}$-$GWM540_{14}$	No	0.028*	No	−0.052**
$GWM136_{44}$-$GWM095_{14}$	No	−0.035	No	0.062***
$GWM332b_{43}$-$GWM429_{13}$	No	0.080*	0.020	−0.104****
$GWM124_{48}$-$GWM540_{14}$	No	0.024*	−0.005	−0.052**

*$p < 0.95$, **$p < 0.01$, ***$p < 0.001$, ****$p < 0.00005$ (χ^2 test)
[a] No LD because at least one allele did not exist

chromosomes were significantly (p < 0.01–0.00005) larger than the 5% expected by chance (Li et al. 2000c). In the Karst subpopulation, the percentage of significant LDs was in most cases significantly (χ^2 test, p < 0.05–0.00005) higher than in the other subpopulations (Table 5.34a). Sample size may have affected the significant LDs. However, the sample size at Karst ($N = 17$) was much smaller than that at Ridge ($N = 37$) and similar to that in the Valley ($N = 14$). This may suggest that the significant difference in the percentages of LDs did not result from different sample sizes. We found that 94 and 96% of LDs were significantly (at $p < 0.05$ and $p < 0.01$ levels) different between the two extreme, opposite Valley and Karst subpopulations, namely these LDs were habitat-*specific*. Many habitat-*unique* LDs were also found at different pairs of loci; some examples are presented in Table 5.34b. Random drift and founder effects may also affect the LDs. In order to determine the significance of the observed habitat *specificity* or *uniqueness* of LDs, a permutation test was performed. Out of 5,000 random permutations, no cases showed more Karst-*specific* or -*unique* LDs than the real data, suggesting high significance ($p < 0.0002$) of the observed Karst *specificity* and *uniqueness* for LDs. After exclusion of the rare alleles (observed in ≤5 individuals in the whole population), the LD Karst *specificity* or *uniqueness* remained the same based on the 5,000 random permutations. This result suggested that the Karst-*specific* or -*unique* LDs did not arise from chance events, e.g., random drift, but were produced by microniche selection on allele associations (Li et al. 2000c).

The proportions of significant LDs between loci within genomes A and B, between A and B, and within chromosomes were similar in the Valley and Ridge subpopulations. However, a significant ($\chi^2 = 15.01$, df = 1, $p < 0.0001$) difference was found within genomes A (46.5%) and B (24.9%, Table 5.34a) in the Karst subpopulation (Table 5.27; Li et al. 2000c). In the North subpopulation, more significant LDs were observed between genomes A and B (10.4%) than within genome A (6.2%) or genome B (7.7%). These results may suggest that the effects of genetic factors (genome, linkage itself) on LDs were not significant in the Valley and Ridge subpopulations, but that the genome effect (A vs B) very strongly influenced LDs in the Karst subpopulation.

5.4.2.11
SSR Multilocus Structure

The observed and expected variances of *K*, the number of heterozygous loci between two randomly chosen gametes in a (sub)population (Brown et al. 1980), were estimated (Table 5.35). The difference between the observed and expected variances was significant ($p < 0.05$) in the four subpopulations. The observed variance of *K* in the Karst subpopulation was the highest (39.83), and larger than those in North (25.88), Valley (25.29), and Ridge (20.64). The standardized index of the intensity of multilocus structure, X(2), was also obviously higher (8.50) in the Karst, which is relatively more humidity-variable from mesic to xeric than the other habitats, suggesting that a higher level of multilocus association exists in the spatially extremely heterogeneous (wider niche width) Karst than in the other habitats. For a regional trend of increasing allozymically derived X(2) towards mesic north Israel, see Nevo and Beiles (1989). The local trends of multilocus association, X(2), mirror regional trends across Israel: higher values associated with increasing humidity and spatial ecological heterogeneity (Li et al. 2000c).

Table 5.35. Genome organization based on the 23 polymorphic microsatellite loci for the subpopulations of *T. dicoccoides* from the four habitats at the Ammiad microsite, Israel. (Li et al. 2000c)

Habitat	Expected variance of *K*	Observed variance of *K*	Upper limit of 95%	X(2)
North	4.249	25.876*	8.605	5.091
Valley	4.846	25.288*	8.380	4.219
Ridge	4.579	20.640*	6.638	3.507
Karst	4.193	39.833*	6.968	8.500
Overall	4.606	21.692*	6.060	3.710

K Number of heterozygous loci in two randomly chosen gametes, *X(2)* index of the intensity of the multilocus structure, *$p < 0.05$

5.4.3 Parallel Microgeographic Patterns of Genetic Diversity and Divergence Revealed by Allozyme, RAPD, and Microsatellites in *Triticum dicoccoides* at Ammiad, Israel

Does the noncoding part of the genome, comprising more than 95% of the entire genome, represent "junk", i.e., nonfunctional DNA, or is it sensitive to environmental stresses and selection and at least partly related to regulatory functions? Many studies have compared allozyme (coding) and RAPD (largely noncoding) diversities (e.g., Heun et al. 1994; Yu and Nguyen 1994; Isabel et al. 1995; Maab and Klaas 1995; Peakall et al. 1995; Ayres and Ryan 1997, 1999; Le Corre et al. 1997; Owuor et al. 1997; Fahima et al. 1999; Keane et al. 1999; Li et al. 1999). Only a few studies have compared allozyme and SSR (largely noncoding) diversity (e.g. Barker et al. 1997). No study to date has compared and contrasted allozyme, RAPD, and SSR genetic diversities in the same organism and ecologically subdivided microscale site.

We used wild emmer wheat as a model organism to compare genetic structures in protein-*coding* and primarily *noncoding* DNA regions on a microgeographic scale at the Ammiad microsite in northern Israel (Li et al. 2000d). We found *locally* similar patterns of genetic diversities at allozyme, RAPD, and SSR loci, suggesting that natural selection may adaptively affect both *coding* (structural) and *noncoding* (partly regulatory) genomic regions, as we earlier found on a *regional* scale in Israel and Turkey (Fahima et al. 1999, 2001). The implications of these findings for genetic conservation are of paramount importance, because the *noncoding* genome regions, still considered by many as "junk" DNA, may in fact play a substantial role in the regulation and control of *coding* genomic regions. As in the regional analysis (Fahima et al. 2001), at Ammiad the correlation among allozymes, RAPDs, and SSRs is high, but sometimes displays opposite trends (i.e., high allozyme and low SSR diversity), suggesting some independence of the two parts of the genome, and possibly *differential selection* pressures on each.

The levels of genetic diversity were compared by means of 35 allozyme, 60 RAPD, and 25 microsatellite (SSR) markers for 75–175 individuals of tetraploid wild emmer wheat, *T. dicoccoides*, collected in 1993 from the microgeographic microsite of Ammiad, north of the Sea of Galilee, Israel. This microsite included, as described above, four major habitats that showed highly significant differentiation in ecological factors, particularly with respect to rock cover, proximity, and height, and surface soil moisture in the early growing season of *T. dicoccoides* (Anikster and Noy-Meir 1991a). Higher within-subpopulation genetic diversity was found in the primarily *noncoding* DNA regions (RAPD and SSR) rather than in the protein-*coding* (allozymes) regions. However, much larger gene differentiation (*Gst*) among the subpopulations was observed in the protein-*coding* allozymes than in the RAPDs and SSRs (*Gst* = 0.428, 0.291, and 0.235, respectively). This differential *Gst* pattern between the *coding* and *noncoding* parts of the genome may again suggest their *partly independent* selection regimes despite the high inbreeding rate. This hypothesis calls for critical testing.

Larger genetic distance was found at SSR loci, followed by allozyme and RAPD loci. The subpopulations in *drier* habitats tend to have higher allozyme, RAPD, and SSR diversities (*He*); the relatively wet Karst subpopulation showed only about half the *He* of the other relatively drier habitats. The subpopulations with larger differences in soil moisture between habitats tend to show larger genetic distances at allozyme, RAPD, and SSR loci. These results may suggest that *climatic selection* through aridity stress acts on both structural protein-*coding* and presumably partly regulatory, *noncoding* DNA regions, resulting in microscale adaptive patterns, although this study cannot absolutely reject hitchhiking effects. These results have profound implications for genetic conservation both in situ and ex situ. It is plausible that the *noncoding* regulatory elements of the genome may be more polymorphic and more sensitive to environmental stress than the coding and usually more evolutionarily conserved elements. Below are the main results (Li et al. 2000d).

5.4.3.1
Patterns of Genetic Diversity Among the Four Subpopulations

Allele frequencies in the four subpopulations significantly varied at 13 allozyme loci, 38 RAPD loci, and 21 SSR loci. The genetic divergence was both qualitative and quantitative. All three types of markers showed that *habitat-specific* alleles predominated in a specific habitat. At allozymic loci, alleles $Nadh\text{-}2A_{null}$, $Nadh\text{-}2B_a$, $Pgi\text{-}B_b$ and $Mdh\text{-}1A_a$ were North-, Valley-, Ridge- ,and Karst-*specific*, respectively. RAPD alleles $OPA15_{870}$, $OPA15_{620}$, $OPA15_{500}$, and $OPT16_{500}$ were North-, Valley-, Ridge- and Karst-*specific*, respectively. At SSR loci, alleles $GWM124_{17}$, $GWM429_9$, and $GWM099_7$ were *specific* to the North, Ridge, and Karst habitats, respectively. At SSR loci, some alleles were present in only one habitat (*habitat-unique* allele), for example, $GWM060_{22}$ and $GWM136_{67}$ were only present in the Ridge subpopulation, and six SSR Karst-*unique* alleles, $GWM018_{36}$, $GWM060_{17}$, $GWM095_{22}$, $GWM136_{58}$, $GWM162_{21}$, and $GWM218_{26}$, were almost fixed in the Karst subpopulation (frequency of most alleles: 0.82) (Li et al. 2000d).

5.4.3.2
Genetic Diversity over All Four Subpopulations

The levels of genetic diversity were expressed in number of alleles per locus (*A*), proportion of polymorphic loci (*P*), and gene diversity (*He*). Over the whole population, the *A* and *P* for allozymes, RAPDs, and SSRs were $A = 1.7$, 1.9, 5.7; $P = 0.54$, 0.95, 0.92, respectively. Over the four subpopulations, the *A* and *P* of SSRs were larger than these of RAPDs and allozymes (Fig. 5.15). In the Karst subpopulation, the *A* and *P* of allozymes and RAPDs were lower than in the other three subpopulations. Decreased sample size could have

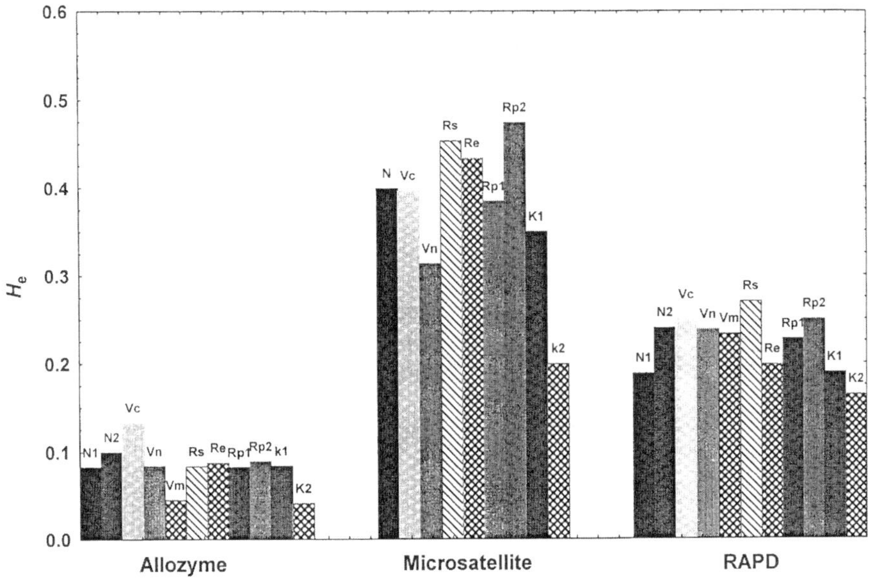

Fig. 5.15. Histogram of average gene diversity (*He*) in the 11 subpopulations of *T. dicoccoides* from the 11 subhabitats revealed by 35 allozyme, 60 RAPD, and 25 SSR loci. *N1* Upper North, *N2* Middle North, *Vn* Narrow Valley, *Vc* Main Valley center, *Vm* Main Valley margins, *Rs* Ridge south, *Re* Ridge east, *Rp1* Ridge shoulder, *Rp2* Ridge top, *K1* Upper Karst, *K2* Lower Karst. (from Nevo 2001a, based on Li 2000 and Li et al. 2000d)

reduced the *A* and *P* of allozymes and RAPDs in the North and Valley subpopulations, but not in the Ridge and Karst subpopulations.

We estimated all genetic indices (*A*, *P*, and *He*) using different sample sizes and again for the same 75 genotypes (equal sample size), since sample size may affect mainly *A* and *P*, but to some extent also *He*. Under different sample sizes and the identical genotypes and sample size, SSR diversities were larger than the RAPD and allozyme diversities in the four subpopulations, and the *He* values were obviously smaller in the Karst than in the other three subpopulations. These results demonstrated that different types of molecular markers revealed *similar patterns but different levels of genetic diversity*. With the same genotypes, the *He* values of allozyme, RAPD, and SSR loci in the four subpopulations were positively correlated (Pearson's $r = 0.80$, $N = 4$, $p > 0.05$ possibly due to the small *n*). These results suggest similar patterns of within-subpopulation genetic diversity in both protein-*coding* and *noncoding* DNA regions, but possibly independent selection on the *coding* and *noncoding* parts of the genome. Increased sample sizes changed only a few of the *He* values of allozymes and RAPDs, particularly in the Ridge and Valley subpopulations, and in the overall population (Table 4 in Li et al. 2000d). Among the North ($N = 7$), Valley ($N = 14$) and Ridge ($N = 37$) subpopulations, the *He* values of allozymes were almost the same ($He = 0.102$–0.105), but the Karst

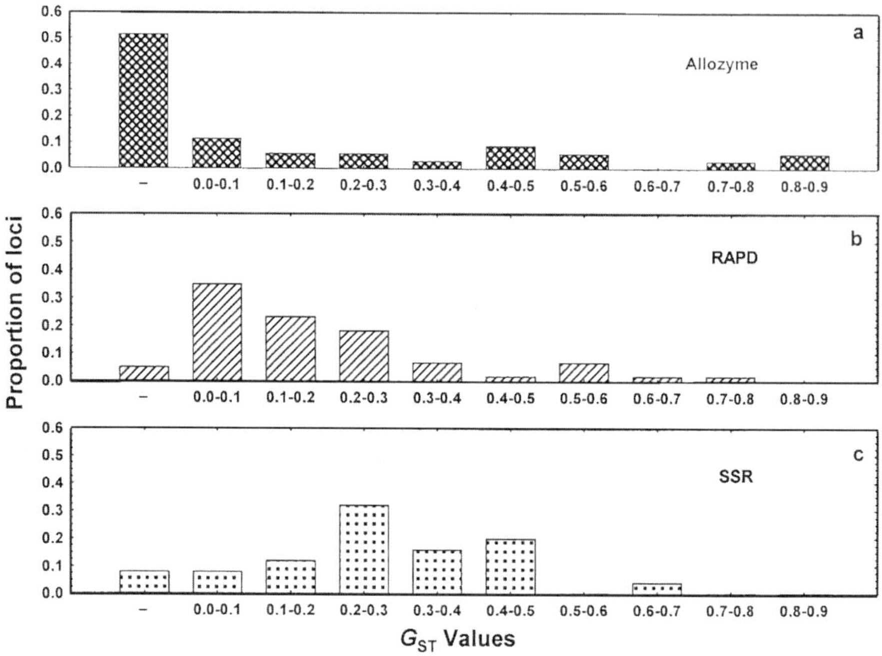

Fig. 5.16. Histogram showing the proportional distribution of *Gst* values among the four subpopulations of *T. dicoccoides* at 35 allozyme, 60 RAPD, and 25 microsatellite loci. (Li et al. 2000d) – no Gst value, i.e. monomonphic loci.

subpopulation (N = 17) showed the lowest *He* (0.048). A similar pattern was found for RAPD *He*. The results indicated that the effect of sample size was small and in both genetic systems the direction of the *He* values of allozyme and RAPD diversity of the four subpopulations displayed the same direction. Once again, the similarity of the *coding* and *noncoding* genomes is obvious, though each may be somewhat differentially correlated with environment.

The four habitats significantly ($p < 0.0001$, Noy-Meir et al. 1991a) varied in moisture content of the soil surface (0–5 cm) after the early rains of the growing season of wild emmer wheat. Such soil moisture influenced seed germination and plant density. The values of *He* for allozyme, RAPD, and SSR markers seem to be *negatively correlated* with soil moisture (Fig. 5.18): *The drier the habitat, the higher the gene diversity.* The relatively wet Karst subpopulation showed the lowest allozyme, RAPD, and SSR diversity.

5.4.3.3
Partition of Genetic Diversity Within and Among the Four Subpopulations

Different types of molecular markers revealed variable levels of relative gene differentiation (*Gst*) among subpopulations (Fig. 5.16). Over all polymorphic

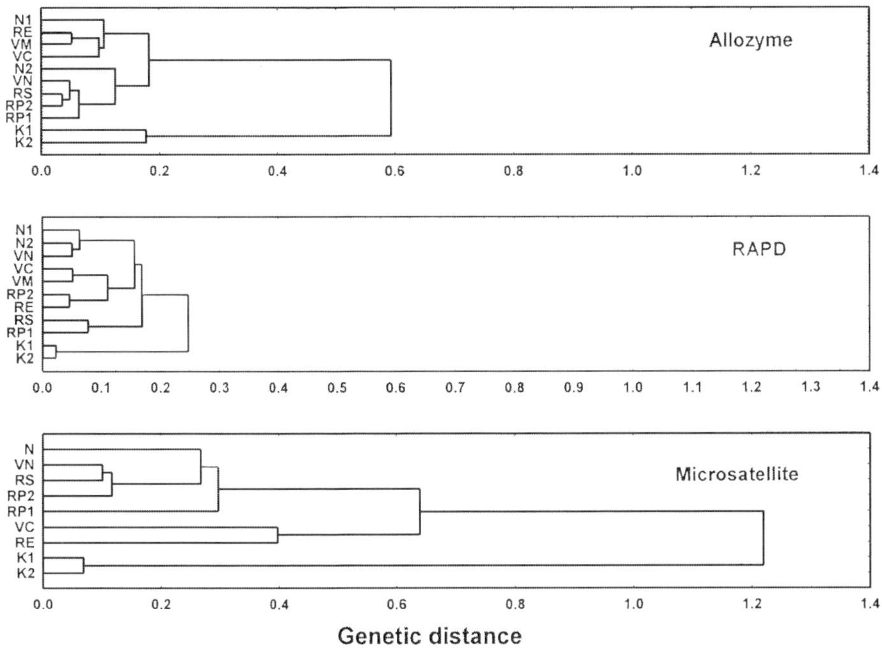

Fig. 5.17. Dendrogram of relationship among 11 subpopulations using Nei's unbiased genetic distance (Nei 1978) and the UPGMA method of clustering, which was based on 35 allozymic loci, 60 RAPD loci, and 25 microsatellite loci. (based on Li 2000, and Li et al. 2000d). For abbreviations see Fig 5.15.

loci with the same genotypes, the highest average level of *Gst* among the four subpopulations was revealed by allozymes (*Gst* = 0.437), followed by SSRs (*Gst* = 0.291) and RAPDs (*Gst* = 0.236). A similar pattern was observed with different sample sizes for the three marker systems. These results indicated that genetic differentiation among the four habitats was larger at the protein level than at the DNA level. At the DNA level, differentiation was larger in the tandem-repeated DNA regions (SSRs) than in the randomly amplified DNA regions (RAPDs).

The three molecular marker systems showed significant differential *Gst* values among the four subpopulations [Kruskal-Wallis $H_{(2, 94)} = 12.78$, $p < 0.01$)]. Pairwise comparison was further conducted by a Mann-Whitney *U* test for *Gst* values between each two markers. The result showed significant differences in the *Gst* values between RAPD and the other two marker systems ($p < 0.01$) (Fig. 5.16). Once again, this interhabitat differential of the three molecular markers may suggest their partly independent selection regimes, challenging the hypothesis of clonal genotypic selection due to high inbreeding and hitchhiking to one selected marker. This hypothesis calls for further critical tests.

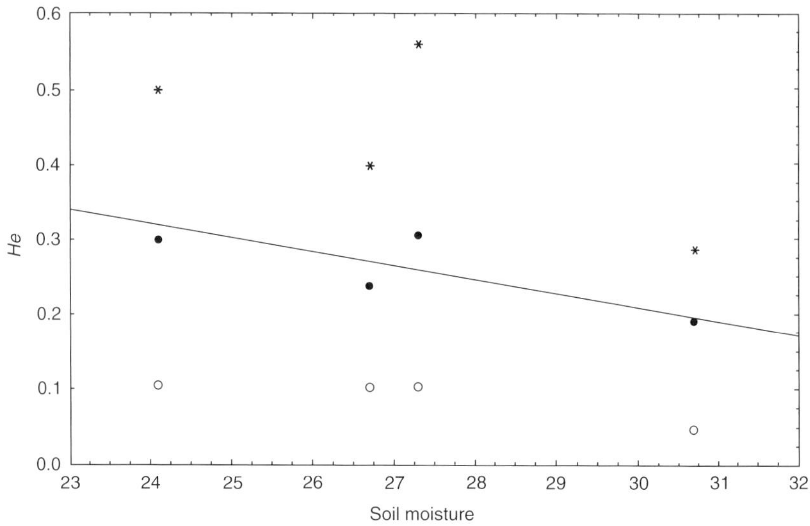

Fig. 5.18. The negative trend (r = –0.28, $p > 0.05$) between gene diversity (*He*) of allozymes ○, RAPD ●, and SSR * markers and surface soil moisture (0–5 cm) after early rains (Oct.–Nov.) in wild emmer wheat at Ammiad in the growing season (Oct.–May). (Li 2000)

5.4.3.4
Genetic Distance

Based on the data of the same genotypes, genetic distances among the four subpopulations were estimated by means of unbiased genetic distance (Nei 1978) for the three marker systems (Li et al. 2000d). The largest genetic distances among the four subpopulations were found at the SSR region. Very similar genetic distances were observed at allozyme and RAPD loci. The dendrograms were constructed by the UPGMA clustering method for the four subpopulations (for 11 subpopulations see Fig. 5.17). The maximal genetic distances were between Karst and the other three subpopulations in all three marker systems. Positive correlation was found among the three matrices of genetic distance between allozymes and RAPDs (Spearman $r_s = 0.89$, $N = 6$, $p < 0.05$), between allozymes and SSRs ($r_s = 0.94$, $N = 6$, $p < 0.01$), and between RAPDs and SSRs ($r_s = 0.77$, $N = 6$, $p > 0.05$). This positive correlation indicates a similar pattern of microgeographic genetic divergence in the protein-*coding* and DNA largely *noncoding* regions. Rather than being "junk" DNA, these regions may be subjected to *natural selection* and function in the *regulation* and control of the plant's physiology, reacting *similarly, but partly independently*, to *environmental aridity* stress.

Among the four subpopulations, gene diversity (*He*) tends to increase with the difference in soil-moisture content (Fig. 5.18). The largest differences in soil moisture existed between the Karst and Valley, and North habitats. The genetic distance between the three subpopulations (ridge, valley, north) and Karst subpopulation appeared to be the largest at all three molecular systems, allozymes, RAPDs, and SSR loci (Fig. 5.17).

Since SSR may evolve as predicted by the stepwise mutation model, while allozymes and RAPDs do not, use of Nei's genetic distance (1978) may bias the estimate of SSR genetic distance. Hence, the stepwise, weighted genetic distance (Shriver et al. 1995) was also used to estimate SSR genetic distances among the four subpopulations (see Fig. 5.17 and Li 2000). The two types of genetic distances significantly corresponded ($r_s = 0.94$, $N = 6$, $p > 0.01$), suggesting that the two measurements of genetic distance at SSR loci showed a parallel matrix.

5.5 Microsatellite Diversity Correlated with Ecological-Edaphic and Genetic Factors in Three Microsites of Wild Emmer Wheat in North Israel

We compared and contrasted the effects of internal (genetic) and external (ecologic) factors on allelic diversity at 27 dinucleotide microsatellite (SSR) loci in the aforementioned three Israeli natural populations of *T. dicoccoides* at Ammiad, Tabigha, and Yehudiyya, north of the Sea of Galilee (Li et al. 2000b) (Fig. 5.7). The results demonstrated that SSR diversity is correlated with the interaction of ecological and genetic factors. Genetic factors, including genomes (A vs B), chromosome, motif, and locus, affected ARN, variance (σ^2) in repeat number, and number of alleles (NA) of SSRs, but the significance of some factors varied among populations. Genome effect on SSR variation may result from different motif types, particularly compound (or imperfect) vs perfect motifs, which may be related to the different evolutionary histories of genomes A and B. Ecological factors significantly affected SSR variation. Soil-*unique* and soil-*specific* alleles were found in the two edaphic groups dwelling on terra rossa (TR) and basalt (BA) soils across *macro-* and *micro*geographic scales. The largest contributions of genetic and ecological effects were found for diversity of ARN and NA, respectively. Multiple regression indicated, albeit indirectly, that replication slippage and recombination could be important mutational mechanisms, but their significance varied among motifs (see Sect. 5.5.6). Edaphic stresses may affect the probability of replication errors and recombination intermediates, and then control diversity level and divergence of SSRs (Li et al. 2000b, 2001d). The results may indicate that SSR diversity is *adaptive*, channeled by *natural edaphic selection* and influenced by both internal and external factors and their interactions. Detailed results are presented in the following sections.

Table 5.36. Motifs and chromosomal locations of 28 microsatellite (simple sequence repeats, SSR) loci of wheat used in the study. (Li et al. 2000b)

Locus[a]	SSR motif	Chromosomal location[b]	Chromosomal distance (D)[c]
GWM018 ...	$(CA)_n GA(TA)_k$	1BS	5.7
GWM060 ...	$(CA)_n$	7AS	52.3
GWM095 ...	$(AC)_n$	2AS	10.8
GWM099 ...	$(AC)_n$	1AL	95.3
GWM120 ...	$(CT)_n(CA)_k$	2BL	23.7
GWM124 ...	$(CT)_n(GT)_k$imp[d]	1BL	82.2
GWM136 ...	$(CT)_n$	1AS	38.8
GWM162 ...	$(CA)_n AA(CA)_k$	3AL	91.0
GWM169 ...	$(GA)_n$	6AL	49.6
GWM186 ...	$(GA)_n$	5AL	38.4
GWM218 ...	$(CT)_n$	3AS	30.4
GWM219 ...	$(GA)_n$imp	6BL	40.5
GWM251 ...	$(CA)_n$	4BL	28.9
GWM294 ...	$(GA)_n TA(GA)_k$	2AL	16.1
GWM332 ...	$(GA)_n$	7AL	53.0
GWM340 ...	$(GA)_n$	3BL	131.4
GWM361 ...	$(GA)_n$imp	6BS	11.4
GWM368 ...	$(AT)_n$	4BS	8.7
GWM389 ...	$(CA)_n(GT)_k$	3BS	118.4
GWM408 ...	$(CA)_n TA(CA)_n(TA)_k$	5BL	73.8
GWM415 ...	$(GA)_n$imp	5AS	25.4
GWM429 ...	$(CT)_n$	2BS	29.0
GWM459 ...	$(GA)_n$	6AS	62.5
GWM537 ...	$(CA)_n(TA)_k$	7BS	43.6
GWM540 ...	$(CT)_n(CC)(CT)_k$	5BS	9.3
GWM577 ...	$(CA)_n(TA)_k$	7BL	105.0
GWM601 ...	$(CT)_n$	4AS	8.1
GWM637 ...	$(CA)_n$	4AL	50.1

[a] Locus names from Röder et al. (1995, 1998)
[b] Chromosome number, followed by A for genome A or B for genome B, followed by S for short arm or L for long arm
[c] Distances from the centromere (in cM) estimated according to the maps of (Röder et al. 1998); the additional map information from Peng et al. (2000b) was used to estimate the distance of GWM218
[d] Imperfect

5.5.1 Divergence Among the Three Populations of *Triticum dicoccoides*

The SSR alleles were labeled by means of repeat numbers of motifs (Table 5.36). The distributions of alleles differed in the three populations (Li et al. 2000b). The NA per locus was significantly lower in the A (= Ammiad) (mean NA = 5.6) than in the T (= Tabigha) (mean NA = 7.7) and Y (= Yehudiyya) (mean NA = 6.7) populations at most SSR loci. The proportions of polymorphic loci were 0.923, 0.963, and 0.926 in the A, T, and Y populations, respectively. The observed SSR diversity (*He*, Nei 1973) in the T population (*He* = 0.709) was significantly higher than that in the A population (*He* = 0.596; Wilcoxon's signed-rank test, $z = 2.085$, $p < 0.05$). The *He* in the Y population (0.674) was between that in the A and in the T populations. This presumably reflects the two soil types (TR and BA) in the T population vs only one soil type in the A (TR) and Y (BA) populations (Li et al. 2000b).

Among the three populations, the ARN was significantly different at 23 polymorphic loci (Kruskal-Wallis test, $p < 0.05–0.00005$, Table 5.37). The Wilcoxon's signed-rank test showed that the ARN over many loci was significantly larger in the Y population than in the other two populations ($z = 2.623$, $p < 0.01$). The variance in repeat number (σ^2) significantly differed at 25 loci among the three populations (Levene's test, $p < 0.05–0.00005$). At 15 loci, the A population showed much higher σ^2 values than did the T and Y populations. Over all loci considered simultaneously, the σ^2 of the A population was significantly greater (Wilcoxon's signed-rank test, $p < 0.01$) than those of the T ($z = 2.677$) and Y ($z = 2.785$) populations (Table 5.37). This may reflect the larger and topographically more variable area of A (250,000 m^2) compared to T and Y (each about 1,000 m^2).

5.5.2 Divergence Between Terra Rossa and Basalt Groups

Between the TR and BA groups (Li et al. 2000b), significant SSR diversity was observed at most of the loci assayed (Table 5.38). ARN was significantly ($p < 0.01–0.00005$) higher in the BA than in TR group at 12 SSR loci. An opposite trend of ARN was observed at eight other loci between the two edaphic groups. The values of σ^2 were significantly larger at 15 loci in the TR than in the BA group ($p < 0.01–0.00005$). The opposite pattern of σ^2 was observed at only five other loci. Pairwise comparison (Wilcoxon signed-rank test) showed that the σ^2 in repeat number over all 27 SSR loci was significantly ($z = 2.53$, $p < 0.05$) higher in the TR ($\sigma^2 = 40.15$) than in BA group ($\sigma^2 = 20.76$) (Table 5.38).

In the TR and BA groups, 364 alleles were observed in the 28 SSR loci. Only 175 alleles were shared between the two edaphic groups (Table 5.38). After exclusion of rare alleles (observed in ≤14 individuals based on 0.05 probability of chance), eight BA-*specific* and seven TR-*specific* alleles were found (χ^2 test, $p < 0.05$). Among these soil-*specific* alleles, one TR-*unique* allele and one BA-*unique* allele were found. Permutation tests were conducted to determine the effects of random sampling. For this purpose, genotypes within each pair of soil groups (micro- only in T; and macrogeographic in A, T, Y subdivided into TR and BA) were randomly shuffled, with maintenance of the initial sample sizes. Out of 10^6 randomly shuffled sets, in no case did the number of soil-*specific* alleles exceed four. Thus, the significance of the observed 15 (BA + TR) soil-*specific* alleles was at least

Table 5.37. Microsatellite (SSR) diversity of 27 loci in Ammiad (A), Tabigha (T), and Yehudiyya (Y) populations of *T. dicoccoides*. (Li et al. 2000b) *ARN* Average repeat number, σ^2 variance in repeat number, *NA* number of alleles, *He* observed SSR diversity, *K-W test* Kruksal Wallis test; $*p < 0.05$, $**p < 0.01$, $***p < 0.0005$, $****p < 0.00005$

Locus	ARN			K-W test *H*(2, *N*)	σ^2			Leven's Test *F*(2, >200)	NA			*He*		
	A	T	Y		A	T	Y		A	T	Y	A	T	Y
GWM018 ...	24.3	33.1	36.9	168.3****	37.561	15.381	2.232	90.388****	7	7	5	0.626	0.804	0.734
GWM060 ...	12.6	22.0	24.0	70.2****	67.592	27.571	20.941	25.751****	6	9	5	0.784	0.754	0.626
GWM095 ...	15.7	16.7	17.8	51.3****	10.672	4.549	2.931	75.005****	6	6	7	0.485	0.797	0.789
GWM099 ...	15.6	8.1	10.4	16.2**	66.536	4.772	33.836	60.453****	4	3	5	0.497	0.410	0.646
GWM120 ...	32.5	33.3	33.1	4.6	0.550	6.431	2.324	11.892****	4	10	6	0.592	0.717	0.706
GWM124 ...	40.2	52.7	52.8	56.2****	245.650	25.083	27.180	78.369****	7	12	9	0.771	0.878	0.841
GWM136 ...	48.8	60.1	69.9	59.4****	233.511	69.036	71.190	6.344**	10	11	8	0.836	0.740	0.826
GWM162 ...	25.0	23.0	22.0	33.3****	13.419	3.104	2.949	61.287****	5	6	4	0.714	0.802	0.447
GWM169 ...	31.0	22.5	23.0	169.0****	35.201	4.033	6.727	27.085****	12	8	10	0.857	0.773	0.808
GWM186 ...	16.8	23.3	23.2	16.2**	118.551	23.006	13.412	14.094****	9	9	6	0.814	0.780	0.722
GWM218 ...	10.9	22.4	23.5	82.8**	64.783	41.856	20.425	32.554****	6	12	9	0.798	0.878	0.813
GWM219 ...	28.6	24.2	26.1	3.7	30.574	25.320	59.455	20.123****	6	6	8	0.703	0.725	0.744
GWM251 ...	25.1	20.5	25.7	8.6*	7.631	41.393	3.228	524.267****	4	9	6	0.324	0.688	0.677
GWM294 ...	15.6	19.8	23.4	128.7****	13.375	9.871	14.096	1.454	7	9	13	0.704	0.860	0.901
GWM332 ...	18.0	20.4	18.2	184.0****	0.766	2.091	0.414	52.414****	3	5	3	0.665	0.690	0.578
GWM340 ...	33.3	27.6	32.9	100.4****	9.479	36.357	9.648	34.493****	4	12	6	0.607	0.817	0.815
GWM361 ...	19.0	21.8	22.3	181.9****	0.000	1.142	1.651	14.082****	1	6	5	0.000	0.696	0.474
GWM368 ...	26.8	22.9	26.6	17.2*	23.794	11.107	59.261	82.275****	3	7	11	0.426	0.769	0.868
GWM389 ...	30.3	32.1	32.7	43.5****	20.803	7.683	5.295	6.400**	7	8	6	0.687	0.791	0.739
GWM408 ...	34.2	26.6	27.4	21.5****	120.012	11.709	8.755	47.003****	7	6	7	0.805	0.761	0.773
GWM415 ...	21.9	19.3	19.0	173.5****	3.215	1.237	0.000	93.800****	3	2	1	0.452	0.154	0.000
GWM429 ...	13.7	15.4	12.8	57.1****	12.957	22.646	2.914	79.561****	3	6	5	0.654	0.619	0.476
GWM459 ...	28.2	25.7	24.6	29.6****	1.540	13.156	21.101	13.146****	5	9	11	0.713	0.853	0.870
GWM537 ...	32.8	34.7	36.8	25.1****	66.049	31.742	20.808	7.179**	8	16	6	0.782	0.901	0.728
GWM540 ...	20.4	17.8	18.7	15.3**	9.000	1.596	5.055	63.007****	8	4	8	0.797	0.645	0.771
GWM577 ...	–	31.8	32.6	0.3	–	57.940	61.018	5.157*	–	9	10	–	0.838	0.818
GWM601 ...	17.0	17.0	17.0	0.0	0.000	0.000	0.000	0.000	1	1	1	0.000	0.000	0.000
Mean	24.5	25.7	27.1		46.661	18.512	17.376		5.6	7.7	6.7	0.596	0.709	0.674

Table 5.38. Comparison of microsatellite (SSR) diversity in terra rossa (TR) and basalt (BA) soil groups in three microsites in north, Israel. (Li et al. 2000b)

Locus	ARN		K-W test $H(1, N)$	σ^2		Hartley *F*-test $F(151, 184)$	NA				*He*		χ^2 (df)
	TR	BA		TR	BA		Total	TR	BA	Shared	TR	BA	
GWM018 ...	28.6	35.3	57.89****	51.79	5.56	9.31****	14	4	6	4	0.764	0.853	175.44**** (13)
GWM060 ...	17.8	23.4	42.51****	65.34	28.31	2.31****	14	5	4	5	0.817	0.752	180.43**** (13)
GWM095 ...	15.6	17.7	83.55****	7.17	3.01	2.38****	10	2	1	7	0.560	0.837	200.27**** (9)
GWM099 ...	11.6	9.6	0.90	51.67	24.20	2.14****	9	4	3	2	0.735	0.559	122.58**** (9)
GWM120 ...	32.6	33.4	7.48**	2.40	4.84	2.01****	11	0	3	8	0.737	0.808	47.13**** (10)
GWM124 ...	45.6	53.2	40.70****	161.42	27.52	5.87****	18	4	3	11	0.837	0.892	156.50**** (17)
GWM136 ...	55.8	63.4	20.61****	183.49	104.86	1.75**	21	6	4	11	0.846	0.857	79.94**** (20)
GWM162 ...	24.3	22.1	40.03****	8.69	2.74	3.17****	8	2	0	6	0.838	0.641	119.96**** (7)
GWM169 ...	25.2	19.1	47.01****	58.83	7.53	7.81****	20	10	2	8	0.901	0.810	121.47**** (19)
GWM186 ...	20.5	23.0	3.79	89.67	11.67	7.68****	18	9	4	5	0.863	0.851	177.44**** (17)
GWM218 ...	16.3	23.4	39.10****	89.22	22.98	3.88****	20	4	4	10	0.882	0.874	186.19**** (17)
GWM219 ...	25.3	26.1	2.12	29.87	49.24	1.65**	14	4	6	4	0.826	0.780	120.02**** (14)
GWM251 ...	24.2	22.6	30.90****	25.23	29.09	1.15	12	3	4	5	0.737	0.779	142.59**** (11)
GWM294 ...	18.2	21.7	42.58****	20.93	15.88	1.32	17	3	4	10	0.857	0.899	160.39**** (16)
GWM332 ...	19.3	19.1	1.70	2.35	2.48	1.06	6	0	1	5	0.829	0.745	174.90**** (9)
GWM340 ...	29.9	31.1	3.80	32.33	26.56	1.22	14	0	5	9	0.814	0.894	109.96**** (13)
GWM361 ...	20.5	22.0	46.29****	3.10	1.35	2.30****	8	1	2	5	0.663	0.524	193.77**** (7)
GWM368 ...	23.8	24.8	0.49	15.15	40.31	2.66****	21	1	12	8	0.790	0.860	57.74**** (12)
GWM389 ...	31.5	32.3	16.00****	16.91	4.72	3.59****	13	4	2	7	0.830	0.799	105.36**** (12)
GWM408 ...	30.4	26.4	14.72****	60.56	10.26	5.90****	18	4	9	5	0.807	0.840	152.62**** (12)
GWM415 ...	20.8	19.0	99.72****	3.90	0.09	44.80****	3	1	0	2	0.518	0.011	100.38**** (2)
GWM429 ...	16.6	13.8	49.30****	17.21	14.11	1.22	9	2	3	4	0.715	0.561	90.56**** (8)
GWM459 ...	26.4	25.3	10.93***	10.41	17.68	1.70***	14	0	5	9	0.822	0.905	104.71**** (13)
GWM537 ...	33.3	36.6	20.50****	39.47	38.61	1.02	21	6	7	8	0.887	0.877	186.69**** (20)
GWM540 ...	18.3	18.1	4.40*	5.50	3.62	1.52**	11	2	1	8	0.857	0.719	81.09**** (10)
GWM577 ...	28.7	33.3	14.18***	31.53	63.39	2.01**	21	0	13	8	0.665	0.865	51.21**** (12)
GWM601 ...	17.0	17.0	0.00	0.00	0.00	–	1	0	0	1	0.000	0.000	–
Overall	25.1	26.4	$z = 1.53$	40.15	20.76	$z = 2.53$*	364	81	108	175	0.755	0.733	

ARN Average repeat number, σ^2 variance in repeat number, *NA* number of alleles, *He* observed SSR diversity, *K-W test* Kruksal Wallis test; $^*p < 0.05$, $^{**}p < 0.01$, $^{***}p < 0.0005$, $^{****}p < 0.00005$

$p < 10^{-6}$. Moreover, in no case did the random permutations produce any soil-*unique* allele in the foregoing sense. This result suggests with high confidence the *nonrandomness* of the revealed soil-*specific* population differentiation. Moreover, the fact that two out of 15 soil-*specific* alleles appeared to be soil-*unique* may indicate that these two alleles are not only markers linked to the selected loci, but are the selected targets themselves (or sites within selected genes).

SSR diversity (*He*, Nei 1973) was estimated for the two edaphic groups (Table 5.38). Although the overall *He* was similar between the TR (0.755) and BA (0.733) groups, significant differences in allele composition were observed at some loci. For example, at five loci (GWM099, GWM162, GWM361, GWM415, and GWM429), the *He* values based on significantly different allele components (according to a χ^2 test for homogeneity of allele frequencies at each locus, $p < 0.00005$) were larger in the TR than in BA group; the differences in *He* ranged from 0.139 to 0.501. At two other loci (GWM095 and GWM537) allele composition was significantly ($p < 0.00005$) different, and *He* values were higher in the BA than in the TR group; the differences ranged from 0.200 to 0.277 (Table 5.38) (Li et al. 2000b).

5.5.3 Genetic Effects on SSR Diversity

The ARN scores in genome B were higher than those in genome A over all three populations (Table 5.39). Interestingly, the ARNs for each genome were very similar among the three populations; values for genome A were 20.8–23.8 and for genome B 28.2–29.6 (Li et al. 2000b). The NA was higher in genome B than in genome A for all three populations. Analysis of variance (MANOVA) indicated that the effects of genome on SSR variation (ARN, NA, and σ^2) differed among the three populations (Table 5.40). Genome significantly ($p <$ 0.05–0.00005) affected ARN in the A and T populations, and NA in the T and Y populations (Table 5.40). Chromosome also affected ($p <$ 0.05–0.00005) ARN, NA, and σ^2 of the A population, ARN and NA of the T population, and ARN and σ^2 of the Y population (Table 5.40). Chromosome 1 (both 1A and 1B) showed the largest ARN and σ^2 in all three populations (Table 5.39). The interaction of genome × chromosome significantly ($p <$ 0.05–0.0005) affected the NA in all three populations, and ARN and σ^2 in the A and Y populations, respectively. This study included seven types of motifs. Motif types significantly ($p <$ 0.05–0.00005) affected the ARN of all populations (Table 5.40). Compound SSRs, such as $(CA)_n(TA)_k$ and $(CT)_n(GT)_k$, had higher ARN, σ^2, and NA than simple SSRs [$(CA)_n$, $(GA)_n$, or $(CT)_n$] (Table 5.40). A locus effect was significant ($p <$ 0.0005–0.00005) for the ARN, NA, and σ^2 in all populations (Table 5.40). Locus GWM136 showed the largest ARN (48.8–69.9) and σ^2 (69.0–233.6) in the three populations (Table 5.37). Locus GWM099 showed the smallest ARN (8.1–15.6). The lowest σ^2 ($\sigma^2 = 0$) and NA (NA = 1) were found in GWM601. The largest NA was found at GWM169 in the A population, at

Table 5.39. Microsatellite (simple sequence repeats, SSR) diversity in different genomes, chromosomes, and motifs across 27 SSR loci in the Ammiad (A), Tabigha (T), and Yehudiyya (Y) populations of *T. dicoccoides*. (Li et al. 2000b)

	ARN				σ^2				NA			
	A	T	Y	Mean	A	T	Y	Mean	A	T	Y	Mean
Genome												
A	20.8	22.7	23.8	22.4	42.960	14.538	14.734	24.077	3.0	5.6	5.3	4.6
B	28.2	28.2	29.6	28.7	35.174	19.498	18.161	24.278	3.5	6.8	6.5	5.6
Chromosome												
1	31.1	38.6	42.5	37.3	141.185	28.266	32.245	67.552	3.2	6.8	5.6	5.2
2	20.5	21.3	21.8	21.2	2.844	11.057	5.393	6.431	2.8	6.6	7.0	5.5
3	24.5	26.4	27.8	26.2	18.130	21.818	9.546	16.498	3.4	8.4	5.8	5.9
4	21.6	19.5	21.6	20.9	5.110	12.322	14.154	10.529	1.5	3.5	4.4	3.1
5	23.6	21.8	21.3	22.2	55.422	8.666	6.294	23.461	4.0	4.4	5.3	4.5
6	26.2	23.6	24.2	24.7	13.363	9.834	21.228	14.808	3.5	6.4	7.5	5.8
7	24.1	27.2	27.9	26.4	37.416	27.202	25.266	29.960	4.3	7.4	5.9	5.9
Motif												
$(CT)_n$	23.4	26.5	27.8	25.9	73.394	26.971	19.907	40.091	3.1	5.8	5.3	4.7
$(CA)_n$	18.3	18.1	20.0	18.8	21.709	15.154	12.461	16.441	2.5	5.1	4.8	4.1
$(GA)_n$	23.3	22.8	23.7	23.3	19.418	12.531	13.464	15.138	3.3	6.2	6.4	5.3
$(CA)_n(TA)_k$	30.9	31.6	33.4	32.0	59.296	26.092	22.431	35.940	4.3	7.5	6.6	6.1
$(CT)_n(GT)_k$	33.0	42.4	42.7	39.4	99.611	15.795	16.043	43.816	4.5	8.8	6.8	6.7
$(CT)_n(CA)_k$	32.6	33.3	33.1	33.0	0.425	6.040	2.320	2.928	3.0	8.5	6.0	5.8
$(AT)_n$	26.6	23.0	26.8	25.4	11.572	11.103	53.375	25.350	1.5	6.5	10.0	6.0

ARN Average repeat number, σ^2 variance in repeat number, *NA* number of alleles

Table 5.40. The *F* ratio of MANOVA for effects of genetic factors on microsatellites (simple sequence repeats, SSR), diversity in natural populations of wild emmer wheat *T. dicoccoides* from the microsites in Ammiad, Tabigha, and Yehudiyya populations, Israel. (Li et al. 2000b)

SSR Variation[a]	Genome	Chromosome	Genome × chromosome	Motif	Locus
Ammiad					
df_1, df_2	1, 88	6, 88	6, 88	6, 95	25, 76
ARN	21.651****	3.025**	4.315***	4.949***	18.305****
σ^2	0.207	5.228***	0.890	1.693	4.597****
NA	1.984	3.504**	3.966**	2.288	3.092***
Tabigha					
df_1, df_2	1, 40	6, 40	6, 40	6, 47	26, 27
ARN	4.465*	3.386**	0.461	4.566***	96.269****
σ^2	0.978	1.717	1.332	1.085	4.594***
NA	4.634*	4.448**	3.212*	1.555	9.926****
Yehudiyya					
df_1, df_2	1, 40	6, 40	6, 40	6, 47	26, 27
ARN	3.865	3.847**	0.470	2.844*	114.132****
σ^2	0.474	2.716*	3.059*	1.473	7.957****
NA	4.987*	1.886	5.896***	1.588	37.717****
Overall					
df_1, df_2	1, 196	6, 196	6, 196	6, 196	31, 178
ARN	8.070**	13.532****	7.188****	12.183****	31.691****
σ^2	2.774	7.899****	3.151**	2.030	3.539****
NA	20.835****	2.173*	2.465**	2.972*	4.022****

Variation included: average repeat number (ARN), variance in repeat number (σ^2), number of alleles (NA); *$p < 0.05$, **$p < 0.01$, ***$p < 0.001$, ****$p < 0.00005$

GWM537 in the T population (overall maximum NA), and at GWM294 in the Y population (Table 5.37).

5.5.4 Effects of Genetic and Environmental Factors on SSR Variation

Multivariate analysis of variance (MANOVA) was carried out to measure the effects of interactions among genetic (genome, chromosome, motif) and environmental (population and soil type) factors on SSR diversity (Li et al. 2000b). The results showed that the main effects of genetic and ecological factors and most of their interactions were significant ($p < 0.05$–0.00005) for ARN and NA. Motif seemed to influence the significance of these main effects and their interactions for the ARN and NA. When motif effect was ignored, the *F* ratios became smaller. But if motif effect was taken into account (as a covariant), the *F* ratios and significance level became larger and higher (Table 5.41) for the ARN and NA. This result suggests that motif modifies genetic and ecological effects on the ARN and NA.

Table 5.41. The *F* ratio of MANOVA for effects of genetic and environmental factors on microsatellites (SSR), diversities in three microsites of wild emmer wheat. (Li et al. 2000b)

Effect[a]	df_1, df_2	SSR Variation					
		ARN		σ^2		NA	
		With[b]	Without[c]	With[b]	Without[c]	With[b]	Without[c]
1	1, 166	19.799****	1.926	1.607	1.936	0.078	3.985**
2	6, 166	23.064****	17.103****	2.147*	2.162*	7.166****	6.395****
3	2, 166	8.433***	5.304**	1.706	1.758	49.877****	45.156****
4	1, 166	4.977*	3.144	0.906	0.927	18.819****	16.771****
1 × 2	6, 166	9.521****	6.108****	0.698	0.699	5.059***	4.737***
1 × 3	2, 166	22.618****	14.338****	0.262	0.251	2.282	2.019
1 × 4	1, 166	8.970**	6.630	0.001	0.001	1.229	1.347
2 × 3	12, 166	3.248***	2.154*	1.833*	1.864*	3.242***	2.874**
2 × 4	6, 166	0.915	0.597	0.882	0.891	0.892	0.878

$^*p < 0.05$, $^{**}p < 0.01$, $^{***}p < 0.0005$, $^{****}p < 0.00005$

[a] *1* Genome, *2* chromosome, *3* population, *4* edaphic niche: terra rossa and basalt. The interactions of tri- and tetra-cofactors could not be analyzed due to unbalanced design

[b] With motif as a covariant

[c] Without any covariant

5.5.5 Contribution of Genetic and Environmental Factors to SSR Variation

The relative contributions of ecological (populations and edaphic niches) and genetic factors (genome, chromosome, and motif) to SSR diversity were estimated according to the proportion of sums of squares (SS) of ecological and genetic factors to the total sum of squares arising from all the factors (Table 5.42) (Li et al. 2000b). For the diversity of ARN, the relative contribution of genetic factors was high (43.0%), the ecological factors contributed 4.4%, and the interaction of genetic and ecological factors contributed 19.7%. For the diversity of σ^2, the interaction genetics × ecological factors contributed 12.9%, the genetic factors contributed 8.7%, and the ecological factors contributed 2.0%. For the variation of NA, however, the ecological contribution reached 29.1%, the genetic contribution was only 18.0%, and the contribution of the interaction of genetics × ecology was 12.3% (Table 5.42). The ecological effect was stronger for NA than for ARN and σ^2, but the genetic effect appeared to be stronger for ARN; the interaction of ecological and genetic factors seemed to be similar for ARN, σ^2, and NA. The total contributions of the ecological and genetic effects investigated were 67.1, 23.6, and 59.4% for ARN, σ^2, and NA, respectively, suggesting that other factors may also be responsible for SSR diversity.

Table 5.42. Relative contributions of genetic (genome, chromosome, and motif) and ecological (population and microniche) factors to microsatellite (SSR) diversity according to percentage of sums of squares of ecological and genetic factors to the total sums of squares. (Li et al. 2000b)

Effect	Relative Contribution (%)		
	ARN	σ^2	NA
Ecology[a]	4.4	2.0	29.1
Genetics	43.0	8.7	18.0
Ecology × genetics	19.7	12.9	12.3
Total	67.1	23.6	59.6

[a] Ecological effect included populations and edaphic effects (terra rossa and basalt)

5.5.6 Determination of Mutational Mechanisms and Edaphic Effect on SSR Diversity

Multiple regression analysis was used to estimate the coefficients of determination of mutational mechanisms (replication slippage and unequal crossing over) and edaphic effect for SSR diversity (Table 5.43). The ARN and locus distance from the centromere (*D*) were used to estimate indirectly the importance of replication slippage and recombination based on previous results obtained by Levinson and Gutman (1987), Wolff et al. (1991), and Innan et al. (1997). The products of the factors were used to measure the effects of interactions between them. The results suggest that edaphic factors may affect SSR diversity, either directly (selection) or indirectly (e.g., by influencing mutational mechanisms or owing to hitchhiking) (Table 5.43). Over all SSRs assayed in this study, replication slippage seems to be the most important mutational mechanism and it might be affected by edaphic stresses. Both soil × ARN and soil significantly affected the NA ($R^2 = 0.247$, $p < 0.00005$). ARN, soil × ARN, and soil were significantly correlated to σ^2 in repeat number ($R^2 = 0.183$, $p < 0.00005$). The observed coefficients of determination (R^2) to SSR diversity across all motifs seemed quite low, suggesting a motif-*specific* mechanism.

Next, multiple regression was conducted for each motif except $(AT)_n$ and $(CT)_n(CA)_k$ due to too few observations. Different mechanisms and R^2 were observed among the five motifs (Table 5.43). The interaction of soil × ARN significantly affected NA of $(GA)_n$ SSRs ($R^2 = 0.277$, $p < 0.00005$), suggesting that edaphic type may influence replication slippage by changing ARN and then modifying the NA. For $(CA)_n$ SSRs, soil × ARN and *D* × ARN were significantly ($R^2 = 0.297$, $p < 0.01$) correlated to NA, indicating effects from

Table 5.43. The number of alleles (NA) and variance in microsatellite repeat number (σ^2) in different motifs of *T. dicoccoides* dependent on the mutational mechanisms (see footnote) and edaphic factors. (Li et al. 2000b)

Motif	No. of cases	Step	Variables	NA		Step	Variables	σ^2	
				r	R^2			*r*	R^2
All	212	1	Soil × ARN	0.306****	0.230****	1	ARN	0.315****	0.118****
		2	Soil	0.151*	0.247****	2	Soil × ARN	–0.212**	0.169****
						3	Soil	0.129*	0.183****
$(GA)_n$	72	1	Soil × ARN	0.526****	0.277****		–	–	–
$(CA)_n$	40	1	Soil × ARN	0.545***	0.340***		–	–	–
		2	*D* × ARN	–0.170	0.297**				
$(CT)_n$	40	1	Soil × *D*	0.498**	0.228**	1	*D* × ARN	0.568***	0.240**
						2	Soil × ARN	–0.394*	0.358***
$(CA)_n(TA)_k$	28	1	Soil × *D*	0.569**	0.313**	1	*D* × ARN	0.694***	0.295**
		2	ARN	0.294	0.373**	2	*D*	–0.648***	0.604****
						3	Soil × ARN	–0.492*	0.703****
						4	ARN	0.484*	0.719****
						5	Soil	0.436*	0.772****
$(CT)_n(GT)_k$	16	1	*D* × ARN	0.695****	0.459**	1	*D* × ARN	–0.681**	0.336**
		2	*D*	–0.299	0.507**	2	*D*	–0.648**	0.615**

Two motifs were excluded from this analysis due to small sample size; *ARN* average repeat number, *D* locus distance from the centromere (cM). *ARN* and *D* were used to indirectly estimate importance of replication slippage and recombination, respectively. *r* Coefficient of partial correlation with significance of *t* test, R^2 coefficient of multiple determination with significance of *F* test for multiple correlation, – no significant factors entered the multiple regression; $^*p < 0.05$, $^{**}p < 0.01$, $^{***}p < 0.001$, $^{****}p < 0.00005$

the interactions of edaphic factors and mechanisms. The interaction soil × *D* significantly affected NA of $(CT)_n$ loci ($R^2 = 0.228$, $p < 0.01$), suggesting that edaphic stresses may affect chromosomal recombination and then change the NA. The interactions *D* × ARN and soil × ARN could significantly affect the σ^2 of $(CT)_n$ loci ($R^2 = 0.358$, $p < 0.001$). For $(CA)_n(TA)_k$ loci, soil × *D* and ARN were significantly correlated to the NA ($R^2 = 0.373$,p < 0.01); *D* × ARN, *D*, soil × ARN, ARN, and soil could determine 77.2% variation in σ^2 of $(CA)_n(TA)_k$ SSRs ($R^2 = 0.772$, $p < 0.00005$). In $(CT)_n(GT)_k$ loci, *D* × ARN and *D* significantly affected NA ($R^2 = 0.507$, $p < 0.01$) and σ^2 ($R^2 = 0.615$, $p < 0.01$). These results may indicate motif-*specificity* of mechanisms generating SSR diversity and mechanisms maintaining population diversity (habitat-*specific* and soil-*specific* selection affecting differential clone selection).

In summary, strong SSR diversity was found among the three populations and two edaphic groups of *T. dicoccoides* across macro- and microgeographic scales. SSR diversity was correlated with both ecological and genetic factors. The revealed patterns may indicate that the relative importance of different mutational mechanisms (replication slippage and recombination) in generating new alleles at SSR loci vary among motifs. Ecological factors may affect the level and the pattern of SSR diversity directly by modulating mutation mechanisms, through natural selection for the maintenance of favorable mutants at SSR loci or through linkage to other selected loci and LDs caused by high selfing rates. At any rate, SSRs may be far from representing neutral genetic diversity, as indicated by the correlation of SSR variation with ecological factors and the observation of *specific* and, particularly, *unique* edaphic alleles. We believe that the idea of "junk DNA" needs to be replaced by the idea of fine-tuned regulation in diverse ecologies and environmental stresses. This idea should be critically tested in model organisms facing sharp ecological contrasts at microscales (Li et al. 2000b).

Part III

Genetic Resources of Wild Emmer for Wheat Improvement

Chapter 6

Genetic Variation in Agronomic Traits

6.1 Quantitative Morphological and Physiological Variations in Phenotypes and Genotypes

Wild emmer wheat in Israel varies not only in genetic polymorphisms of allozymes and DNA markers, but also in quantitative traits of agronomic importance (Nevo 1983a, 1986a, 1989a, 1995, 2001b; Nevo et al. 1984b; Peng et al. 2000c, 2001). These include agronomic traits, disease resistance, protein content, ecological tolerances to drought, salt, herbicides, and photosynthetic yield. These traits, described below, are economically significant and should be conserved and utilized in wheat crop improvement. Our results substantiated earlier claims that wild emmer in Israel displays remarkable morphological and physiological variation both within and between populations in each of the agronomic characters studied (Table 2 and Fig. 1 in Nevo et al. 1984b; Fig. 6.1). These traits are derived by natural selection as adaptive traits at both *macro*- and *micro*geographic scales. A detailed quantitative trait loci (QTL) mapping for 11 agronomic traits appears below and is elaborated on in Peng et al. (2001).

6.1.1 Genetic Variation in Phenotypic Agronomic Traits

Populations of *T. dicoccoides* in Israel, originating from diverse habitats and tested for allozyme (Nevo et al. 1982) and disease resistance polymorphisms (Moseman et al. 1984; Nevo et al. 1985), were compared and contrasted for performance in agronomically important phenotypic traits (Nevo et al. 1984b and Fig. 6.1). The compared traits involved ten variables including germination, earliness, biomass, and yield variables. The field experiments were conducted in 1980, 1981, and 1982 in two relatively standardized and contrasting environments: mesic (Mount Carmel, Haifa) and xeric (Avedat Farm and Sede Boqer, in the northern Negev desert). The experimental design involved 12 population quadrangles at Avedat Farm in 1980 and rows of randomized genotypes of five populations in both Haifa (Mediterranean climate) and Sede Boqer (desert climate) in 1981 and 1982.

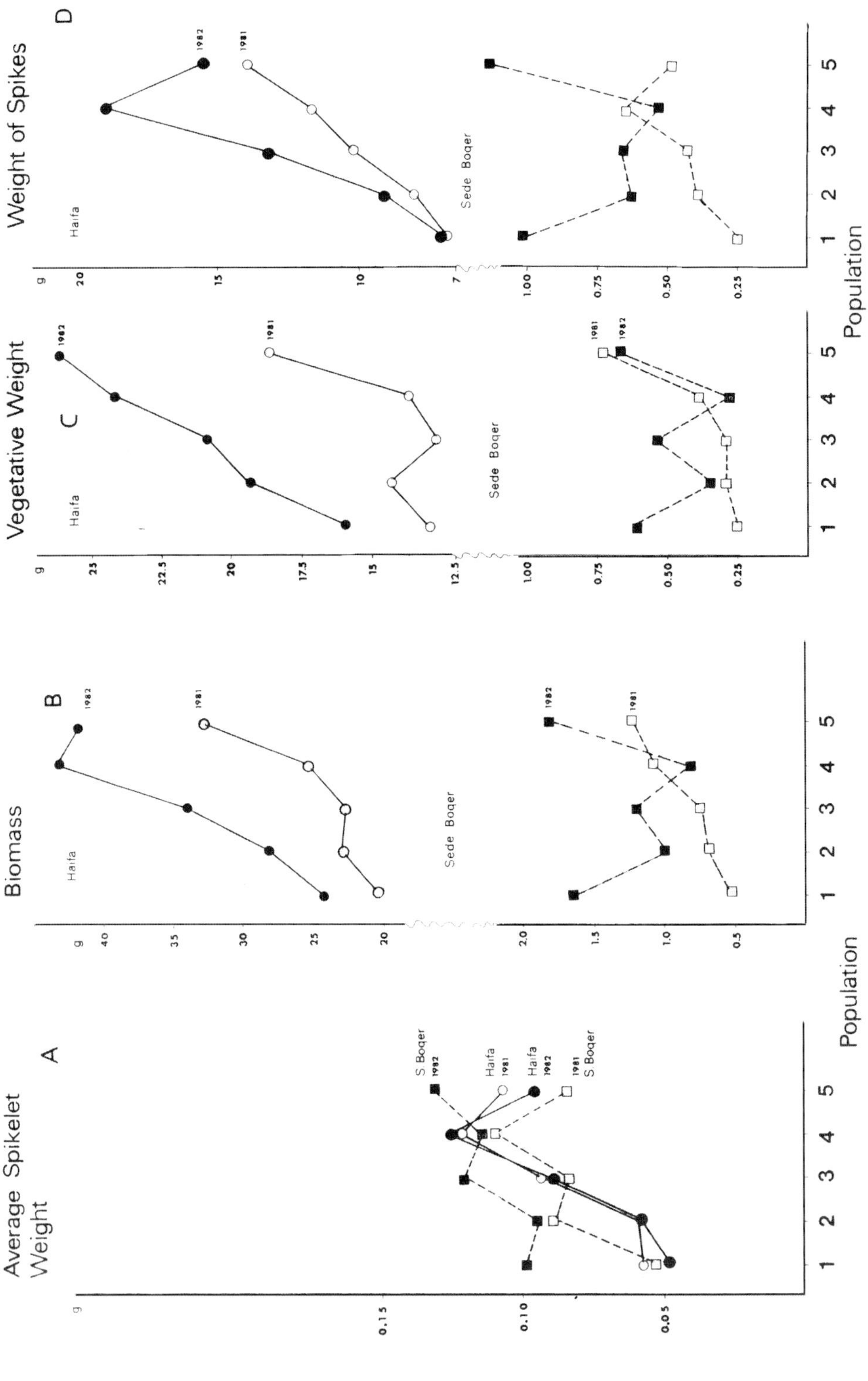
Average Spikelet Weight
A
g
0.15
0.10
0.05
S Boqer 1982
Haifa 1981
Haifa 1982
1981 S Boqer
Population
1
2
3
4
5
Biomass
B
Haifa
Sede Boqer
g
40
35
30
25
20
2.0
1.5
1.0
0.5
1982
1981
Vegetative Weight
C
Haifa
Sede Boqer
g
25
22.5
20
17.5
15
12.5
1.00
0.75
0.50
0.25
Weight of Spikes
D
Haifa
Sede Boqer
g
20
15
10
7
1.00
0.75
0.50
0.25
Population

The results indicated that the characters studied are partly genetically determined. Striking genotypic and phenotypic variation was found between and, at least for some characters, also within populations in each site, whereas remarkable environmental variation including genetic-environmental interaction was found between the mesic and the xeric sites (Fig. 6.1).

Interest in the progenitors of crop species should not be restricted to their wealth of resistance genes against pathogens acquired over long periods of host-pathogen coevolution. *T. dicoccoides* harbors a wealth of additional genetic traits of economic significance, such as earliness, high yield, and high biomass. Apart from their direct importance in breeding, they can also improve breeding indirectly. When parental genotypes of the progenitor are used in breeding, it is desirable that they also combine the genetic capacity for high yield and biomass to avoid yield depression in the resulting crosses (see the QTLs of 11 agronomic traits below, also detailed in Peng et al. 2001). In addition, with the QTL mapping approach it is possible to identify valuable cryptic alleles, i.e., beneficial alleles that are otherwise hidden in a sea of agronomically deleterious ones (Peng 2000; Peng et al. 2001).

6.2 Heat Production in Wild Cereals

We calorimetrically compared and contrasted heat production in seedlings incubated at 5 and 24 °C, testing genotypes from cold and warm Israeli populations of the wild progenitors of barley (*Hordeum spontaneum*) and wheat (*T. dicoccoides*) (Nevo et al. 1992b). The wild barley sample involved 14 accessions, seven from cold and seven from warm localities. The wild emmer wheat sample involved 12 accessions, six from cold and six from warm localities. Our results (Fig. 6.2) indicated that: (1) in both wild cereals, heat production was significantly higher at 5 °C than at 24 °C; (2) interspecifically, wild barley generated significantly more heat than wild wheat at both 5 and 24 °C; (3) intraspecifically, at 24 °C wild barley from warm environments generated significantly more heat than wild barley from cold environments. We hypothesized that both the *inter*- and *intra*specific differences in heat production evolved adaptively by natural selection in accordance with the genetic niche width variation hypothesis (Van Valen 1965). These differences presumably enhance biochemical processes, hence growth, thereby leading to the shorter annual cycle of barley than that of wheat. They may also explain the wider range of the wild and cultivated gene pools of barley as compared with those of wheat. We proposed that shortening the growth period through the uti-

Fig. 6.1A–D. Graphic representation of the population means of five agronomic characters in wild wheat, *Triticum dicoccoides*, in Israel. **A** Average spikelet weight; **B** biomass; **C** vegetative weight; **D** weight of spikes. ❍ Haifa, 1981; ● Haifa, 1982; ❑ Sede Boqer, 1981; ■ Sede Boqer, 1982. Populations were arranged from marginal (*1* Mt. Hermon, *2* Kokhav Hashahar, *3* Bat Shelomo) to central (*4* Yehudiyya, *5* Qazrin). (Nevo et al. 1984b)

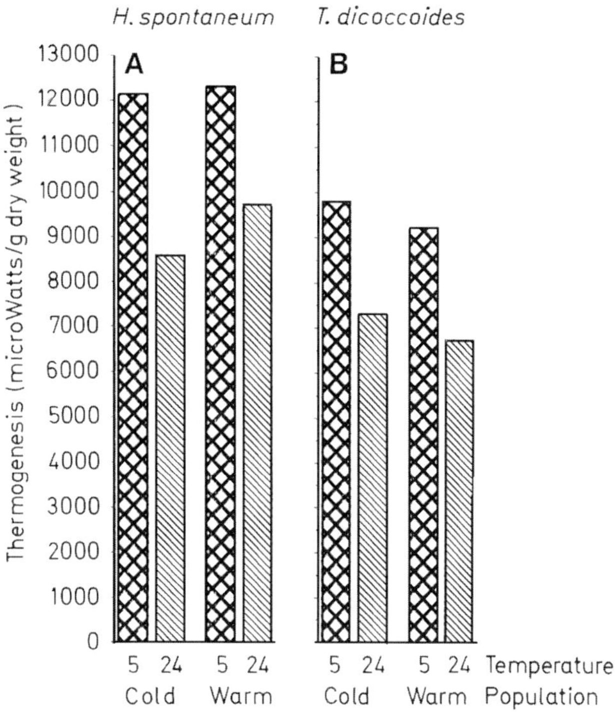

Fig. 6.2A.B. Heat production in wild barley (*Hordeum spontaneum*) (**A**) and wild emmer wheat (*Triticum dicoccoides*) (**B**) from cold and warm populations incubated at 5 and 24 °C. (Nevo et al. 1992a)

lization of heat production gene(s) is feasible in breeding by classical methods and/or modern biotechnology.

6.3 Abiotic Stress Tolerance and Variation in Physiological Performance: Salt and Drought Tolerance

Natural populations of *T. dicoccoides* are characterized by abundant genetic variation in resistance to physiological stresses, such as salinity and drought. This is highlighted by the tremendous variation both between and within populations observed in our extreme desert station at Sede Boqer (Table 2 and Fig. 2 in Nevo et al. 1984b). In our Sede Boqer experiments, under a water regime of 130 mm and highly saline soil conditions, some populations and superior genotypes have proved outstanding in both growth and yield despite these extremely dry ecological conditions. Thus, the genetic resources of the progenitor display striking morphological and physiological traits that are of potential economic importance for wheat improvement (Fig 6.1). We conducted critical salt tolerance tests on these populations; these are described below.

6.3.1 Salinity Tolerance

Salt tolerance was tested in the progenitors of the cultivated cereals wild barley (*H. spontaneum*) and wild emmer wheat (*T. dicoccoides*) from Israel (Nevo et al. 1993b; Table 6.1). Plants from five populations of *T. dicoccoides* from the eastern Samaria steppes, Mt. Hermon, and Mt. Carmel were grown in 175 and 250 mM NaCl. Likewise, five selected populations of *H. spontaneum* from the Mediterranean coastal plain and northern Negev desert were grown in 250 and 350 mM NaCl. We found superior genotypes of *T. dicoccoides* ripening at 250 mM (=40% seawater) and of *H. spontaneum* ripening at 350 mM NaCl (=60% seawater). We are now proceeding with both genetic and physiological studies aimed at chromosomally locating salt tolerance genes and unraveling the mechanism(s) of salt resistance in these wild cereals.

In wild barley, *Hordeum spontaneum*, the progenitor of all cultivated barleys (Nevo 1992), we found that chromosome 4 controlled potential water use efficiency ($\delta^{13}C$) (Handley et al. 1994), and that shoot $\delta^{15}N$ is correlated with salt stress (Handley et al. 1997). Likewise, we found amplified fragment-length polymorphism (AFLP) associations with salt tolerance (Pakniyat et al. 1997) and SSR associations with drought tolerance (Forster et al. 1997). It is expected that similar, though less extreme, salt and drought resistances will also be found in *T. dicoccoides* as indicated below by the genetic polymorphisms in ^{22}Na uptake. We identified genes with positive effects of salt tolerance on chromosomes 4 and 5 of *H. vulgare* and $1H^{ch}$, $4H^{ch}$, and $5H^{ch}$ of *H. chilense*. Thus, the salt tolerance of barley could be improved by gene transfer from chromosomes 4 and 5 of salt-tolerant species such as *H. spontaneum*, not only within the genus *Hordeum* (Forster et al. 1990), but also from wild barley to cultivated wheat.

6.3.1.1 Genetic Polymorphisms in ^{22}Na Uptake

We also found variation in ^{22}Na uptake in wild emmer, which may provide salt tolerance resources for wheat improvement (Nevo et al. 1992a). Polymorphisms in ^{22}Na uptake relative to the uptake into *T. durum* cv. Langdon (relative uptake Langodon, RUL), were found among 230 accessions of tetraploid *T. dicoccoides* in Israel. The accessions were collected in 15 xeric margins of the Israeli range (Table 6.2). The results indicated that: (1) there is widespread genetic variation both within and primarily between populations in ^{22}Na uptake. (2) Remarkably, the xeric eastern and southern populations, characterized by a drier and hotter climate, displayed significantly lower values of

Table 6.1. Salt tolerance of genotypes from five populations of wild barley, *Hordeum spontaneum*, and wild wheat, *Triticum dicoccoides*, in Israel (from Nevo et al. 1988b)

Population	Genotype	Lower salt concentration					Higher salt concentration					Control average (g)	Performance[g]
		Plant condition			Average[d] (g)	Maximum[e] (g)	Plant condition			Average[d] (g)	Maximum[e] (g)		
		I[a]	II[b]	FS[c]			I[a]	II[b]	FS[c]				
Wild barley, *Hordeum spontaneum*													
Mehola	30	9	8	3	14.25	17.30	5	4	3	6.43	17.89	160.18	1
Mehola	25	8	8	3	11.84	18.32	6	6	3	3.38	5.83	186.59	
Ceasarea	24	8	8	4	12.93	19.28	7	6	2	2.72	5.23	122.89	
Ceasarea	56	8	7	1	7.78	10.46	5	4	2	3.13	7.20	122.90	
Akhziv	42	7	4	3	16.72	29.36	4	2	1	2.30	3.35	103.61	2
Akhziv	26	9	7	3	10.56	26.09	6	2	1	0.75	3.03	155.56	
Ha-Makhtesh Ha-Gadol	9	9	7	2	2.62	4.54	7	4	1	1.67	3.77	180.34	
Ha-Makhtesh Ha-Gadol	6	9	8	3	15.10	18.86	6	2	0	0.95	2.45	149.22	3
Atlit	49	8	6	0	10.68	15.03	3	2	0	0.62	2.48	100.25	
Atlit	10	6	6	0	5.76	11.70	3	0	0	0	0	130.53	
Wild wheat, *Triticum dicoccoides*													
Gilboa	28	9	7	2	44.95	89.19	9	3	1	2.36	5.53	71.79	1
Gitit	51	9	2	2	20.04	58.60	9	1	0	2.25	5.35	165.83	
Gitit	1	9	8	2	23.46	30.22	6	1	1	0.76	1.67	–[f]	
Bet-Oren	1	9	5	0	23.74	45.37	8	1	0	1.96	4.73	145.42	
Gilboa	40	9	8	2	14.90	22.98	7	0	0	1.57	2.60	69.91	2
Hermon	13	9	4	1	15.22	32.70	6	1	0	1.24	3.80	–	
Bet-Oren	14	9	5	0	9.76	17.80	7	0	0	1.98	4.88	45.66	
Kokhav-Hashahar	25	8	5	0	14.35	22.60	7	0	0	1.64	4.33	–	
Hermon	22	9	2	0	5.18	13.02	6	1	0	0	0	–	3
Kokhav-Hashahar	32	8	4	1	4.30	17.20	6	0	0	0	0	–	

[a] Plant condition at 4 weeks after salinization on a scale of 0–9; 0 = dead, 9 = excellent
[b] Plant condition at 3 months after sowing
[c] Flowering and seed ripening, number of flowering containers
[d] Average of total plant dry weight including dead plants
[e] Maximum plant weight achieved by genotype
[f] Control plants were removed from the greenhouse because of excessive growth
[g] Phenotypic performance: 1 superior, 2 intermediate, 3 inferior

Table 6.2. Frequency distribution of ^{22}Na uptake in 15 populations of *T. dicoccoides* relative to ^{22}Na uptake in Langdon in Israel and two ecogeographic regions. (Nevo et al. 1992b)

	N	Relative ^{22}Na uptake compared to Langdon (RUL)			
		0–0.25 (%)	0.26–0.48 (%)	0.49–0.89 (%)	0.90–1.52 (%)
Population no.					
16	15	2 (13.3)	6 (40.0)	7 (46.7)	0 (0.0)
17	7	1 (14.3)	3 (42.9)	2 (28.6)	1 (14.3)
18	17	7 (41.2)	8 (47.1)	2 (11.8)	0 (0.0)
19	18	3 (16.7)	9 (50.0)	6 (33.3)	0 (0.0)
20	12	3 (25.0)	5 (41.7)	4 (33.3)	0 (0.0)
21	7	2 (28.6)	5 (71.4)	0 (0.0)	0 (0.0)
23	18	7 (38.9)	8 (44.4)	3 (16.7)	0 (0.0)
24	16	1 (6.3)	1 (6.3)	7 (43.8)	7 (43.8)
25	15	0 (0.0)	3 (20.0)	4 (26.7)	8 (53.3)
26	16	9 (56.3)	6 (37.5)	1 (6.3)	0 (0.0)
27	18	3 (16.7)	4 (22.2)	10 (55.6)	1 (5.6)
28	14	3 (21.4)	2 (14.3)	5 (35.7)	4 (28.6)
29	20	0 (0.0)	2 (10.0)	14 (70.0)	4 (20.0)
32	18	0 (0.0)	9 (50.0)	8 (44.4)	1 (5.6)
33	19	1 (5.3)	5 (26.3)	13 (68.4)	0 (0.0)
Total	230	42 (18.3)	76 (33.0)	86 (37.4)	26 (11.3)
Marginal region					
Xeric (east and south)	94	25 (26.6)	44 (46.8)	24 (25.5)	1 (1.1)
Mesic (west)	136	17 (12.5)	32 (23.5)	62 (45.6)	25 (18.4)

Significance by χ^2: $p < 0.000005$

relative ^{22}Na uptake as compared with the mesic western populations (0.38 vs 0.61 RUL) (Table 6.3). (3) Ecological factors and allozyme markers, either singly or in combination, explained up to 81% of the variance in RUL and, therefore, appear to be good guidelines for predicting the ecogeographic location and allozymic constitution of specific genotypes that have low ^{22}Na uptake and which presumably are associated with high salt tolerance. We concluded that *T. dicoccoides* harbors genetic resources for salt tolerance that can be utilized in wheat improvement by both classical plant breeding and by modern biotechnology.

Some variation for salt tolerance is available in wheat, *Triticum aestivum* (Kingsbury and Epstein 1984), and the potential exists for small improvements by conventional breeding. Genes for salt tolerance have been located in the D genome of wheat (Gorham et al. 1987; Shah et al. 1987). *Aegilops squarrosa,* the D genome donor of bread wheat, and other D genome-carrying species, therefore, widen the gene pool. The tolerance of *Aegilops* species is not as great as that of more distant wheat relatives, particularly some members of the genus *Thinopyrum. Thinopyrum bessarabicum* (syn. *Agropyron*

Table 6.3. Mean ^{22}Na uptake in 15 populations of *Triticum dicoccoides* in **A** Israel and **B** two ecogeographic regions. (Nevo et al. 1992b)

A. No	Population		*N*	RUL ± SD
16	Mt. Gilboa		15	0.441 ± 0.212
17	Mt. Gerizim		7	0.559 ± 0.321
18	Gitit		17	0.317 ± 0.201
19	Kokhav-Hashahar		18	0.400 ± 0.161
20	Taiyiba		12	0.403 ± 0.175
21	Sanhedriyya		7	0.310 ± 0.118
23	J'aba		18	0.301 ± 0.195
24	Amirim		16	0.806 ± 0.267
25	Nahef		15	0.844 ± 0.329
26	Ahihud		16	0.268 ± 0.128
27	Nesher		18	0.571 ± 0.245
28	Bet-Oren		14	0.651 ± 0.355
29	Daliyya		20	0.759 ± 0.178
32	Yabad		18	0.493 ± 0.140
33	Givat-Koach		19	0.544 ± 0.122
Mean				0.518 ± 0.279
Total			230	
Significance by ANOVA		F		11.783
		df		14/215
		p		<0.00005
B. Marginal region				
Xeric (east, south)			94	0.378 ± 0.025
Mesic (west)			136	0.615 ± 0.283
Significance by ANOVA		F		48.220
		df		1/228
		p		<0.00005

RUL ^{22}Na uptake relative to Langdon control

junceum L.) is native to sand dunes of the Black Sea (former USSR) and is highly tolerant to prolonged exposure to salt. Since it is diploid (2n = 2x = 14), genes are confined to just seven chromosomes of the E genome. It can also be hybridized with wheat (Alonso and Kimber 1980; Forster and Miller 1985). For these reasons *Th. bessarabicum* was chosen as a source of salt tolerance genes for introduction into wheat (Forster et al. 1990). Our salt tolerance tests of *T. dicoccoides* suggest that some of its genotypes and populations (Nevo et al. 1993b) may be good candidates for improving salt tolerance in wheat cultivars (Table 6.1). Likewise, salt tolerance genes of wild barley, which is superior to wheat in salt tolerance, can be transferred for improving salt tolerance in wheat.

6.3.2
Drought Tolerance

Populations of wild emmer wheat, *T. dicoccoides*, in Israel, originating from diverse habitats, were compared and contrasted for their performance in dry and humid environments, regarding agronomically important phenotypic traits. The traits involved comparison of ten variables with respect to germination, earliness, biomass and yield variables. The field experiments were conducted in 1980, 1981, and 1982 in two relatively standardized and contrasting environments: mesic (Mount Carmel, Haifa) and xeric (Avedat Farm, and Sede Boqer, in the northern Negev desert), where drought resistance was tested. The experimental design involved 12 population quadrangles at Avedat Farm in 1980, and rows of randomized genotypes of five populations in both Haifa (Mediterranean climate) and Sede Boqer (desert climate) in 1981 and 1982 (Nevo et al. 1984b).

The results indicate that the characters studied are partly genetically determined. Striking genetic variation was found between and, at least in some characters, also within populations in each site, whereas remarkable environmental variation, including genetic-environmental interaction, was found between the mesic and the xeric sites (Fig. 6.1). We concluded that natural populations of wild emmer wheat in Israel vary not only in genetic polymorphisms of allozymes and disease resistance, but also in quantitative traits of agronomic importance, including drought tolerance. These traits are economically significant and should be conserved and utilized in wheat crop improvement. We currently evaluate them physiologically, and genetically map drought resistance QTLs (Peng 2000; Peng et al. 2001) and candidate genes.

6.4
Herbicide Response

Crop species can show differential responses to successful, widely used herbicides. Some varieties are unaffected by application, whilst others show symptoms of damage ranging from a slight reduction in vigor to complete plant death (Snape et al. 1991a). The elucidation of the genetic control of such responses is important for developing strategies for breeding for herbicide resistance within crop species and also in understanding the modes of action of the herbicides and the evolution of resistance in weed species.

The responses of wild populations of *T. dicoccoides* from different ecogeographic areas of Israel, to three herbicides, difenzoquat, chlorotoluron and metoxuron, commonly used on cultivated wheat, were studied (Snape et al. 1991b). Although cultivated wheats were polymorphic in their response to difenzoquat, all families of all populations of the wild species were resistant. These species, were, however, polymorphic in their response to both chlorotoluron and metoxuron. In addition, there appeared to be differences

Table 6.4. Responses of wild populations of wild emmer wheat to application of twice the recommended dose of chlortoluron (1–9 scale). (Snape et al. 1991b)

Population	Experiment 1 (winter sowing 1987)				Experiment 2 (spring sowing 1987)			
	Number of families tested	Mean response	Range	Classification of families	Number of families tested	Mean response	Range	Classification of families
Amirim	8	7.3	6–9	All S[1]	2	8.9	8–9	All S
Bat-Shelomo	16	7.3	3–9	6R: 10S	9	8.0	3–9	1R: 8S
Bet-Meir	6	8.2	6–9	All S	6	8.0	8–9	All S
Daliyya	8	6.2	3–9	4R: 4S	3	8.8	8–9	All S
Kokhav-Hashahar	16	7.2	3–9	8R: 8S	10	8.0	2–9	2R: 6S: 2 Seg
Mt. Dov (Mt Hermon)	12	6.3	2–9	5R: 7S	7	8.8	7–9	All S
Rosh-Pinna	10	4.9	2–9	8R: 2S	15	6.5	2–9	8R: 6S: 1 Seg
Sanhedriyya	5	4.4	2–9	4R: 1S	6	4.0	2–9	1R: 5S
Tabigha	12	7.1	3–9	3R: 9S	19	7.7	3–9	13R: 6S
Yabad	6	4.5	3–5	All R	7	5.7	2–9	5R: 1S: 1 Seg
Yehudiyya	14	5.8	2–9	7R: 7S	11	6.6	2–9	5R: 5S: 1 Seg
total								86R: 117S: 5 Seg

[1] S, R, and Seg stand for susceptible, resistance and segregating, respectively

between populations in the frequencies of resistant and susceptible morphs for these herbicides (for chlorotoluron, see Table 6.4). There was also a close correspondence between the responses of individual families to chlorotoluron and metoxuron (Table 5 in Snape et al. 1991b), which suggests a common genetic control. The implications of these findings for understanding the evolution of herbicide resistance and for developing strategies for breeding for resistance in the cultivated species were thus highlighted.

We demonstrated that the scores and frequencies of chlorotoluron (CT) and metoxuron (MX) resistant and susceptible phenotypes of *T. dicoccoides* correlated with ecological factors and allozyme markers (Nevo et al. 1992c). Some isozyme markers located on chromosome 6B (e.g. *Adh*, *Est*-4 and *Got*), which also harbors the CT and MX resistance genes, are good genetic markers for herbicide resistance breeding. Significant correlations between herbicide and photosynthetic characters suggest that the evolution of herbicide resistance polymorphisms may be related to the process of photosynthesis in nature and predated domestication of cultivated wheat. Later (Sect. 8.2) we compare and contrast the genetic mapping of CT resistance in cultivated and wild emmer wheat (Krugman et al. 1997)

6.5 Grain Protein Quality and Quantity

World food assessments indicate shortages in, and biased distribution of, high-quality food. The primary deficiency is protein, and more than half of the people in the world do not consume enough high-quality protein for a nutritionally balanced diet. The ongoing population explosion in less developed countries, with 10 billion people expected in 2050, and currently a billion people starving, primarily in developing countries, highlights the seriousness of the problem. Protein deficiency is most serious for the young and may cause permanent brain damage. The current serious protein shortage threatens to become a real nutritional disaster in the near future, primarily since in Asia and Africa about 80% of the protein of human diets is supplied by plants and only about 20% is from animal sources. This dictates an active search for additional, genetically based high-quality protein sources in nature. The urgent need to increase high-quality protein sources is exacerbated by the major problems affecting cultivated crops with respect to the drastic reduction in genetic diversity (Plucknett et al. 1983, 1987). As indicated earlier, the amount of storage grain proteins in wheat cultivars is 7–22%, but 17–28% in *T. dicoccoides* in nature and 24–43% under greenhouse conditions (Levy 1987). Furthermore, in wild emmer the high-protein content is under distinctly strict genetic control (Grama et al. 1983). We have shown that molecular markers and ecological factors (Nevo et al. 1986a) can predict high-quality and -quantity protein parameters. Geographic variation of protein and seed characters of wild emmer wheat in Israel and the associations with ecological and allozyme markers were tested in an attempt to derive predictive guidelines for conservation and utilization in breeding programs (Nevo et al. 1986a). The

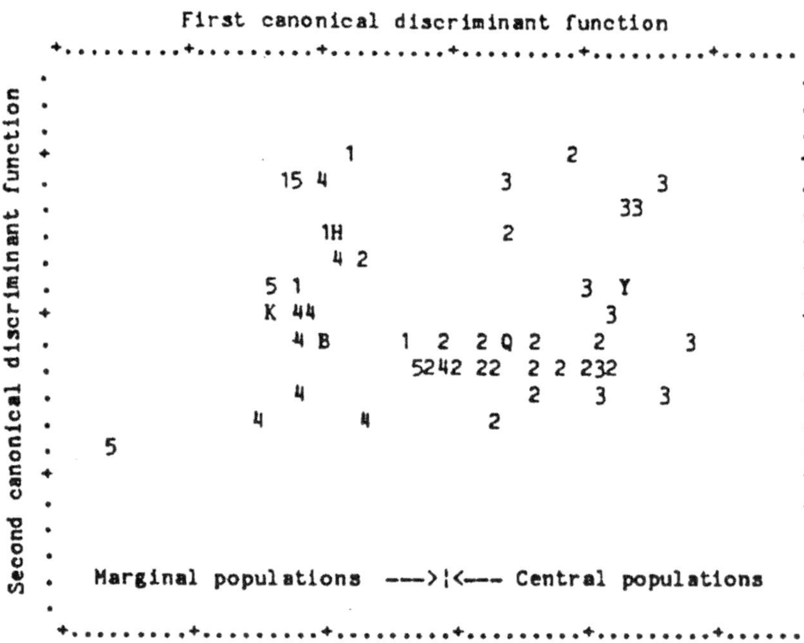

Fig. 6.3. Discriminant analysis between five populations of wild emmer by three seed characters: kernel weight, filling score, and protein percentage. *H*, *1* Hermon; *Q*, *2* Qazrin; *Y*, *3* Yehudiyya; *B*, *4* Bat Shelomo; K, 5 Kokhav Hashahar (*letters* are group centroids) (Nevo et al. 1986a)

study involved 46 genotypes of wild emmer from five populations in Israel, two central and three marginal. These populations were also tested for allozymic variation (Nevo et al. 1982).

The results indicate that protein percentage, kernel, and protein weight (the product of the former two values) vary both within, but particularly between, populations (Table 6.5b, Fig. 6.3). Notably, the three marginal populations (Fig. 6.3) exhibited high protein content but low kernel weight, hence low protein weight as compared with the two central populations, which displayed lower protein percentage but high kernel weight, hence higher protein weight. Three-variable combinations of climatic factors explained $R^2 = 0.70$ of the variance in kernel weight and $R^2 = 0.60$ of the variance in protein weight. Likewise, three-variable combinations of allozyme genotypes significantly explained the spatial variances in protein percentage, kernel, and protein weight ($R^2 = 0.60$, 0.69, and 0.54, respectively) (Tables 6.5a, 6.6).

We concluded, together with others (Gerechter-Amitai and Grama 1977; Avivi 1978, 1979; Avivi et al. 1983; Grama et al. 1983; Levy 1987), that natural populations of wild emmer in Israel contain a large number of yet untapped genes for elite protein and high seed weight. These could be effectively screened and utilized for producing high-quality and -quantity protein wheat

Table 6.5a The ecogeographic background[a] of five populations of wild wheat, *T. dicoccoides*, in Israel. Climatic values were taken from the Atlas of Israel (1970) and from multiple-year records of the Meteorological Service of Israel

Locality	Ecogeographic variables																	
	Long.	Lat.	Alt.	Tm	Ta	Tj	Td	Tdd	Rn	Rd	Hu14	Huan	Dw	Sh	Th	Trd	Ev	So[b]
Mt. Hermon	35.73	33.30	1300	11	21	3	18	6	1400	66	48	60	60	80	–	0	150	1
Qazrin	35.67	33.02	350	18	26	10	16	12	530	50	43	58	58	50	–	60	155	5
Yehudiyya	35.70	32.93	200	19	27	11	16	12	550	47	42	58	58	50	–	100	160	5
Bat-Shelomo	35.02	32.60	75	20	26	13	13	10	650	55	58	68	77	40	−10	30	150	2
Kokhav-Hashahar	35.34	31.95	600	20	28	12	16	12	400	40	45	59	30	80	−30	25	165	1

[a] *Long.* Longitude, *Lat.* latitude (in decimals), *Alt.* altitude (in m), *Tm* mean annual temperature, *Ta* mean August temperature, *Tj* mean January temperature, *Td* mean seasonal temperature difference, *Tdd* mean day-night temperature difference, *Trd* mean number of tropical days, *Sh* mean number of Sharav days (i.e., hot and dry days), *Ev* mean annual evaporation, *Dw* mean number of dew nights in summer, *Rn* mean annual rainfall (in mm), *Rd* mean number of rainy days, *Hu14* mean humidity at 14:00 h, *Huan* mean annual humidity, *Th* Thornwaite's moisture index. [b] *So* Soil type: *1* terra rossa, *2* rendzina, *5* basalt

Table 6.5b. Population means and ranges of protein percentage, kernel weight, protein weight, and filling scores. (Nevo et al. 1986b)

Locality	No. of accessions	Protein percentage		Kernel weight (1000 seeds)		Protein weight (1000 seeds)		Filling scores (mean ± SE)
		(%) ± SE	Range	(%) ± Se	Range	(%) ± SE	Range	
Mt. Hermon	6	21.75 ± 1.36	16.6 ± 26.0	20.63 ± 1.79	16.0–28.8	4.41 ± 0.28	3.49–5.53	1.83 ± 0.28
Qazrin	17	19.92 ± 0.38	17.4 ± 22.9	33.85 ± 1.05	23.5–40.0	6.73 ± 0.22	4.58–8.14	1.18 ± 0.10
Yehudiyya	10	19.67 ± 0.31	17.9 ± 20.8	37.60 ± 1.51	28.0–44.0	7.39 ± 0.31	5.40–8.98	1.50 ± 0.15
Bat-Shelomo	9	22.82 ± 0.49	20.3 ± 24.1	23.86 ± 1.38	18.0–31.5	5.44 ± 0.31	3.65–6.93	1.39 ± 0.14
Kokhav-Hashahar	4	22.10 ± 1.62	19.8 ± 26.7	20.88 ± 3.14	16.5–30.0	4.54 ± 0.55	3.27–5.97	1.38 ± 0.24
Total	46	20.86 ± 0.33	16.6 ± 26.7	29.86 ± 1.18	16.0–44.0	6.3 ± 0.21	3.27–8.98	1.39 ± 0.07
Significance		**		***		***		c

***$p < 0.001$, **$p < 0.01$, [c]$p < 0.1$

Table 6.6. Coefficients of multiple determination (R^2) with protein and seed characters as dependent variables and allozyme genotypes of *T. dicoccoides*, in Israel, as independent variables. Stepwise model. (Nevo et al. 1986b)

Traits	Variables		
A. All genotypes			
Protein percentage	*Pept-2bb*	+ *Est-2Acc*	+ *Acph-3cc*
	0.262**	0.433***	0.598***
Kernel weight	*Hk-aa*	+ *Est-5Baa*	+ *Pept-1Baa*
	0.254**	0.552***	0.685***
Protein weight	*Hk-aa*	+ *Est-5Baa*	+ *Pept-1Baa*
	0.224**	0.484***	0.539***
B. Positively correlated genotypes[a]			
Protein percentage	*Est-2Bcc*	+ *Pept-2aa*	+ *Adh-1Baa*
	0.198*	0.337*	0.458**
Kernel weight	*Hk-bb*	+ *Ipol-Acc*	+ *Est-2Bbb*
	0.254*	0.549***	0.610***
Protein weight	*Gluc-cc*	+ *Ipol-Acc*	+ *Hk-bb*
	0.243*	0.426**	0.495**

***$p < 0.001$, **$p < 0.01$, *$p < 0.05$

[a] employing allozyme genotypes associated with high levels of the three seed parameters

cultivars by means of effectively following ecological factors and molecular markers as predictive guidelines in screening natural populations of wild emmer wheat. Several programs to utilize these high-protein genetic resources are underway in studies being carried out by Prof. Feldman (Weizmann Institute, Israel), Dr. Joppa (Fargo, ND), and Dr. Dubcovsky (Davis, CA).

6.5.1 The Genetics of Grain Protein Content

It is of paramount importance to note that, while most of the variation in protein content in cultivated wheats is environmental, and the genetic determination amounts to only about 5% (Johnson et al. 1973, 1975), the opposite is true in wild emmer wheat, *T. dicoccoides*. Protein values range in seeds of wild emmer collected in nature from 13.9 to 25.5% and in seeds grown in nurseries from 15.4–26.9%. The correlation factors for protein values in seeds collected in nature and grown in nurseries are significantly high, $R^2 = 0.487–0.644$ (Grama et al. 1983), indicating that, in wild emmer grain, protein potential is under distinctly strict genetic control (see more on protein genetic control in Sect. 6.12.4).

6.5.2 Utilization of High Grain Protein from Wild Emmer

A decline in genetic diversity of the world's crops has accelerated as a result of domestication, particularly the breeding of modern cultivars. This has also been the fate of grain protein content, which has declined in cultivated wheats. It has become increasingly evident that wild relatives of crops harbor rich reservoirs of genetic diversity. These have been especially useful for breeding programs designed to improve cultivated crops, including wheats. In the development of high-quality bread wheats, multiple factors must be taken into account based on the common and diversified genetic resources derived from the wild relatives. Wild emmer wheat is superbly suitable for selecting diverse sources of genetic diversity for resistance against multiple diseases, pests, environmental stresses, and for elite agronomic traits and high grain protein content and weight. This is due to its long evolution in the Fertile Crescent in general and particularly in Israel, which appears to be its center of diversity and where most of the ecological spectrum of wild emmer traits is found. Among other adaptive traits, wild emmer has evolved a high grain protein content and kernel size, possibly to overcome competitive situations in central dense populations, as well as efficient and rapid germination and seedling growth in ecological margins. All these diverse genetic adaptations to the varied ecologies occupied by wild emmer are far from being fully tapped for the development of elite wheat cultivars.

We have outlined ecological and allozymic predictive guidelines that could maximize effective selection of elite wild germplasm for powdery mildew, leaf-rust, and stripe-rust resistances (Nevo et al. 1985), for agronomic elite traits (Nevo et al. 1984b), and also for protein percentage and seed characters (kernel weight and protein weight) (Nevo et al. 1986a). Furthermore, optimal multiple-character genotypes of wild emmer, including genotypes featuring disease resistance, high protein quality and quantity, and elite agronomic traits, can conceivably be either identified in nature by means of allozyme markers or generated by crossings. Indeed, such efforts have begun and have proved successful at the Volcani Center in Israel, where cultivated wheats were improved by the transfer of several desirable genes from wild emmer (Gerechter-Amitai and Grama 1974, 1977; Grama et al. 1983). These crosses, involving resistance to rusts coupled with high grain protein content, successfully combined in a cultivar a variety of superior quality traits derived from both wild progenitors and/or cultivated crops. Simultaneously, the negative traits inherited from the wild parent were eliminated. Thus, it is possible to retain the desired characters of the cultivated wheats (e.g., non-brittleness of the rachis, good threshability, good baking quality and high yield capacity) along with those of the wild progenitor (e.g., effective disease resistance, high grain protein content, and other elite agronomic characters). The predictive value of ecological factors and allozyme markers can greatly help to maximize the conservation, exploitation, and utilization of the rich genetic resources of wild emmer.

6.5.3 Wheat Storage Proteins and Glutenin DNA Diversity in *Triticum dicoccoides* in Israel

Glutenin and gliadin prolamin storage proteins comprise most of the wheat grain endosperm. Glutenin consists of approximately 20% high-molecular-weight (HMW) glutenin subunits and 80% low-molecular-weight (LMW) subunits. The biochemistry and genetics of endosperm proteins in bread wheat have been extensively studied (e.g., Kasarda et al. 1976; Konzak 1977; Payne and Rhodes 1982; Payne et al. 1980, 1981a,b, 1984; Galili and Feldman 1983, 1985; Nevo and Payne 1987; Levy and Feldman 1987, 1988; Gepts 1990; Felsenburg et al. 1991) due to their importance in bread baking (Wall 1979).

6.5.3.1 Utilization of Glutenin Diversity in Bread-Making Quality

Glutenins, like other wheat grain proteins, play a key role in food processing such as the manufacture of bread, biscuits, breakfast cereals, and pasta products (Payne and Rhodes 1982). A major effort of plant breeders is to develop elite bread wheats with improved bread-making quality. Significant associations were found for certain HMW subunits of glutenin and bread-making quality (Payne et al. 1981b). These results have been confirmed (Moonen et al. 1982) and expanded (Payne et al. 1984). Dough strength is the limiting factor in the bread-making process in western European wheats. This character is controlled by the elasticity of glutenin, which in turn is determined by its specific composition of LMW and HMW subunits.

One approach to improving the quality of wheat cultivars for bread making is by utilizing genes from: (1) land races of primitive agriculture (Payne et al. 1984) and (2) the wild relatives of bread wheat (Law and Payne 1983). Wild emmer could provide a host of genetic variants of the endosperm storage proteins for improving bread-making quality (Nevo and Payne 1987; Felsenburg et al. 1991; Pagnotta et al. 1995), as it does for other agronomic traits (Nevo 1983a, 1989a, 1995, 2001b; Feldman 1979). The range of allelic variation in the glutenin loci of wild emmer is remarkable, and all of it could be readily transferred to bread wheat for testing by plant breeders, in an attempt to improve baking quality.

6.5.3.2 Genetic Differentiation of High Molecular Weight Glutenin Subunits

The diversity of HMW glutenin subunits in the tetraploid wild progenitor of wheat, *T. dicoccoides,* was studied electrophoretically in 231 individuals representing 11 populations of wild emmer from Israel (Nevo and Payne 1987). The results showed that: (1) the two HMW glutenin loci, *Glu-A1* and *Glu-B1*, were rich in variation, having 11 and 15 alleles, respectively (Figs. 6. 4a–c and

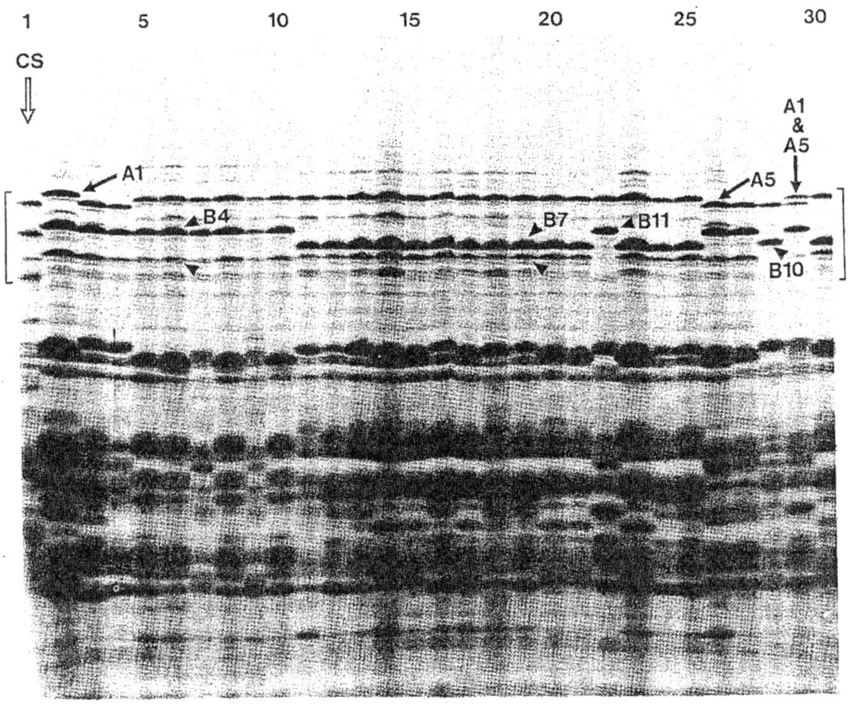

a

Fig. 6.4. **a** SDS-PAGE of grain proteins found in a population of wild emmer from Kokhav Hashahar, near Jerusalem. The region of the gel containing the HMW glutenin subunits is marked by *brackets*. The proteins thought to be coded by the *Glu-A1* and *Glu-B1* loci are prefixed with letters *A* and *B*, respectively. The HMW glutenin subunits of bread-wheat cultivar Chinese Spring (*CS*) are separated in *lane 1*. They are, with increasing mobility, *2* (chromosome 1D), *7 + 8* (1B), and *12* (1D). (From Nevo and Payne 1987) **b** SDS-PAGE of grain proteins in two wild emmer populations: *lanes 1–15* Qazrin, *lanes 16–23* Yehudiyya. (From Nevo and Payne 1987) **c** SDS-PAGE of a range of wild emmers to show the extent of allelic variation in HMW glutenin subunits. Grains were taken from populations Qazrin (*lane 1*), Yehudiyya (*lanes 2, 7, 8,* and *10*), Tabigha (*lane 12*), Bet-Meir (*lane 14*), Rosh-Pinna (*lane 11*), Bat-Shelomo (*lane 9*), Mt. Hermon (*lane 15*), Sanhedriyya (*lanes 3–6*, and *13*), and Taiyiba (*lane 16*). (Nevo and Payne 1987)

6.5). (2) Genetic variation in HMW glutenin subunits was often severely restricted in individual populations, supporting an island population genetic model (Tables 6.7a,b and 6.8). (3) Significant correlations were found between glutenin diversity and the frequencies of specific glutenin alleles and physical (climate and soil) and biotic (vegetation) variables (Table 6.8). Our results suggest that at least part of the glutenin polymorphisms in wild emmer can be accounted for by environmental factors. Furthermore, the endosperm of wild emmer contains many allelic variants of glutenin storage proteins that are not present in bread wheat and could be utilized in breeding varieties with improved bread-making qualities.

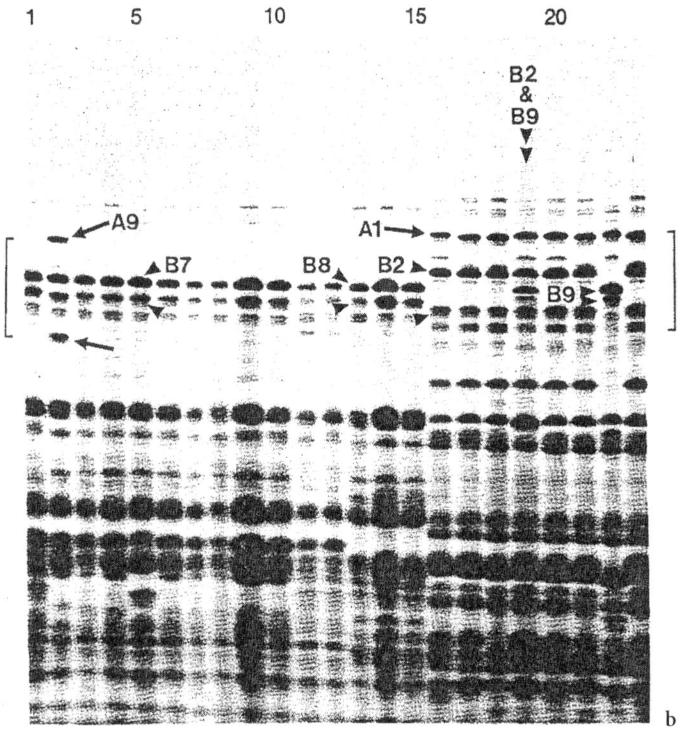

Fig. 6.4. *Continued* (Fig. 6.4c see page 198)

Three main features characterize glutenin differentiation in natural populations of wild emmer in Israel (Nevo and Payne 1987): (1) extensive variability, (2) extreme population differentiation, and (3) correlation with ecological factors. First, allele diversity is remarkable in the glutenin subunits and is similar to that of hordein storage proteins in wild barley, *H. spontaneum* (Nevo et al. 1983), in which 15 hordein-1 phenotypes and 16 hordein-2 phenotypes have been described in Israel. In line with this extremely high hordein diversity, we found 11 *Glu-A1* and 15 *Glu-B1* alleles in the present study (Figs. 6.4a–c, 6.5). Similarly, our unpublished studies also revealed extensive allelic variation in the gliadin proteins of wild emmer. The high levels of allele diversities of hordein and glutenin storage proteins contrast sharply with the relatively lower levels of allozyme allele diversity per locus in each population of wild barley and wild emmer in Israel. The mean number of alleles per locus per population, A, for allozymes in wild barley (Nevo et al. 1979) and wild emmer (Nevo et al. 1982) were 1.48 and 1.33 for 28 and 50 loci, respectively. By contrast, $A = 2.5$ for the two glutenin loci in wild emmer. Our estimates of allelic variation in the HMW glutenin subunits of emmer is also clearly an underestimation. The proteins have only been fractionated in one dimension and, therefore, different subunits of very similar molecular

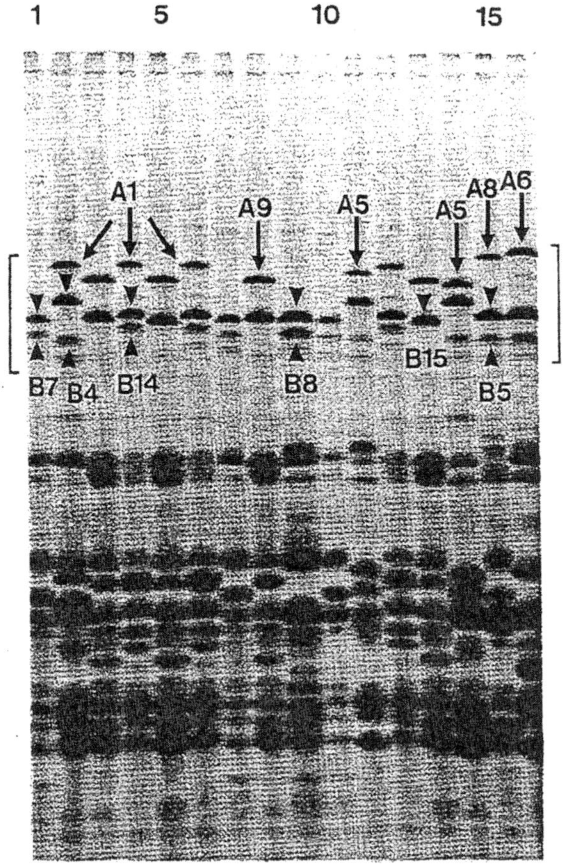

c

Fig. 6.4. *Continued*

weight will not be resolved. A good example of this is the subunits labeled Al in Fig. 6.4a–c. A careful examination reveals two alternative proteins of very similar mobility. As only exceptional separations showed this polymorphism, the subunits were not subclassified. However, different proteins of more similar molecular weights would never be resolved by this technique.

6.5.3.3 Glutenin DNA Diversity

Following our study of the diversity of HMW glutenin subunits in wild emmer (Nevo and Payne 1987), we also analyzed glutenin DNA diversity (Pagnotta et al. 1995).

Germplasm of *T. dicoccoides* collected from different environments in Israel was evaluated by using the PCR products as molecular markers

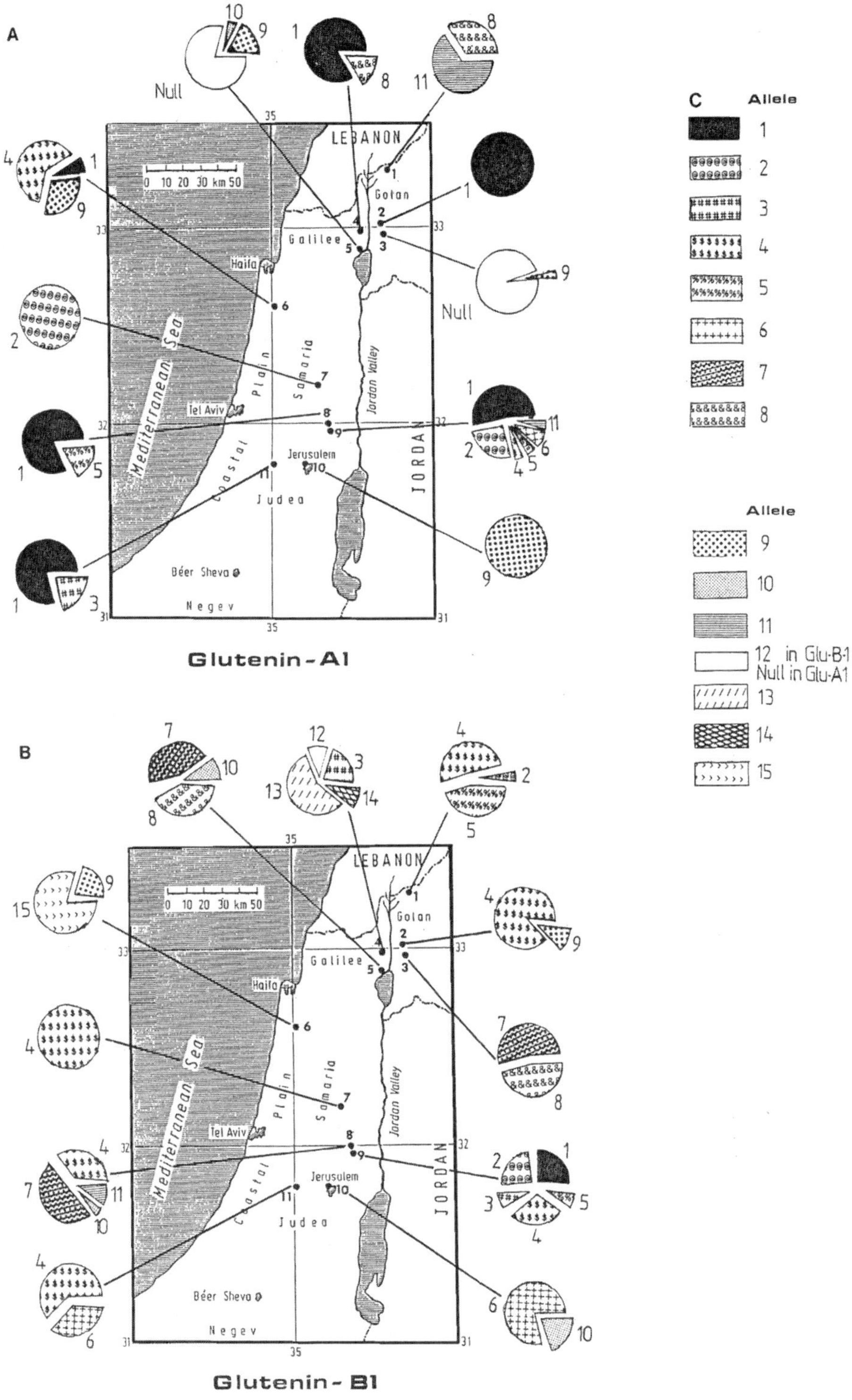

Fig. 6.5A–C. Pie diagrams displaying the percentage of the 11 alleles of *Glu-A1* and 15 alleles of *Glu-B1* in 11 populations of wild emmer wheat, *Triticum turgidum dicoccoides*, and their geographic location in Israel. Population numbers as in Table 6.7b **A** *Glu-A1*; **B** *Glu-B1*; **C** key to diagrams. Alleles are numbered as in Table 6.7a.

Table 6.7a. Allele frequencies and gene diversity (*He*)[a] at two glutenin loci in 11 populations of *T. dicoccoides* in Israel. (Nevo and Payne 1987)

Locality[b]	1	2	3	4	5	6	7	8	9	10	11	Mean
N	20	26	23	19	24	26	2	30	23	19	19	231
Locus allele												
Glu-A1												
1	0	1.000	0	0.842	0	0.077	0	0.817	0.522	0	0.789	0.413
2	0	0	0	0	0	0	1.000	0	0.261	0	0	0.035
3	0	0	0	0	0	0	0	0	0	0	0.211	0.017
4	0	0	0	0	0	0.615	0	0	0.043	0	0	0.074
5	0	0	0	0	0	0	0	0.183	0.043	0	0	0.028
6	0	0	0	0	0	0	0	0	0.087	0	0	0.009
8	0.350	0	0	0.158	0	0	0	0	0	0	0	0.043
9	0	0	0.043	0	0.167	0.308	0	0	0	1.000	0	0.139
10	0	0	0	0	0.042	0	0	0	0	0	0	0.004
11	0.650	0	0	0	0	0	0	0	0.043	0	0	0.061
null	0	0	0.957	0	0.792	0	0	0	0	0	0	0.177
He	0.455	0	0.083	0.266	0.344	0.521	0	0.299	0.647	0	0.322	0.268
Glu-B1												
1	0	0	0	0	0	0	0	0	0.261	0	0	0.026
2	0.050	0	0	0	0	0	0	0	0.261	0	0	0.030
3	0	0	0	0.211	0	0	0	0	0.087	0	0	0.026
4	0.500	0.885	0	0	0	0	1.000	0.367	0.304	0	0.623	0.281
5	0.450	0	0	0	0	0	0	0	0.087	0	0	0.048
6	0	0	0	0	0	0	0	0	0	0.789	0.368	0.095
7	0	0	0.522	0	0.458	0	0	0.500	0	0	0	0.165
8	0	0	0.478	0	0.417	0	0	0	0	0	0	0.091
9	0	0.115	0	0	0	0.192	0	0	0	0	0	0.035
10	0	0	0	0	0.125	0	0	0.033	0	0.211	0	0.035
11	0	0	0	0	0	0	0	0.100	0	0	0	0.013
12	0	0	0	0.105	0	0	0	0	0	0	0	0.009
13	0	0	0	0.579	0	0	0	0	0	0	0	0.048
14	0	0	0	0.105	0	0	0	0	0	0	0	0.009
15	0	0	0	0	0	0.808	0	0	0	0	0	0.091
He	0.545	0.204	0.499	0.598	0.601	0.311	0	0.604	0.756	0.332	0.465	0.447

[a] The index of gene diversity (*He*) for each locus is recorded below each set of frequencies
[b] Population numbers refer to those listed in Table 6.7b

(Pagnotta et al. 1995). Two pairs of primers were used for PCR amplification of the DNA tracts coding for the HMW and LMW glutenin subunits. Analyses revealed great variability between and within populations, indicating the high value of this germplasm for future breeding programs to improve protein quality in wheat (Tables 6.9, 6.10).

We also showed that the DNA diversity of HMW and LMW glutenins correlated with environmental physical and biotic factors (climate, soil and pathogen resistance) and is associated with allozymic variation (Nevo et al. 1995). We concluded that glutenin DNA, like glutenin protein diversity, was *nonrandomly* distributed and could be more optimally sampled in nature for use in future breeding programs to improve bread quality.

The potential genetic resources of wild emmer include genes coding for disease resistance against viruses (wheat soil-borne mosaic virus), fungi

Table 6.7b. Glutenin diversity based on two loci in 11 populations of *Triticum dicoccoides* in Israel. (Nevo and Payne 1987)

Locality	Sample size (*N*)	Mean no. of alleles per locus (*A*)	Mean proportion of loci: Polymorphic per population (*P*-5%)	Mean proportion of loci: Heterozygous per individual (*H*) Mean	Heterozygous per individual (*H*) SE	Gene diversity (*He*)
1 Mt. Dov	20	2.5	1.0	0.0	0.0	0.500
2 Qazrin	26	1.5	0.5	0.038	0.038	0.102
3 Yehudiyya	23	2.0	0.5	0.0	0.0	0.291
4 Rosh-Pinna	19	3.0	1.0	0.0	0.0	0.432
5 Tabigha	24	3.0	1.0	0.0	0.0	0.472
6 Bat-Shelomo	26	2.5	1.0	0.0	0.0	0.416
7 Mt. Gerizim	2	1.0	0.0	0.0	0.0	0.0
8 Kokhav Hashahar	30	3.0	1.0	0.017	0.017	0.452
9 Taiyiba	23	5.5	1.0	0.0	0.0	0.701
10 Sanhedriyya	19	1.5	0.5	0.0	0.0	0.166
11 Bet-Meir	19	2.0	1.0	0.0	0.0	0.399
Mean	231	2.5	0.773	0.006	0.002	0.357
Range		1.0–5.5	0–1	0–0.038		0–0.701

Table 6.8. Coefficients of multiple regressions (R^2) of glutenin diversity and allele frequencies as the dependent variables and geographic, climatic, ecological, edaphic, and plant formation variables in 10 populations of *T. dicoccoides* as independent variables (see abbreviations in Table 6.19). (Nevo and Payne 1987)

	N	Stepwise model		
Glu-A1 1	10	Tdd 0.172 ns	Tdd Alt 0.327 ns	
Glu-A1 9	10	Hu14 0.254 ns	Hu14 Sh 0.450 ns	Hu14 Sh So 0.760*
He of *Glu-A1*	10	So 0.306[c]	So Pl 0.660*	So Pl Ev 0.828**
Glu-B1 7	10	Ta 0.467*	Ta Tm 0.657*	Ta Tm Pl 0.747*
Glu-B1 10	10	Sh 0.267 ns	Sh Alt 0.512[c]	Sh Alt Rd 0.753*
He of *Glu-B1*	10	Dw 0.332[c]	Dw Pl 0.508[c]	
He of *Glu-A1* and *B1*	10	Pl 0.323[c]	Pl Lt 0.531[c]	

Tdd Mean day-night temperature difference, *Hu14* mean humidity at 14:00 h, *Alt* altitude, *Sh* mean number of Sharav days (i.e., hot and dry days), *So* soil type, *Ta* mean August temperature, *Tm* mean annual temperature, *Dw* mean number of dew nights in summer, *Pl* climax plant community, *Lt* latitude, *Ev* mean annual evaporation, *Rd* mean number of rainy days

$^*p < 0.05$, $^{**}p < 0.01$, $^{c}p < 0.10$; *ns* $p > 0.10$

Table 6.9. Frequencies and distribution of glutenin DNA PCR patterns in 25 populations of *T. dicoccoides* in Israel and Turkey. Pattern number in *parentheses*. (Pagnotta et al. 1995)

Locality	Pattern frequencies	
	LMW glutenin	HMW glutenin
Mt. Hermon	(1) 0.200; (2) 0.500; (13) 0.300	(1) 0.200; (2) 0.800
Qazrin	(3) 0.800; (4) 0.100; (25) 0.100	(3) 1.000
Yehudiyya	(5) 0.700; (6) 0.300	(4) 0.300; (5) 0.100; (6) 0.400; (7) 0.200
Gamla	(7) 0.200; (8) 0.500; (9) 0.100; (10) 0.200	(8) 0.100; (9) 0.100; (10) 0.700; (11) 0.100
Rosh-Pinna	(1) 0.800; (11) 0.100; (12) 0.100	(1) 0.100; (10) 0.300; (12) 0.100; (14) 0.400
Tabigha	(6) 0.100; (14) 0.900	(1) 0.500; (15) 0.500
Mt. Gilboa	(5) 0.100; (16) 0.500; (17) 0.100; (18) 0.300	(16) 0.900; (17) 0.100
Mt. Gerizim	(1) 1.000	(3) 0.100; (6) 0.100; (18) 0.800
Gitit	(1) 0.100; (2) 0.200; (4) 0.300; (7) 0.300; (16) 0.100	(3) 0.100; (4) 0.700; (7) 0.100; (19) 0.100
Kokhav-Hashahar	(1) 0.556; (7) 0.111; (10) 0.222; (16) 0.111	(2) 0.111; (7) 0.222; (20) 0.667
Taiyiba	(4) 0.400; (19) 0.200; (20) 0.400	(7) 0.500; (21) 0.500
Sanhedriyya	(2) 0.700; (4) 0.100; (21) 0.100; (22) 0.100	(2) 0.100; (3) 0.300; (4) 0.300; (7) 0.100; (21) 0.100; (22) 0.100
Bet-Meir	(7) 0.300; (8) 0.200; (16) 0.500	(2) 0.800; (6) 0.200
Jaba	(1) 0.300; (2) 0.300; (19) 0.400	(2) 0.500; (4) 0.300; (6) 0.200
Amirim	(2) 0.100; (5) 0.200; (7) 0.500; (10) 0.100; (14) 0.100; (23) 0.100	(2) 0.100; (3) 0.600; (4) 0.200; (24) 0.100
Nahef	(10) 0.900; (24) 0.100	(3) 0.600; (4) 0.300; (6) 0.100
Ahihud	(5) 0.636; (23) 0.364	(6) 0.900; (25) 0.100
Nesher	Not recorded	(8) 0.700; (16) 0.300
Bet-Oren	(7) 0.800; (21) 0.200	(22) 0.800; (26) 0.200
Daliyya	(22) 0.900; (26) 0.100	(4) 1.000
Bat-Shelomo	(5) 0.100; (7) 0.300; (20) 0.600	(3) 0.600; (4) 0.100; (6) 0.100; (10) 0.100; (22) 0.100
Kabara	(7) 1.000	(11) 0.100; (16) 0.100; (18) 0.800
Yabad	(2) 0.600; (20) 0.400	(24) 1.000
Givat-Koach	(1) 0.111; (2) 0.778; (20) 0.111	(4) 0.778; (27) 0.111; (28) 0.111
Diyarbakir (Turkey)	(1) 0.100; (2) 0.600; (27) 0.300	(28) 0.400; (29) 0.600

(powdery mildew, yellow or stripe rust, and stem rust), salt and herbicides, and genes elevating protein quantity and quality, amino acid levels, amylases, heat production, and photosynthetic yields, among others (see Nevo 1983a, 1989a, 1995). Employment of these genes (e.g., against powdery mildew, Reader and Miller 1991) in commercial cultivars is just beginning, and yellow-rust resistance and high protein level will follow soon. But even these are just the start; most of the genetic resources of wild emmer wheat (as well as other

Table 6.10. Genetic DNA variation in *T. dicoccoides* (detected by PCR) of low- and high-molecular-weight glutenin. Based on bands regarded as 19 loci in 23 populations in Israel and one in Turkey. (Pagnotta et al. 1995)

Locality	Sample size (*N*)	Mean no. of alleles per locus (*A*)	Polymorphism of population (*P*-5%)	Gene diversity	
				He	SE
Mt. Hermon	10	1.158	0.158	0.069	0.038
Qazrin	10	1.211	0.211	0.055	0.027
Yehudiyya	10	1.263	0.263	0.101	0.042
Gamla	10	1.421	0.421	0.120	0.037
Rosh-Pinna	10	1.579	0.579	0.143	0.034
Tabigha	10	1.105	0.105	0.038	0.029
Mt. Gilboa	10	1.421	0.421	0.140	0.045
Mt. Gerizim	10	1.158	0.158	0.045	0.026
Gitit	10	1.421	0.421	0.125	0.039
Kokhav-Hashahar	9	1.368	0.368	0.135	0.045
Taiyiba	10	1.263	0.263	0.135	0.053
Sanhedriyya	10	1.526	0.526	0.156	0.039
Bet-Meir	10	1.316	0.316	0.112	0.039
Jaba	10	1.316	0.316	0.151	0.053
Amirim	10	1.421	0.421	0.131	0.041
Nahef	10	1.211	0.211	0.057	0.030
Ahihud	11	1.158	0.158	0.045	0.028
Bet-Oren	10	1.211	0.211	0.071	0.032
Daliyya	10	1.053	0.053	0.010	0.010
Bat-Shelomo	10	1.421	0.421	0.170	0.051
Kabara	10	1.158	0.158	0.030	0.016
Yabad	10	1.053	0.053	0.027	0.027
Givat-Koach	9	1.421	0.421	0.105	0.031
Diyarbakir (Turkey)	10	1.263	0.263	0.123	0.049
Mean	239	1.287	0.287	0.096	0.010

Triticum and *Aegilops* species) have yet to be exploited, including the LMW and HMW glutenins described here.

6.6 Amino-Acid Resources in *Triticum dicoccoides*: Polymorphisms and Predictability by Ecology and Isozymes

Amino acid content was measured in 109 genotypes from 22 populations of wild emmer wheat, *T. dicoccoides*, across its ecological range in Israel. The results are given in Table 6.11 (Nevo and Beiles 1992). Plants were grown outdoors on mesic Mt. Carmel in a standardized environment and without irrigation and fertilization. *T. dicoccoides* was high in lysine and isoleucine contents as compared to six other species of wild wheats, but low in threo-

Table 6.11. Means and standard deviations of essential amino acid contents of *T. dicoccoides* in 22 populations in Israel, partitioned into three ecogeographic regions. (Nevo and Beiles 1992)

	N[b]	Lysine (LYS) Mean + SD	Isoleucine (ISO) Mean + SD	Methionine (MET) Mean + SD	Leucine (LEU) Mean + SD	Threonine (THR) Mean + SD	Proline (PRO) Mean + SD	Arginine (ARG) Mean + SD	Glutamic[a] (GLU) Mean + SD
A. Central populations									
5. Qazrin	5 (2 + 3)	2.90 + 0.44	4.02 + 0.21	0.91 + 0.24	7.11 + 0.37	1.86 + 0.36	8.77 + 1.68	4.81 + 0.73	38.69 + 2.89
7. Yehudiyya	4 (3 + 1)	2.65 + 0.16	3.96 + 0.13	0.92 + 0.15	7.27 + 0.21	2.36 + 0.04	10.73 + 1.90	4.40 + 0.25	38.24 + 0.98
9. Rosh-Pinna	4 (2 + 2)	2.66 + 0.20	3.92 + 0.40	1.06 + 0.29	6.92 + 0.48	2.02 + 0.27	8.87 + 1.86	4.56 + 0.68	38.95 + 3.12
11. Tabigha	6 (3 + 3)	2.66 + 0.24	3.82 + 0.41	1.08 + 0.17	6.90 + 0.68	2.28 + 0.22	9.79 + 1.48	4.46 + 0.49	39.57 + 3.09
Mean	19	2.72 + 0.29	3.92 + 0.30	1.00 + 0.21	7.04 + 0.48	2.13 + 0.31	9.53 + 1.74	4.56 + 0.55	38.93 + 2.56
B1. Xeric-cold marginal (northern) populations									
1. Mt. Hermon	5 (2 + 3)	2.68 + 0.31	4.14 + 0.48	1.72 + 0.92	7.42 + 0.74	2.13 + 0.17	7.75 + 1.68	4.66 + 0.54	38.02 + 3.98
B2. Xeric-warm marginal (eastern and southern) populations									
16. Mt. Gilboa	5 (2 + 3)	2.73 + 0.14	3.91 + 0.53	1.02 + 0.32	7.30 + 1.29	2.13 + 0.18	10.67 + 0.76	4.37 + 0.43	38.28 + 3.56
17. Mt. Gerizim	5 (2 + 3)	2.70 + 0.20	3.85 + 0.33	1.25 + 0.37	7.00 + 0.27	2.25 + 0.14	9.79 + 1.02	4.25 + 0.20	39.85 + 1.57
18. Gitit	5 (2 + 3)	2.46 + 0.12	4.07 + 0.27	1.35 + 0.37	6.67 + 0.18	2.07 + 0.24	8.82 + 1.57	4.02 + 0.52	40.92 + 1.49
19. Kokhav-Hashahar	6 (3 + 3)	2.59 + 0.23	3.66 + 0.39	0.72 + 0.48	6.79 + 0.58	2.14 + 0.22	8.17 + 1.37	4.02 + 0.32	43.38 + 4.13
20. Taiyiba	5 (2 + 3)	2.87 + 0.22	3.75 + 0.50	0.93 + 0.44	7.23 + 0.60	2.05 + 0.17	8.76 + 1.63	5.05 + 0.39	39.08 + 1.77
21. Sanhedriyya	4 (2 + 2)	2.77 + 0.28	3.79 + 0.22	1.17 + 0.09	6.29 + 0.18	1.97 + 0.08	8.35 + 1.84	4.52 + 0.46	40.86 + 4.72

22. Bet-Meir	5 (3 + 2)	2.59 + 0.17	4.02 + 0.46	1.28 + 0.27	7.26 + 0.75	2.20 + 0.14	10.16 + 2.19	4.23 + 0.61	37.93 + 3.60
23. J'aba	5 (2 + 3)	2.78 + 0.38	4.56 + 0.39	1.53 + 0.26	7.52 + 0.38	2.08 + 0.42	9.60 + 1.55	4.75 + 1.06	37.16 + 4.18
Mean	40	2.68 + 0.24	3.95 + 0.45	1.14 + 0.41	7.02 + 0.69	2.12 + 0.22	9.29 + 1.63	4.39 + 0.61	39.60 + 3.48
Mean of B1 + B2	45	2.68 + 0.25	3.97 + 0.45	1.21 + 0.51	7.07 + 0.70	2.12 + 0.21	9.11 + 1.69	4.42 + 0.60	39.42 + 3.52
C. Mesic marginal (western) populations									
24. Amirim	4 (2 + 2)	2.85 + 0.78	3.90 + 0.09	1.23 + 0.42	7.11 + 0.33	2.24 + 0.14	9.08 + 1.87	4.72 + 1.03	35.08 + 1.29
25. Nahef	5 (2 + 3)	2.74 + 0.14	3.81 + 0.23	0.98 + 0.33	6.52 + 0.33	1.85 + 0.40	10.10 + 1.45	4.17 + 0.29	40.10 + 2.87
26. Ahihud	5 (2 + 3)	2.54 + 0.23	4.13 + 0.18	1.17 + 0.26	7.15 + 0.46	1.91 + 0.47	8.97 + 1.88	4.45 + 0.34	39.97 + 2.87
27. Nesher	5 (2 + 3)	2.63 + 0.29	3.79 + 0.28	1.29 + 0.17	7.14 + 0.21	2.41 + 0.06	8.27 + 0.76	4.23 + 0.37	39.27 + 2.28
28. Bet-Oren	5 (2 + 3)	2.59 + 0.31	4.08 + 0.17	1.23 + 0.27	6.88 + 0.51	2.36 + 0.18	9.33 + 2.01	4.34 + 0.58	39.33 + 1.60
29. Daliyya	5 (2 + 3)	2.57 + 0.27	4.34 + 0.14	1.33 + 0.18	7.37 + 0.14	2.44 + 0.35	10.66 + 1.43	4.80 + 0.93	35.66 + 3.34
30. Bat-Shelomo	5 (3 + 2)	2.37 + 0.25	3.96 + 0.22	1.24 + 0.10	7.07 + 0.55	2.08 + 0.11	9.27 + 1.20	4.08 + 0.30	40.27 + 2.27
32. Yabad	6 (2 + 4)	2.79 + 0.32	4.30 + 0.31	1.51 + 0.40	7.13 + 0.37	2.23 + 0.16	8.60 + 1.22	5.09 + 0.77	36.60 + 3.41
33. Givat-Koach	5 (2 + 3)	2.58 + 0.19	3.98 + 0.29	1.23 + 0.13	7.05 + 0.53	2.02 + 0.16	9.25 + 1.35	4.35 + 0.41	40.25 + 2.56
Mean	45	2.63 + 0.33	4.04 + 0.28	1.25 + 0.28	7.05 + 0.43	2.17 + 0.32	9.27 + 1.52	4.48 + 0.64	38.68 + 3.04
Grand mean	109 (50 + 59)	2.67 + 0.29	3.99 + 0.37	1.19 + 0.39	7.05 + 0.56	2.14 + 0.27	9.25 + 1.62	4.47 + 0.61	39.03 + 3.17
Significance by Kruskal-Wallis tests									
Between populations	$p =$	0.319	0.013	0.004	0.036	0.004	0.184	0.133	0.173
Between regions	$p =$	0.542	0.395	0.005	0.423	0.615	0.332	0.519	0.356

[a] Glutamic includes both glutamic acid and glutamine
[b] *N* of first (1986) and second (1987) batches in *parentheses*

nine and proline. Significant correlations were found among lysine and the other essential amino acids. Significant differences between Israeli populations of *T. dicoccoides* were found for isoleucine, methionine, leucine, and threonine, but not for lysine, arginine, proline, and glutamic acid. Significant differences between regions were only found for methionine. Ecological factors and allozyme markers appear to be good guidelines for predicting the ecogeographic location of specific elite amino acid genotypes, either singly or in combination. *T. dicoccoides* may be used in the future as a promising genetic resource for genetically improving the nutritional value of cultivated wheats.

6.6.1 Amino Acids in Wild Emmer

The study of amino acid composition in wild emmer wheat, *T. dicoccoides,* the progenitor of most wheats, suggests that it is a good source for improving the nutritive value of cultivated wheats. This holds particularly true if specific, unique populations and genotypes are selected as donors for breeding programs, since significant geographic variation occurs between populations across the ecological range of *T. dicoccoides* in Israel regarding both single and combinations of essential amino acids. These single and combinatorial syndromes can be predicted by ecology and allozyme markers, thus optimizing sampling strategies for elite amino acids and combination with other agronomically important characters. Predictabilities based on ecology and allozyme markers for diverse agronomic traits were reviewed by Nevo (1987).

The poor nutritive value of seed storage proteins of cereals, primarily of prolamines, was briefly reviewed in the Introduction of Nevo and Beiles (1992) and has been extensively reviewed by Porceddu et al. (1983 and Table 1 therein) and Barrett (1985). Notably, three major consequences result from unbalanced amino acid profiles. First, the requirement for the limiting amino acid becomes higher, as was the case with lysine when wheat gluten was added to the diet of mice in order to raise the amount of basic amino acids (Munaver and Harper 1959). Second, if one amino acid is in excess, it may depress the utilization of another, structurally similar amino acid. For example, excess leucine in maize depressed the utilization of isoleucine (Harper et al. 1955), while an excess of arginine in millet depresses the utilization of lysine (Ganapaty and Chitre 1970). Third, an excess of one amino acid may contribute to the limited digestibility of all proteins. For example, an excess of proline vs lysine and arginine limits the digestibility of cereal proteins (Kakade 1974). The genetic control of the content, amino acid composition, and processing properties of proteins in wheat was reviewed by Konzak (1977).

Specific populations and allozyme genotypes of *T. dicoccoides* in Israel are rich in lysine and the other essential amino acids, such as threonine, isoleucine, and methionine, and there are also populations with the desirable low levels of leucine and arginine (Table 6.11). Balanced profiles of amino acids are also found, where the search for elite single accessions can be conducted, potentially resolving the problem of better digestibility. We conclude that optimal profiles of amino acids, either high or low in the content of single amino acids and/or in overall balanced profiles, occur in natural populations of *T. dicoccoides* and may be used in wheat breeding.

6.7 Amylases

Amylases are of fundamental importance in wheat germination and growth. α- and β-Amylase isozyme diversity was studied electrophoretically by thin-layer polyacrylamide gel isoelectrofocusing in *T. dicoccoides* (Nevo et al. 1993a). We analyzed 225 plants from 23 populations encompassing the ecological spectrum of *T. dicoccoides* in Israel. The results were as follows: (a) band and multilocus genotype polymorphisms abounded and varied within and between the four amylase components, malt, green (α-amylases), dry, and germinating seeds (β-amylases) (Fig. 6.6, Table 6.12). (b) The number of bands of malt, green, dry, and germinating seeds was 20, 6, 11, and 13, generating 40, 6, 51, and 51 patterns, or multilocus genotypes (MGPs), respectively. The MGPs varied widely within and between populations, from monomorphic in some populations with a single pattern to highly polymorphic ones (Table 6.13). (c) Mean H_e values for malt, green, germinating, and dry seed amylases were 0.053, 0.055, 0.088, and 0.077, respectively; the mean number of bands per individual was 11.8, 4.4, 7.6 and 4.0, respectively (Table 6.13). (d) The percentages of 50 bands and 148 MGPs (in parentheses) (Table 6.13) were classified into widespread, sporadic, and localized groups: 84.4 (10.8), 8.9 (12.2), 6.7 (77.0), respectively. Notably, 89.2% of the patterns were not widespread, but *sporadic* and *localized*, as was shown earlier for other allozymes (Nevo and Beiles 1989). (e) The mean value of genetic distances among populations (Nei's D) for the four amylase groups was $D = 0.136$, 0.175, 0.288, and 0.307, respectively, not displaying geographic correlates. (f) Most of the α- and β-amylase diversity is between populations ($G_{st} = 68–75\%$). (g) Significant environmental correlates occurred between either bands or patterns and climatic diversity (water and primarily temperature factors). (h) There were significant associations of multilocus amylase bands across Israel (Table 6.14). Likewise, significant linkage disequilibria, LD, occurred within populations, positively correlated with climatic variables, primarily of temperature (Table 6.14). (i) Discriminant analyses correctly classified (95–100%) the 23 wild emmer populations into their ecogeographic region and soil type (Fig. 6.7). (j) Autocorrelation analysis showed no correlation between bands and geographic distance, thereby excluding migration as a major factor of amylase differentiation.

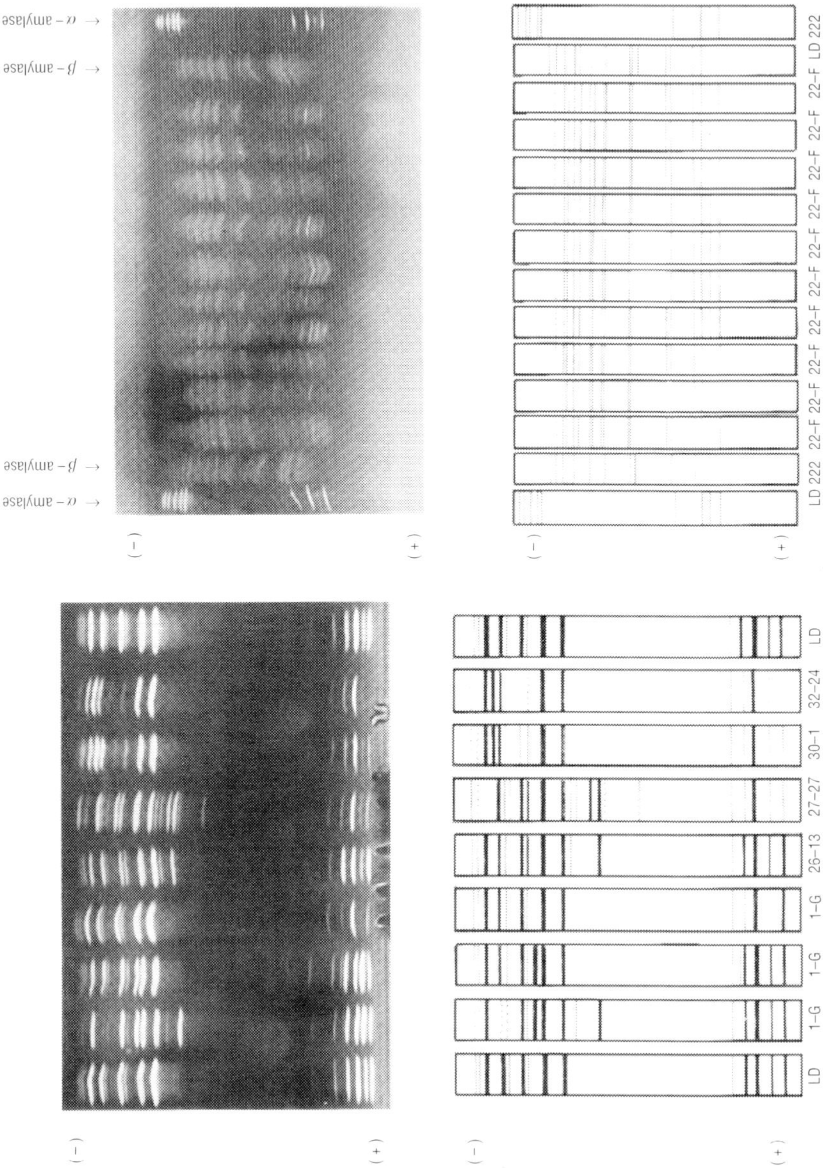

Fig. 6.6a,b. Amylase phenotypes of wild emmer wheat, *Triticum dicoccoides*. **a** α-Amylase, **b** β-amylase. (Nevo et al. 1993a)

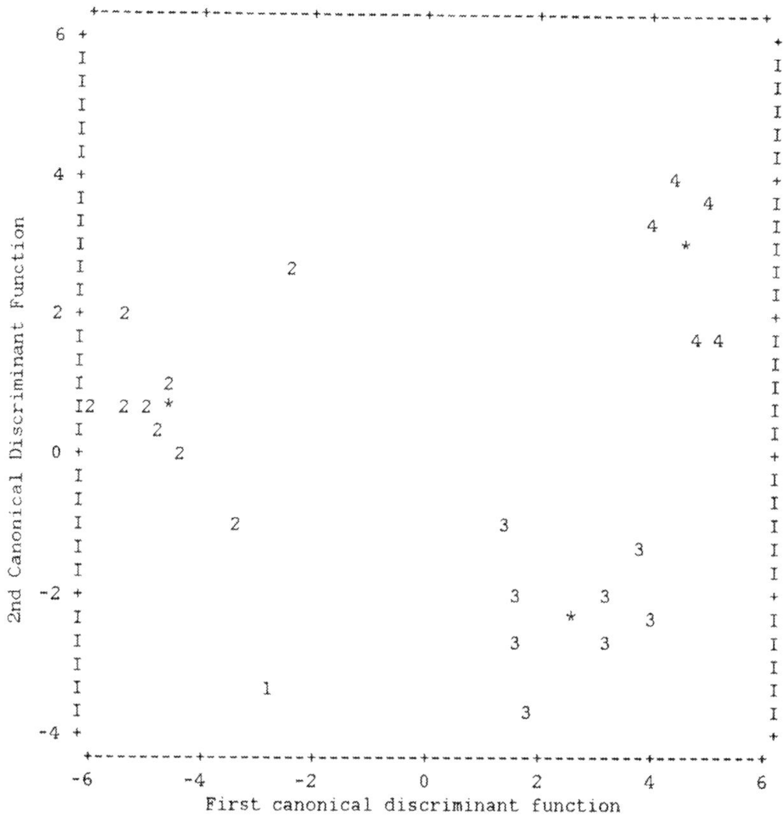

Fig. 6.7. Discriminant analysis of 23 populations of *Triticum dicoccoides* in Israel into four eco-geographic regions by amylases. *1* Cold margin, *2* west margin, *3* east-south margin, *4* central populations, * group centroid. (Nevo et al. 1993a)

These results suggested that diversifying climatic, edaphic, and biotic natural selection rather than stochastic factors or migration is the major evolutionary force driving amylase differentiation at both the single and multilocus levels. Furthermore, wild emmer harbors high levels of α- and β-amylase diversity, detectable both as single bands and multilocus adaptive genetic patterns. These are exploitable both as genetic markers for QTLs and as direct adaptive genetic resources to improve wheat germination and growth through classical breeding and/or biotechnology.

The findings presented here for the multigene families of α- and β-amylases in 23 Israeli populations of wild emmer wheat, *T. dicoccoides*, parallel basically, in population-genetic terms, those regarding the single-gene and multilocus diversity encoded by 42 allozyme gene loci (Nevo and Beiles 1989), as described below.

Table 6.12. Geographic distribution of 26 bands revealed by electrophoretic analysis of α-amylase in 23 populations of *T. dicoccoides* in Israel. (Nevo et al. 1993a)

Population number and name	N	Bands[a]																										Mean number of bands
		Malt																				Green pats						
		23	21	20	20a	19	18	16	13	11	10	Em	6	5	4	3	1'a	1	1a	1b	1c	8	8a	7	5	3	2	
1 Mt. Hermon	10	0	1.0	1.0	0	0	1.0	1.0	0	1.0	1.0	0.70	1.0	0	1.0	1.0	0	0.30	0	0	0	1.0	0.30	1.0	1.0	1.0	1.0	16.5
5 Qazrin	10	0	1.0	1.0	0.10	0	1.0	0.90	0	1.0	1.0	0	1.0	0	1.0	1.0	0	1.0	0	0	0	1.0	0	1.0	1.0	0.70	1.0	17.7
7 Yehudiyya	10	0	1.0	1.0	0	0	1.0	1.0	0	1.0	1.0	0.20	1.0	0	1.0	1.0	0	0.90	0	0	0	1.0	0	1.0	1.0	0.50	1.0	17.5
8 Gamla	9	0	1.0	1.0	0	0.11	1.0	1.0	0	1.0	1.0	0.11	1.0	0	1.0	1.0	0.11	1.0	0.11	0.11	0.11	1.0	0.11	1.0	1.0	1.0	1.0	18.8
9 Rosh-Pinna	9	0	1.0	1.0	0.67	0	1.0	0.33	0	0.33	1.0	0	1.0	0	1.0	1.0	0	0.11	0	0	0	1.0	0	1.0	1.0	0.22	1.0	14.8
11 Tabigha	10	0.70	1.0	1.0	0	0	0.70	1.0	0.90	1.0	1.0	0	1.0	0.70	1.0	1.0	0.10	1.0	0	0	0	1.0	0	1.0	1.0	0.40	1.0	19.2
16 Mt. Gilboa	9	0	1.0	1.0	0.22	0	1.0	0.78	0.44	0.78	1.0	0	1.0	0	1.0	1.0	0	0.89	0	0	0	1.0	0	1.0	1.0	1.0	1.0	18.0
17 Mt. Gerizim	10	0	1.0	1.0	0	0	1.0	0.30	0.20	1.0	1.0	0	1.0	0	1.0	1.0	0	0	0	0	0	1.0	0	1.0	1.0	1.0	1.0	15.5
18 Gitit	10	0	0.70	0.90	0	0.10	0.50	0.70	0.50	1.0	1.0	0.30	1.0	0	1.0	1.0	0	0.20	0	0	0	1.0	0	1.0	1.0	1.0	1.0	16.0
19 Kokhav-Hashahar	10	0	1.0	0.90	0.30	0	1.0	0.70	0	0.70	1.0	1.0	1.0	0	1.0	1.0	0	1.0	0	0	0	1.0	0	0.70	1.0	0.70	1.0	16.9
20 Taiyiba	10	0	1.0	1.0	0.50	0	0.90	0.60	0	0.60	0.80	0	1.0	0	1.0	1.0	0	0.10	0	0	0	1.0	0	0.20	1.0	0.20	1.0	13.8
21 Sanhedriyya	10	0	1.0	1.0	0.70	0	1.0	0.20	0	0.30	1.0	0	1.0	0	1.0	1.0	0	0.90	0	0	0	1.0	0	1.0	1.0	1.0	1.0	17.0
22 Bet-Meir	10	0	1.0	1.0	0.70	0	1.0	0	0	1.0	1.0	1.0	1.0	0	1.0	1.0	0	1.0	0	0	0	1.0	0	0.20	1.0	0.20	1.0	16.1
23 J'aba	10	0	0.70	1.0	0.20	0	1.0	0.80	0	0.80	1.0	0.10	1.0	0	1.0	1.0	0	0.30	0.10	0.10	0.10	1.0	0	0.40	1.0	0.40	1.0	14.5
24 Amirim	9	0	1.0	1.0	0.11	0	1.0	0.89	0.44	0.89	1.0	0	1.0	0	1.0	1.0	0	0.89	0	0	0	1.0	0	1.0	1.0	0.44	0.89	17.3
25 Nahef	10	0	0	1.0	0	0	1.0	0	0	1.0	1.0	0.10	1.0	0	1.0	1.0	0	0.90	0	0	0	1.0	0	0.90	1.0	0.90	1.0	14.1
26 Ahihud (Achihood)	10	0	1.0	1.0	0	0	1.0	1.0	0	1.0	1.0	0	1.0	0	1.0	1.0	0	1.0	0	0	0	1.0	1.0	1.0	1.0	1.0	1.0	19.0
27 Nesher	10	0	1.0	0	0	0	1.0	0	0	1.0	1.0	0	1.0	0	1.0	1.0	0	1.0	1.0	1.0	1.0	1.0	0	0	1.0	0	0	14.0
28 Bet-Oren	10	0	1.0	1.0	0	0.20	1.0	0	0	1.0	1.0	0	1.0	0	1.0	1.0	0	1.0	0	0	0	1.0	0	0.10	1.0	0.10	0.10	13.5
29 Daliyya	10	0	1.0	1.0	0	0	1.0	1.0	1.0	1.0	1.0	0	1.0	0	1.0	1.0	0	0	0	0	0	1.0	0	1.0	1.0	0	1.0	16.0
30 Bat-Shelomo	10	0	1.0	1.0	0.40	0	1.0	0.10	0	0.70	1.0	0	1.0	0	1.0	1.0	0	0.70	0	0	0	1.0	0	1.0	1.0	1.0	1.0	16.4
32 Yabad	10	0	1.0	1.0	0.70	0	1.0	0	0	0.30	1.0	0	1.0	0	1.0	1.0	0	0	0	0	0	1.0	0	1.0	1.0	1.0	0	13.0
33 Givat-Koach	9	0	1.0	1.0	0	0	1.0	0	1.0	1.0	1.0	0	1.0	0	1.0	1.0	0	0	0	0	0	1.0	0	1.0	1.0	0.44	1.0	15.4
Total	225																											
Number of polymorphic populations:		1	2	2	11	3	3	11	5	9	1	6	0	1	0	0	2	12	2	2	2	0	2	6	0	12	2	
Mean frequency		0.03	0.93	0.95	0.20	0.02	0.96	0.53	0.19	0.84	0.99	0.15	1.0	0.03	1.0	1.0	0.01	0.61	0.95	0.95	0.95	1.0	0.06	0.80	1.0	0.62	0.87	
Mean *He*		0.02	0.04	0.02	0.18	0.03	0.05	0.16	0.09	0.15	0.01	0.07	0	0.02	0	0	0.02	0.13	0.02	0.02	0.02	0	0.03	0.08	0	0.20	0.02	

[a] The following bands are duplicates: 1 = 1′, 6 = 7 (in malt), 20 = 22

Table 6.13. Geographic distribution of genetic indices of the bands and multilocus genotypes (MGP) revealed by electrophoretic analysis of α- and β-amylases in 23 populations of *T. dicoccoides* in Israel. (Nevo et al. 1993a)

Population number and name	*N*	α-Amylase										β-Amylase									
		Malt					Green parts					Germinating seeds					Dry seeds				
		MGP no.	*He*	Number of bands	*He*	*P*	MGP no.	*He*	Number of bands	*He*	*P*	MGP no.	*He*	Number of bands	*He*	*P*	MGP no.	*He*	Number of bands	*He*	*P*
1 Mt. Hermon	10	4	0.660	11.2	0.050	0.13	2	0.420	5.3	0.070	0.17	3	0.640	6.8	0.262	0.55	3	0.620	4.5	0.057	0.15
5 Qazrin	10	2	0.180	13.0	0.016	0.09	2	0.420	4.7	0.070	0.17	7	0.820	6.7	0.278	0.73	6	0.760	4.6	0.145	0.38
7 Yehudiyya	10	3	0.460	13.0	0.030	0.13	2	0.500	4.5	0.083	0.17	3	0.640	6.0	0.160	0.36	2	0.180	5.2	0.028	0.15
8 Gamla	9	4	0.519	13.7	0.052	0.26	2	0.198	5.1	0.033	0.17	4	0.617	7.6	0.175	0.55	4	0.519	5.3	0.080	0.31
9 Rosh-Pinna	9	3	0.494	10.6	0.075	0.22	2	0.346	4.2	0.058	0.17	1	0.0	9.0	0.0	0.0	1	0.0	6.0	0.0	0.0
11 Tabigha	10	4	0.640	14.8	0.089	0.26	2	0.480	4.4	0.080	0.17	2	0.180	9.1	0.016	0.09	4	0.660	7.4	0.145	0.54
16 Mt. Gilboa	9	4	0.667	13.0	0.084	0.26	1	0.0	5.0	0.0	0.0	4	0.617	5.7	0.148	0.36	3	0.593	3.3	0.065	0.15
17 Mt. Gerizim	10	3	0.460	10.5	0.032	0.09	1	0.0	5.0	0.0	0.0	5	0.680	6.7	0.238	0.73	6	0.800	3.3	0.275	0.69
18 Gitit	10	10	0.900	11.0	0.150	0.43	1	0.0	5.0	0.0	0.0	1	0.0	6.0	0.0	0.0	3	0.620	5.9	0.103	0.23
19 Kokhav-Hashahar	10	4	0.480	12.5	0.070	0.22	2	0.420	4.4	0.140	0.33	3	0.580	8.7	0.060	0.18	1	0.0	0.0	0.0	0.0
20 Taiyiba	10	6	0.780	10.4	0.117	0.39	2	0.320	3.4	0.107	0.33	4	0.480	8.4	0.091	0.36	3	0.560	2.0	0.049	0.15
21 Sanhedriyya	10	4	0.480	12.0	0.066	0.22	1	0.0	5.0	0.0	0.0	1	0.0	10.0	0.0	0.0	3	0.460	2.5	0.057	0.15
22 Bet-Meir	10	2	0.420	12.7	0.018	0.04	2	0.320	3.4	0.107	0.33	1	0.0	9.0	0.0	0.0	1	0.0	5.0	0.0	0.0
23 J'aba	10	6	0.800	10.7	0.131	0.43	2	0.480	3.8	0.160	0.33	1	0.0	8.0	0.0	0.0	2	0.480	2.4	0.037	0.08
24 Amirim	9	4	0.667	13.0	0.071	0.26	3	0.568	4.3	0.115	0.33	1	0.0	9.0	0.0	0.0	2	0.346	0.8	0.027	0.08
25 Nahef	10	3	0.540	9.3	0.034	0.13	2	0.180	4.8	0.060	0.33	1	0.0	9.0	0.0	0.0	3	0.580	6.3	0.066	0.23
26 Ahihud (Achihood)	10	1	0.0	13.0	0.0	0.0	1	0.0	6.0	0.0	0.0	1	0.0	7.0	0.0	0.0	2	0.500	5.5	0.039	0.08
27 Nesher	10	1	0.0	12.0	0.0	0.0	1	0.0	2.0	0.0	0.0	6	0.760	5.4	0.182	0.45	5	0.720	2.9	0.183	0.38
28 Bet-Oren	10	2	0.320	11.2	0.014	0.04	2	0.180	2.3	0.090	0.50	3	0.580	6.3	0.060	0.18	2	0.500	5.0	0.077	0.15
29 Daliyya	10	1	0.0	12.0	0.0	0.0	1	0.0	4.0	0.0	0.0	3	0.640	6.2	0.073	0.18	2	0.500	5.5	0.039	0.08
30 Bat-Shelomo	10	5	0.760	11.4	0.087	0.22	1	0.0	5.0	0.0	0.0	4	0.480	9.5	0.075	0.27	5	0.680	3.6	0.099	0.38
32 Yabad	10	2	0.420	9.0	0.037	0.09	1	0.0	4.0	0.0	0.0	8	0.860	7.7	0.213	0.55	4	0.740	2.3	0.131	0.31
33 Givat-Koach	9	1	0.0	11.0	0.0	0.0	2	0.494	4.4	0.082	0.17	1	0.0	6.0	0.0	0.0	4	0.667	3.3	0.065	0.15
Mean		3.4	0.463	11.8	0.053	0.17	1.7	0.232	4.4	0.055	0.15	3.0	0.373	7.6	0.088	0.24	3.1	0.499	4.0	0.077	0.21
Total	225																				

P Proportion of polymorphic bands (or loci)

Table 6.14. Linkage disequilibria (LD) of amylases within populations of *T. dicoccoides* in Israel. Only LDs based on frequencies between 0.2 and 0.8 were counted. Significance by χ^2 or Fisher exact tests. (Nevo et al. 1993a)

Population	*N*	Number of polymorphic bands		Linkage disequilibria (LD)				
				Total number	Number significant			
		All	0.2–0.8		<0.05	<0.01	All	%
Mt. Hermon	10	11	11	55	4	7	11	20.0
Qazrin	10	15	13	78	5	3	8	10.3
Yehudiyya	10	9	6	16	0	3	3	20.0
Gamla	9	17	4	6	2	1	3	50.0
Rosh-Pinna	9	4	3	3	1	0	1	33.3
Tabigha	10	14	6	15	3	4	7	46.7
Mt. Gilboa	9	10	8	28	4	3	7	15.6
Mt. Gerizim	10	19	15	105	1	6	7	6.7
Gitit	10	11	9	36	0	1	1	2.8
Kokhav-Hashahar	10	7	5	10	0	4	4	40.0
Taiyiba	10	12	9	36	2	1	3	8.3
Sanhedriyya	10	5	4	6	0	0	0	0.0
Bet-Meir	10	3	3	3	1	0	1	33.3
J'aba	10	11	7	21	1	1	2	9.5
Amirim	9	7	3	3	0	0	0	0.0
Nahef	10	7	1	0	–	–	–	–
Ahihud (= Achihood)	10	1	1	0	–	–	–	–
Nesher	10	10	9	36	12	3	15	41.7
Bet-Oren	10	8	6	15	0	1	1	6.7
Daliyya	10	3	3	3	0	0	0	0.0
Bat-Shelomo	10	11	6	15	0	1	1	6.7
Yabad	10	11	10	45	3	2	5	11.1
Givat-Koach	9	3	3	3	0	0	0	0.0
Total	225							

6.8 The Amylase Multigene Family

6.8.1 Evolutionary Considerations

The amylase multigene families display an "archipelago" pattern of population-genetic structure similar to that of allozymes in wild emmer wheat and wild barley (Nevo and Beiles 1989). The diversity patterns of α- and β-amylases in natural populations of *T. dicoccoides* illustrate the evolutionary driving forces causing amylase differentiation and its potential adaptedness. Out of the potential conceivable forces, i.e., mutation, genetic drift, migration, and natural selection, the last-named appears to be the predominant differentiating and orienting force for the following reasons.

The archipelago pattern of amylase population genetics is *nonrandom*, structured, and correlated with ecology across the range of *T. dicoccoides* in Israel, negating genetic drift as an explanatory model. The predominance of ecogeographically structured polymorphisms (displaying significant climatic and soil correlates) excludes randomness as a major evolutionary factor and strongly suggests the operation of ecologically *diversifying natural selection*. Furthermore, the significant correlation of α- and β-amylases with climatic factors, primarily temperature variables, suggests that many of the selection processes in different temperature regimes operate on MGPs (Nevo et al. 1993a and Table 6.13), in addition to those directly affecting single bands. Finally, the autocorrelation analysis excludes migration as a major differentiating force, again assigning selection an important evolutionary role in amylase differentiation. We conclude that, as in the single-loci allozyme case, and as concluded from both *micro-* and *macro*geographic differentiation (Nevo and Beiles 1989), the amylase multigene families are basically governed by climatic and edaphic diversifying natural selection. This conclusion bears directly on the utilization of amylase variation.

6.8.2 Importance of Amylase Variation for Wheat Improvement

The numerous genes and rich polymorphisms of wild emmer α- and β-amylases may provide unique genetic resources of agricultural importance for improving wheat germination and growth in adverse and stressful environments. The chromosomal location of most of the α- and β-amylases genes is known (McIntosh et al. 1998). Some have been cloned and others from wild emmer are clonable. Most sequence amylase motifs are expected to be common to most of the actively transcribed genes in plants. However, specific sequences are expected in the rich amylase allelic diversity described here in wild emmer, as well as in other genes regulating gibberellic acid (GA3), which regulates amylases. The unique expression and the effects on germination and growth of the wild emmer genes can be tested biotechnologically in *T. aestivum*. It is expected that amylase genes and alleles, and their regulator GA_3 genes, may be unique in some wild emmer populations, particularly in steppic ones. These steppic populations are adapted to rapid germination and quick growth in order to shorten the annual cycle and escape the early hot and dry summer (Nevo and Beiles 1989).

In addition to their direct utilization as described above, α- and β-amylase isozymes can provide useful markers for chromosomal locations of QTLs (Nevo et al. 1993a and references therein). Biochemical loci are superior to morphological loci that affect the gross phenotype since they and their alleles are easily identifiable. The wheat α-amylase genes are useful markers because the products of six individual loci of α-*Amy-1* and three of α-*Amy-2* can be resolved simultaneously, whereas the β-amylase genes are less practical in this respect. We have used amylase probes in our laboratory to map herbicide resistance gene(s) (Krugman et al. 1997) on chromosome 6B of wild emmer wheat.

6.9
Disease Resistance Polymorphisms in *Triticum dicoccoides* (Host-Pathogen Coevolution in the Center of Diversity)

Crop diseases cause loss of yield and prevent a rapid increase in food production. Crop improvement should therefore aim to minimize the risks of epidemic and endemic diseases in an attempt to increase production efficiency without affecting the agronomic properties of existing varieties (Horsfall and Cowling 1977). In general, natural plant ecosystems are not devastated by disease, since, over the course of host-pathogen coevolution, they have developed efficient defense mechanisms. This evolutionary interaction has resulted in a balanced genetic polymorphism, with the highest genetic diversity for virulence in the pathogen and for resistance in the host (Dinoor 1974; Anikster and Wahl 1979; Dinoor and Eshed 1990).

Israel is one of the epicenters where active coevolution of several crop ancestors and their pathogens occurred, including wild barley and wild wheat and their rust and powdery mildew pathogens. In keeping with the theory and genetics of host-parasite coevolution, the highest and broadest genetic immunity against devastation in the progenitors of some crops would be expected in the Fertile Crescent in general, and in Israel in particular. Israel is ideal for the exploitation of resistance genes, which could provide useful material for crop improvement for both the short-term and, more importantly, the long-term solution to disease problems in agriculture (Segal et al. 1980, 1982). This includes clarification of the controversy about specific and general resistance (Anikster and Wahl 1979).

6.9.1
Powdery Mildew, *Erysiphe graminis tritici*

Powdery mildew of the wheat *T. aestivum* L., incited by the pathogen *Erysiphe graminis* DC. ex Merat. f. sp. *tritici* (Em. Marchal), is one of the most damaging diseases of wheat on a worldwide basis. It causes significant losses in wheat production in the United States, United Kingdom, New Zealand, Hungary (Large and Doling 1962; Smith and Smith 1974), China (Zhuang and Li 1993) and elsewhere (Zeller and Hsam 1998).

The reactions of 233 *T. dicoccoides* accessions, collected at ten sites in Israel and elsewhere, to infection with cultures of *Erysiphe graminis tritici* were determined (Moseman et al. 1984; Table 6.15). The results indicated that the number of sources of resistance to *E. graminis tritici* obtained from *T. dicoccoides* plants growing wild in Israel and elsewhere is almost unlimited. Of the accessions, 114 (49%), were resistant, and 137 (59%), were resistant or moderately resistant to infection with four cultures of *E. graminis tritici*, which possess the virulence genes corresponding to most of the identified resistance genes in wheat (Table 6.15). Accessions collected at sites with marginal habi-

Table 6.15. Reactions[a] of *T. dicoccoides* accessions collected in Israel to infection with two composites of cultures of *Erysiphe graminis tritici*. (Moseman et al. 1984)

Location	Site no.	Number of accessions	Composite no. 1[a] reactions			Average infection types	Composite no. 2[a] reactions			Average infection types
			R1	M1	S1		R2	M2	S2	
Mt. Hermon	1	14	6	2	6	5.0	3	3	8	6.1
Qazrin	2	35	33	2		2.1	35			1.7
Yehudiyya	3	13	6		7	5.6	5		8	5.7
Rosh-Pinna	4	28	21	5	2	2.5	27		1	1.8
Tabigha	5	32	15	6	11	4.6	18	4	10	4.5
Bat-Shelomo	6	32	5	5	22	6.8	10	4	18	5.8
Kokhav-Hashahar	9	34	24	4	6	3.4	26	2	6	3.1
Taiyiba	10	9	2	5	2	4.2	5		4	4.9
Sanhedriyya	11	18	2	1	15	7.2	3	1	14	6.7
Bet-Meir	12	18	7		11	5.5	5	1	12	5.9
Total		233	121	30	82		137	15	81	
%			51.9	12.9	35.2		58.8	6.4	34.8	

R Infection types 0–3, *M* infection types 4–6, *S* infection types 7–9
[a] Composite no. 1 was Quincy and Molo; composite no. 2 was 127 and ABK

tats, where *T. dicoccoides* grows poorly and has a lower grain weight, were more susceptible than accessions collected at central sites with an optimal habitat for growth of *T. dicoccoides*. The results agreed with those in a previous study with *H. spontaneum* and indicated that, to obtain *H. spontaneum* or *T. dicoccoides* accessions with the highest level of resistance to powdery mildew pathogens, plants should be collected at sites in ecological and geographic regions where those two species occupy optimum habitats and are also exposed to the powdery mildew pathogens. Powdery mildew resistance genes from *T. dicoccoides* have been extensively used in wheat improvement both in Europe and North America.

A powdery mildew resistance gene originating from a wild emmer wheat (*T. dicoccoides*) accession (C20), from Rosh Pinna, Israel, was successfully transferred to hexaploid wheat through crossing and back-crossing (Liu et al. 2001). Genetic analysis indicated that a single dominant gene controls powdery mildew resistance. Segregating BC_1F_2 progenies of the cross 87-1/C20//2*8866 was used for bulked segregant analysis (BSA). PCR was used to generate polymorphic DNA fragments between the resistant and susceptible DNA pools by means of 10-mer random primers, STS primers, and wheat microsatellite primers. Three markers, *Xgwm*159/430, *Xgwm*159/460, and *Xgwm*159/500, were found to be linked with the resistance gene. After evaluation of the polymorphic markers in two segregating populations, the distance between the markers and the mildew resistance gene was estimated to be 5.4–6.0 cM. Using Chinese Spring nullisomic-tetrasomics and ditelosomics, the polymorphic markers and the resistance gene were assigned to chromosome arm 5BS and were physically mapped on the gene-rich regions of fragment length (FL) 0.41–0.43 by means of Chinese Spring deletion lines. As no powdery mildew resistance gene has been reported on chromosome arm 5BS, the mildew resistance gene originating from C20 might be a new gene and has been temporarily designated *MlC20.*

Two powdery mildew resistance genes originating from wild emmer were previously reported and are located on chromosomes 4A (Reader and Miller 1991, see also end of section 6.12.2) and 2BS (Rong et al. 1998). Two powdery mildew resistance genes, Pm2 and Pm23, were assigned to chromosomes 5DS and 5A, respectively (McIntosh et al. 1998). As noted above, no powdery mildew resistance gene has been reported on chromosome 5B or 5BS. Thus, the wild-emmer-derived powdery mildew resistance gene *MlC20* reported by Liu et al. (2001) appears to be a new gene, which could be used to diversify the resistance resources of wheat powdery mildew originating from *T. dicoccoides.*

6.9.2 Leaf Rust, *Puccinia recondita tritici*

There are three species of wheat rust pathogens, *P. recondita* Rob. ex Desm. f. sp. *tritici* (leaf rust), *Puccinia graminis* Pers. f. s. *tritici* (the pathogen of wheat stem rust), and *P. striiformis* West. f. sp. *tritici* (stripe, or yellow rust). The rusts caused by these pathogens are very important diseases affecting worldwide wheat production and as such have attracted the most attention and research interest. While rust diseases are readily controlled by resistant cultivars, virulent pathotypes are a constant threat (McIntosh 1998; McIntosh et al. 1996). During the past several decades, great progress has been made in breeding for resistance to rust diseases. However, these diseases are still severe and threaten worldwide wheat production because of the continuous and rapid evolution of virulence in these pathogens (Roelfes et al. 1992).

Leaf rust, incited by *Puccinia recondita* Rob. ex Desm. f. sp. *tritici,* is one of the most damaging diseases of wheat *T. aestivum* L. in the United States and throughout the world. Reactions of 687 accessions of *T. dicoccoides* from Israel to infection with culture PRTUS6 of *P. recondita* Rob. ex Desm. f. sp. *tritici* were determined (Moseman et al. 1985). Fourteen percent (98 accessions) were at least moderately resistant in the first leaf stage to infection with culture PRTUS6. Nineteen of 34 accessions collected at Bet Meir were moderately resistant, as were 10 of 35 from Qazrin and four of nine from Mt. Gerizim. Resistant accessions were obtained in low frequencies from Bat Shelomo and Mt. Gilboa. Resistant and moderately resistant accessions were obtained among 353 accessions collected at approximately 180 sites throughout the region from Upper and Lower Galilee, Mt. Gilboa, the Judean Mountains, Mt. Hermon, and the Golan Heights. Reactions of 687 accessions of *T. dicoccoides* collected at 192 sites in Israel to infection with culture PRTU56

Table 6.16. Reactions of 687 *T. dicoccoides* accessions, collected at 192 sites in Israel, to infection with culture PRTUS6 of *Puccinia recondita tritici*. (Moseman et al. 1985)

Collector	Reactions[a]					
	R		MR		S	
	No.	%	No.	%	No.	%
Nevo and Zohary	2	0.6	44	13.2	288	86.2
Gerechter-Amitai	4	1.1	48	13.6	301	85.3

[a] Reactions: R (resistant), infection types 0–3; MR (moderately resistant), infection types 4–6; S (susceptible), infection types 7–9

of *P. recondita tritici* are listed in Table 6.16. Additional resistant and moderately resistant accessions of *T. dicoccoides* could probably be obtained by collection from additional sites. The resistant accessions identified in this study were used to develop enhanced hexaploid and tetraploid wheat germplasm resistant to *P. recondita tritici*, *Erysiphe graminis tritici*, and *P. striiformis* West.

6.9.3 Stem Rust, *Puccinia graminis tritici*

Seedling resistance to wheat stem rust *P. graminis tritici* was determined in populations of *T. dicoccoides*, and characterized by means of ecological factors and allozyme genotypes (Fig. 6.8; Nevo et al. 1991b). Reactions to wheat stem rust were studied in 102 single plant accessions of *T. dicoccoides* from ten populations by inoculation with *P. graminis tritici* race 14, isolate GSR-739. Six populations displayed different degrees of response polymorphism, with reactions ranging from high resistance to complete susceptibility, whereas four populations contained only susceptible plants (Tables 6.17, 6.18). In some of the accessions, unexplained intrasib variation in resistance and intraplant variation of infection types were found. Resistance to stem rust was negatively correlated with two ecological factors, altitude and number of *Sharav* (hot-dry) days, which are unfavorable to disease development. Variation in stem-rust response, conceivably maintained by natural selection, was shown to exist in ecogeographic regions where climatic variables enhance development of the fungus. Likewise, allozyme genotypes, single or in multiple loci combinations, appear to be associated either with resistance or susceptibility to rust (Fig. 6.8). Such associations need to be verified by genetic studies to become established as useful markers. Clearly, current genetic mapping effort by ourselves (Peng et al. 2000a–c) and others may directly contribute

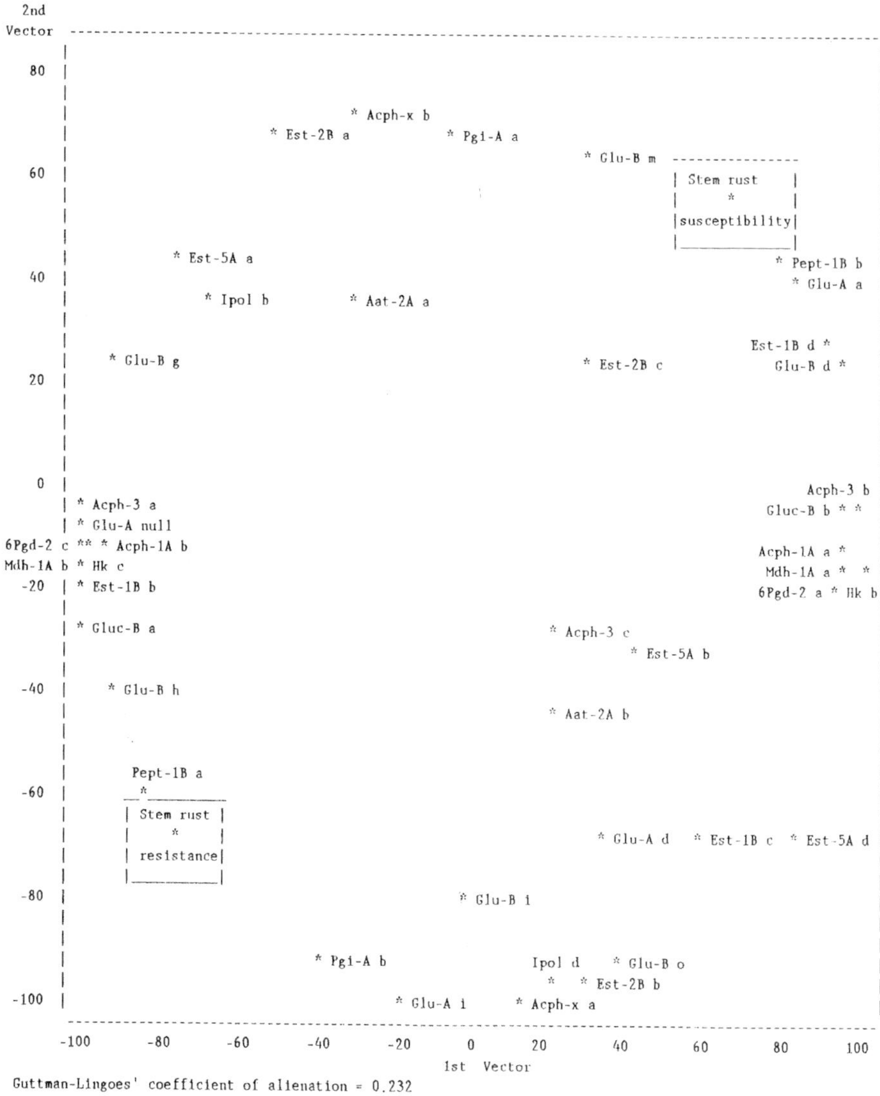

Fig. 6.8. Smallest space analysis (SSA) plotting Spearman correlations among stem-rust response variables and 42 allozyme markers in *T. dicoccoides*. For symbols of enzymatic loci and alleles, see Nevo and Beiles 1989. (Nevo et al. 1991b)

1984)

Population	Accession	N[a]	IT[b]	Mean[c]	Population	Accession	N	IT	N	IT[b]	Mean	Population	Accession	N	IT	N	IT	Mean
Mt. Hermon	1	1	8	8	Tabigha	1		4			4	Kokhav-Hashahar	12	1	9			9
	3	1	8	8		2		4			4		14		8			8
	22	1	7	7		4		4			4		18		8			8
	56	1	8	8		5		3			3		23		8			8
	58	1	8	8		6		2–3			2.5		24		8			8
	60	2	8	8		7		2–3			2.5		25		8			8
Qazrin	5		8	8		8	6	2–3	1	8	3.3		32		9			9
	6		7	7		13	3	5–6	5	7–8	6.8		33		8			8
	21		8	8		14	3	3–7	2	8	6.2		35		8			8
	27		8	8		15		3			3		36		9			9
	43		5–6	5.5		22		2–3			2.5		41		8			8
	56		8	8		23	1	6	1	8	7		45		8			8
	64	1	1–2	1.5		25		2–3			2.5		46		8			8
	79		8	8		28	4	3–7			5		50		8			8
	80		7	7		31	1	6	4	8	7.6	Taiyiba	9		8			8
	82		8	8		37		9			9		14		8			8
	96		8	8		45	1	3–7	6	6–7	6.3		37	1	2–3	2	8	6.2
Yehudiyya	5		3	3		48		9			9		43		8			8
	8	2	3	3		51	1	3	2	3–6	4	Sanhedriyya	4		8			8
	13		2–3	2.5		52		9			9		6		8			8
	15		3	3		54		4			4		12	1	8			8
	34		4	4	Bat-Shelomo	2		2–3			2.5		18		8			8
	37		2–3	2.5		4		2–3			2.5		30		8			8
Rosh-Pinna	3		9	9		5		7			7		44		8			8
	6		2–3	2.5		8		9			9	Bet-Meir	37		8			8
	7		7	7		11		6			6		48		8			8
	9	2	9	9		14		9			9		54		8			8
	19	2	9	9		16	1	8			8							
	21		9	9		22		8			8							
	23		9	9		25		8			8							
	24		9	9		27		2–3			2.5							
	29		9	9		28	1	3	7	8	7.4							
	33		8	8		32	8	3–4	1	8	4							
	42		8	8		34		2–3			2.5							
	43		8	8		36		2			2							
	69	1	9	9		38		2			2							
	70		8	8		42	10	2	1	8	2.5							
						44		6			6							

[a] Number of seedlings if the number was small or when different ITs were recorded in the same accession
[b] If in a seedling more than one IT was scored, the range is presented. If the seedling differed in IT, the two scores are given, each with their N
[c] For accessions with variable scores, the mean score was used in the analysis. The mid-range IT in each seedling entered the weighted mean score of the accession

Table 6.18. Geographic distribution of resistance to *Puccinia graminis tritici* in 102 accessions from ten populations of *T. dicoccoides*. (Nevo et al. 1991b)

Population[a]	Number[b] of accessions with specified reaction					
	IT (*N*)	Resistant 1–3	Intermediate 4–6	Susceptible 7–9	Variable[c]	Nonvariable
Mt. Hermon (1)	6	0	0	6	0	6
Qazrin (5)	11	1	1	9	0	11
Yehudiyya (7)	6	5	1	0	0	6
Rosh-Pinna (9)	14	1	0	13	0	14
Tabigha (11)	21	7.75	7	6.25	8	13
Bat-Shelomo (30)	17	7.5	2.5	7	3	14
Kokhav-Hashahar (19)	14	0	0	14	0	14
Taiyiba (20)	4	0.33	0	3.67	1	3
Sanhedriyya (21)	6	0	0	6	0	6
Bet-Meir (22)	3	0	0	3	0	3
Total	102	22.58	11.50	67.92	12	90
Percent		22.1	11.3	66.6	11.8	88.2

[a] Population numbers in *parentheses* are according to Nevo and Beiles (1989)
[b] The variable phenotypes were added proportionally to each resistance type
[c] The variable category included within-seedling variation (a range of infection types (ITs) on the same seedling) and between-seedling variation (different ITs on different seedlings of the same accession). Within a seedling, a range of one score (i.e., 2–3) was not regarded as variable. In Bat-Shelomo and Taiyiba, only between-seedling variation was recorded. In Tabigha, three within- and seven between-seedling variable phenotypes were recorded

to mapping, cloning, transformation and improvement of susceptible wheat cultivars.

6.9.4 Stripe Rust, *Puccinia striiformis tritici*

Stripe or yellow rust, caused by *P. striiformis* West. f. sp. *tritici*, continues to be a problem in cooler and humid wheat-growing areas at high elevation or higher latitudes (McIntosh 1998). The name of the disease describes the characteristic yellow stripe of uredinia that produce yellow colored urediniospors. This disease usually attacks young wheat plants during the winter and early spring and makes the plants stunted and weakened. In extreme situations, stripe rust can result in up to 100% yield losses (Roelfes et al. 1992). In Israel, stripe rust has become a major threat to wheat production, especially in northern Israel, due to the appearance of new virulent races (Manistersky and Eyal 1994). These could infect wheat cultivars that were resistant to previously occurring races in Israel and cause a considerable decrease in the yield of wheat grown in the Yizreel Valley, Bet Shean Valley, and Hula Valley (Manistersky et al. 1995).

Host resistance is considered to be the most economical and environmentally benign control method (Ma et al. 1997). A majority of the 23 characterized resistance genes to stripe rust (McIntosh et al. 1995) are no longer effective due to virulence factors in the pathogen. In some important high-yielding and widely adapted durum wheats developed at the International Maize and Wheat Improvement Center (CIMMYT), resistance could be conferred by only four to five resistance genes. To globally sustain stripe-rust resistance, the existing resistance gene pool should be widened and diversified (Ma et al. 1997) through the pyramiding process (Roelfes et al. 1992).

From 114 accessions of wild emmer wheat from 11 sites in Israel, known for their allozymic variation (Nevo et al. 1982), individual genotypes were tested for resistance to one isolate of stripe rust, both in the seedling stage in a growth chamber and in the adult plant stage in the field (Nevo et al. 1986b; Peng et al. 1999, 2000a,b). The results indicate that resistance to stripe rust in seedlings and adults was significantly correlated ($r_s = 0.40$, $p < 0.001$). The Mount Hermon population proved uniquely resistant both in the seedling and the adult stages (Table 6.19). Genetic polymorphisms of resistance to stripe rust varied geographically and were predictable by climatic and allozymic markers. Three variable combinations of rainfall, evaporation, and temperature explain significantly 0.40–0.53 of the spatial variance in disease resistance to stripe rust, suggesting the operation of natural selection (Tables 6.19, 6.20). Several allozyme genotypes were significantly associated with disease resistance. We conclude that natural populations of wild emmer wheat in Israel contain large numbers of disease resistance genes (see Fig. 6.9). These populations could be effectively screened and then utilized by phytopathologists for identifying resistant genotypes and producing new resistant cultivars.

6.9.5 Responses of Israeli *Triticum dicoccoides* to Selected Australian Pathotypes of *Puccinia* Species

Seedling responses to one Australian isolate of each of the stripe rust, stem rust, and leaf rust pathogens were determined for 541 accessions of *T. dicoccoides* collected from 23 locations in Israel (The et al. 1993). Resistance to stripe rust was more frequent than resistance to stem rust. Stripe-rust responses showed a wide range of variability indicative of a number of genes for resistance (Table 6.21). Comparisons of the present stem-rust data and those reported for the same accessions tested in Israel (Nevo et al. 1986b) indicated that different genes operated in each country. Only moderately resistant responses to stem rust were obtained. No potentially useful seedling resistance to leaf rust was found.

6.9.6 Resistance to Wheat Soilborne Mosaic Virus (WSBMV)

A fungal vector called *Polymyxa graminis* Ledingham carries wheat soil-borne mosaic virus (WSBMV). The virus particles are carried on or in the fungal zoospores (swimming spores). The fungus invades root hairs of the young

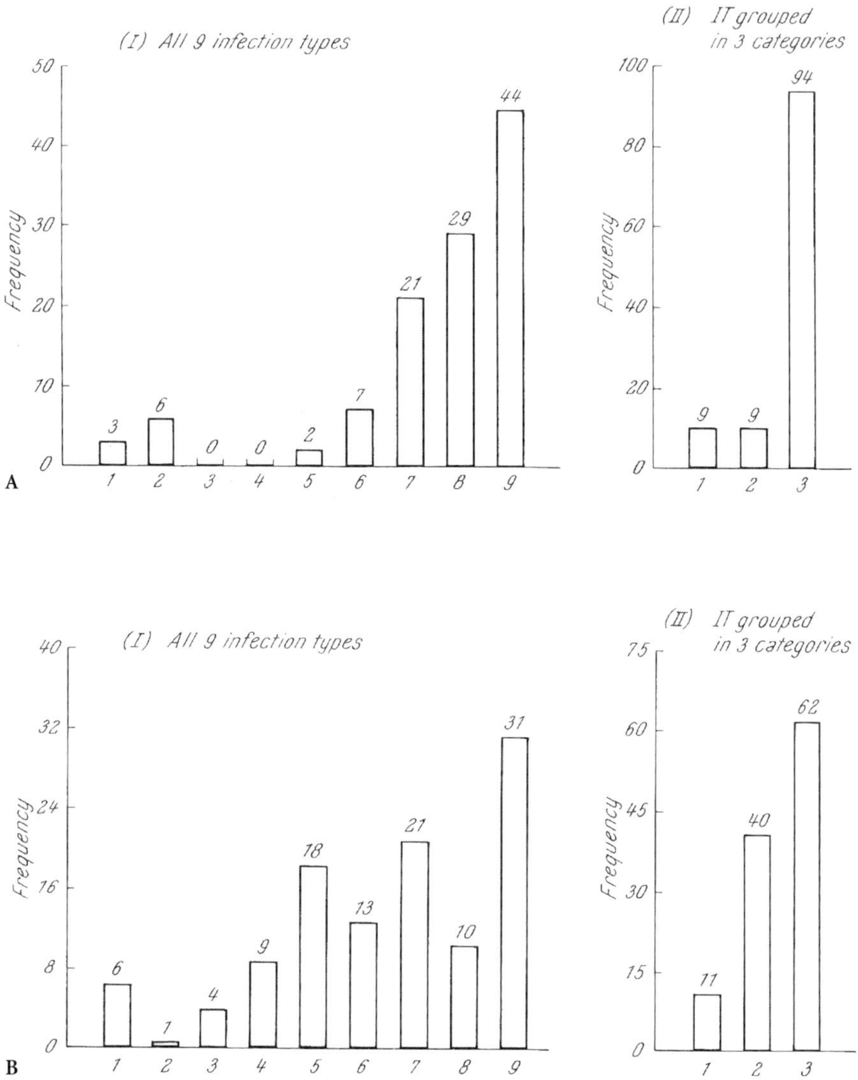

Fig. 6.9A, B. Frequency of infection types of accessions *of T. dicoccoides* infected by stripe rust. A Seedlings (N = 112); B Plants in field nursery (N = 113) (Nevo et al. 1986b)

wheat plants in the fall during periods of high soil moisture and forms dark clusters of resting spores in the wheat roots, which are released to the soil when the roots decay. Since the fungus survives in the soil as resting spores, the disease is always associated with infested soil. Soils may remain infective for many years. Losses are limited to the areas of the field showing symptoms. Within affected areas, losses caused by WSBMV are variable, depending

Table 6.19. The ecogeographic background of 11 populations of wild wheat, *Triticum dicoccoides*, in Israel and their interactions with the pathogen *Puccinia striiformis* expressed as average infection type (IT). (Nevo et al. 1986b)

Locality	Ecogeographic variables																		Seedlings			Adults		
	Lon	Lat	Alt	Tm	Ta	Tj	Td	Tdd	Rn	Rd	Hul4	Huan	Dw	Sh	Th	Trd	Ev	So	No.	IT	±SD	No.	IT	±SD
Mt. Hermon	35.73	33.30	1300	11	21	3	18	6	1400	66	48	60	60	80	–	0	150	1	6	1.25	±0.27	6	1.00	±0.0
Qazrin	35.67	33.02	350	18	26	10	16	12	530	50	43	58	58	50	–	60	155	5	11	6.82	±0.96	10	6.60	±2.07
Yehudiyya	35.70	32.93	200	19	27	11	16	12	550	47	42	58	58	50	–	100	160	5	10	8.10	±0.84	10	7.20	±2.15
Rosh-Pinna	35.52	32.95	700	18	25	9	16	10	697	50	48	58	50	75	−10	35	150	1	16	7.99	±1.46	16	5.63	±1.75
Tabigha	35.53	32.90	0	24	32	15	17	10	436	45	45	57	58	60	−30	120	160	5	23	8.44	±0.84	23	7.57	±2.11
Bat-Shelomo	35.02	32.60	75	20	26	13	13	10	650	55	58	68	77	40	−10	30	150	2	17	7.69	±0.70	17	6.00	±0.71
Gilboa	35.42	32.50	150	21	28	12	16	12	400	44	43	58	40	60	−30	160	165	1	1	9.00		1	8.00	
Kokhav-Hashahar	35.34	31.95	600	20	28	12	16	12	400	40	45	59	30	80	−30	25	165	1	14	6.75	±2.24	15	6.87	±2.03
Taiyiba	35.35	31.92	450	19	26	10	16	12	400	40	44	58	30	80	−10	25	165	1	3	5.83	±3.88	4	6.38	±1.49
Bet-Meir	35.03	31.80	500	19	26	11	15	9	582	44	47	60	61	70	−10	100	160	1	3	9.00	±0.00	3	5.00	±0.00
Sanhedriyya	35.22	31.80	800	17	24	9	15	9	548	44	51	62	44	102	−10	0	155	1	8	8.79	±0.60	8	8.50	±0.93
Total																			112	7.44	±2.07	113	6.46	±2.26
Significance																				***			***	

Climatic values were from the Atlas Israel (1970) and from multiple-year records of the Meteorological Service of Israel
Geographic variables: *Lon* Longitude, in decimals; *Lat* latitude, in decimals; *Alt* Altitude, in meters. Temperature variables: *Tm* Mean annual temperature; *Ta* mean August temperature, *Tj* mean January temperature, *Td* mean seasonal temperature difference, *Tdd* mean day-night temperature difference, *Trd* mean number of tropical days; *Sh* mean number of *Sharav* days, i.e.,. hot and dry days; *Ev* mean annual evaporation. Water availability variables: *Dw* Mean number of dew nights in summer; *Rn* mean annual rainfall, in mm; *Rd* mean number of rainy days; *Hul4* mean humidity at 14:00; *Huan* mean annual humidity; *Th* Thornthwaite's moisture index. Edaphic variables: *So* Soil type; *1* Terra Rossa; *2* Rendzina; *5* Basalt. *No.* Number of accessions; ***$p < 0.001$

Table 6.20. Coefficients of multiple determination (R^2) of dependent variables resistance to stripe rust (IT) with independent climatic variables in 11 populations of wild wheat, *T. dicoccoides*, in Israel (for abbreviations see Table 6.19). (Nevo et al. 1986a)

Stepwise model	Variables			
Seedlings	Rn 0.358***	Rn Ev 0.489***	Rn Ev Trd 0.528***	Rn Ev Tdd Rd 0.583***
Adults	Rn 0.383***	Rn Tdd 0.395***	Rn Tdd Ev 0.403***	Rn Tdd Ev Lon 0.412***

***$p < 0.001$

Table 6.21. Frequencies of *T. dicoccoides* accessions resistant, moderately resistant, moderately susceptible, and susceptible to stripe rust, stem rust, and leaf rust from 23 locations in Israel partitioned into ecogeographic sets. (The et al. 1993)

	Stripe rust					Stem rust				Leaf rust		
	No.	R	MR	MS	S	No.	MR	MS	S	No.	MS	S
Central locations												
Qazrin	24	–	2	1	21	26	1	2	23	26	1	25
Yehudiyya	20	3	10	4	3	23	–	1	22	20	1	19
Gamla	21	–	–	2	19	22	6	1	15	21	–	21
Rosh-Pinna	21	–	7	3	11	23	–	3	20	22	–	22
Tabigha	21	–	7	–	14	25	–	2	23	21	1	20
Total	(107)	(3)	(26)	(10)	(68)	(119)	(7)	(9)	(103)	(110)	(3)	(107)
Xeric-cold marginal (northern) location												
Mt. Hermon	17	17	–	–	–	18	–	3	15	19	2	17
Xeric-warm marginal (eastern and southern) locations												
Mt. Gilboa	18	–	–	1	17	17	–	–	17	17	6	11
Mt. Gerizim	7	–	–	–	7	10	–	3	7	8	–	8
Gitit	23	–	–	–	23	27	–	2	25	27	4	23
Kokhav-Hashahar	13	–	1	–	12	14	–	–	14	12	1	11
Taiyiba	14	–	4	1	9	20	–	–	20	19	1	18
Sanhedriyya	16	–	2	–	14	18	–	–	18	17	–	17
Bet-Meir	11	–	–	–	11	11	–	–	11	10	–	10
J'aba	27	–	6	3	18	27	–	–	27	25	–	25
Total	(146)	(17)	(13)	(5)	(111)	(162)	–	(8)	(154)	(154)	(14)	(140)
Mesic marginal (western) locations												
Amirim	14	–	3	1	10	17	–	9	8	16	–	16
Nahef	15	–	–	15	–	16	–	1	15	16	–	16
Ahihud	28	–	13	5	10	26	–	18	8	28	2	26
Nesher	17	–	6	4	7	16	–	2	14	18	1	17
Bet-Oren	29	1	–	–	28	26	–	1	25	28	6	22
Daliyya	23	–	19	2	2	23	14	–	9	22	–	22
Bat-Shelomo	27	1	12	6	8	27	23	3	1	29	5	24
Yabad	30	–	–	–	30	30	–	–	30	28	2	26
Givat-Koach	17	1	–	–	16	17	–	1	16	17	–	17
Total	(200)	(3)	(53)	(33)	(111)	(198)	(37)	(35)	(126)	(202)	(16)	(186)
Grand total	453	23	92	48	290	479	44	52	383	466	33	433

on the variety of wheat and the weather. Losses are roughly proportional to the length of time during which plants show obvious symptoms. A long cool spring prolongs the symptomatic period and losses may be 30–50%. During a warm spring, losses may be only 10–20% (further information at http://www.ext.nodak.edu/extpubs/plantsci/smgrains/pp646w.htm).

The reaction to WSBMV of 318 plants representing 90 genotypes in 25 populations of *T. dicoccoides* collected in Turkey and Israel were determined by symptomatology and enzyme-linked immunosorbent assay (ELISA) (Hunger et al. 1992). A range of reactions to WSBM existed in the wild emmer collection, although the virus has not been reported as a severe disease in Israel or Turkey. Nine populations contained only WSBM-resistant plants. The other 16 populations contained different ratios of resistant and susceptible plants that ranged from nearly all susceptible to nearly all resistant (Table 6.22). Thus, wild emmer may represent a source of resistance to WSBMV. This could, in turn, help diversify the genetic base for bread-wheat resistance to WSBMV.

6.10 Photosynthetic Characteristics in *Triticum dicoccoides* and Their Predictability by Ecological and Genetic Factors

Crop productivity in general, and specifically grain yield of wheat, is genetically determined by a hierarchy of physiological factors. Central to this hierarchy is the photosynthetic process (Mahon 1983). We tested *T. dicoccoides* for genetic variation (*Vc*) in photosynthetic characteristics residing within and between native populations sampled from three ecogeographic regions of Israel (Fig. 6.10). Our aim was to identify potential sources of high photosynthetic efficiency for future wheat improvements (Carver and Nevo 1990). Accessions sampled in the center of wild emmer distribution (catchment area of upper Jordan Valley) in a relatively narrow geographic range showed the greatest diversity in CO_2 assimilation rate per unit leaf area (A) or per unit chlorophyll (A/Chl). Genetic variation was absent for internal CO_2 concentration (*Ci*) and water-use efficiency (WUE) and was generally lacking for stomatal conductance (*gs*). Leaf area, although quite variable, was not a significant cofactor for assessing genetic potential for photosynthesis. Some accessions within a given population showed ten times more variation in A and A/Chl than populations sampled from different locations in a region. Accessions with the highest photosynthetic efficiency derived from upland steppic populations located in marginal habitats extending in Samarian eastern steppes facing the Jordan rift valley. Some accessions having high photosynthetic capacity ($A = 32\,\mathrm{ml\,m^{-2}\,s^{-1}}$) with no significant reduction in leaf size constitute a potentially valuable genetic resource as yet untapped for genetic improvement of hexaploid (*T. aestivum* L.) wheat. We, therefore, crossed these accessions into hexaploid wheat.

Table 6.22. Reactions of wild emmer wheat populations to wheat soil-borne mosaic virus (WSBMV). (Hunger et al. 1992)

Location site and identification no.[a]		No. of genotypes	No. of plants	Reaction to WSBM[b] (no. of plants)		
				Resistant	Susceptible	Uncertain
Turkey		4	5	5	0	0
Turkey		3	6	6	0	0
Ahihud	26	2	6	2	1	3
Amirim	24	4	16	16	0	0
Bet-Oren	28	4	14	14	0	0
Givat-Koach	33	4	14	0	13	1
Gitit	18	4	17	7	7	3
Jaba	23	4	16	6	4	6
Daliyya	29	4	17	7	8	2
Yabad	32	4	7	7	0	0
Nesher	27	4	18	10	3	5
Nahef	25	5	16	14	2	0
Qazrin	5	3	14	11	3	0
Ammiad	10	3	9	9	0	0
Yehudiyya	7	4	18	17	0	1
Rosh-Pinna	9	2	8	8	0	0
Sanhedriyya	21	3	15	2	13	0
Bet-Meir	22	4	13	6	5	2
Mt. Hermon	1	4	14	6	5	3
Tabigha	11	3	14	2	11	1
Bat-Shelomo	30	4	13	13	0	0
Taiyiba	20	3	8	8	0	0
Kokhav Hashahar	19	4	14	6	4	4
Mt. Gilboa	16	4	18	6	7	5
Mt. Gerizim	17	3	8	7	0	1
Total: 25		90	318	195	86	37

[a] See Fig. 6.10 for further information regarding populations.
[b] Reaction to WSBMV was determined based on the presence or absence of the mosaic symptom in combination with the value obtained from an ELISA conducted on the foliage collected from each seedling. Resistant, no or slight mosaic and an ELISA value <0.100; susceptible, mosaic observed and an ELISA value ≥ 0.100; uncertain, no or slight mosaic and an ELISA value ≥ 0.100, or mosaic present and an ELISA value <0.100

In a twin study (Nevo et al. 1991a), we analyzed, by univariate and multivariate methods, the differentiation of variation in photosynthetic characteristics of the same 107 genotypes from 27 populations of wild emmer in Israel as analyzed by Carver and Nevo (1990). These were distributed in three ecogeographic regions including central, xeric-marginal (northern cold and eastern warm), and mesic-marginal (western) populations. The highest photosynthetic efficiency was displayed by populations of the xeric-marginal region, but most variation for photosynthetic capacity occurred between accessions within ecogeographic regions and populations. Genotypes and

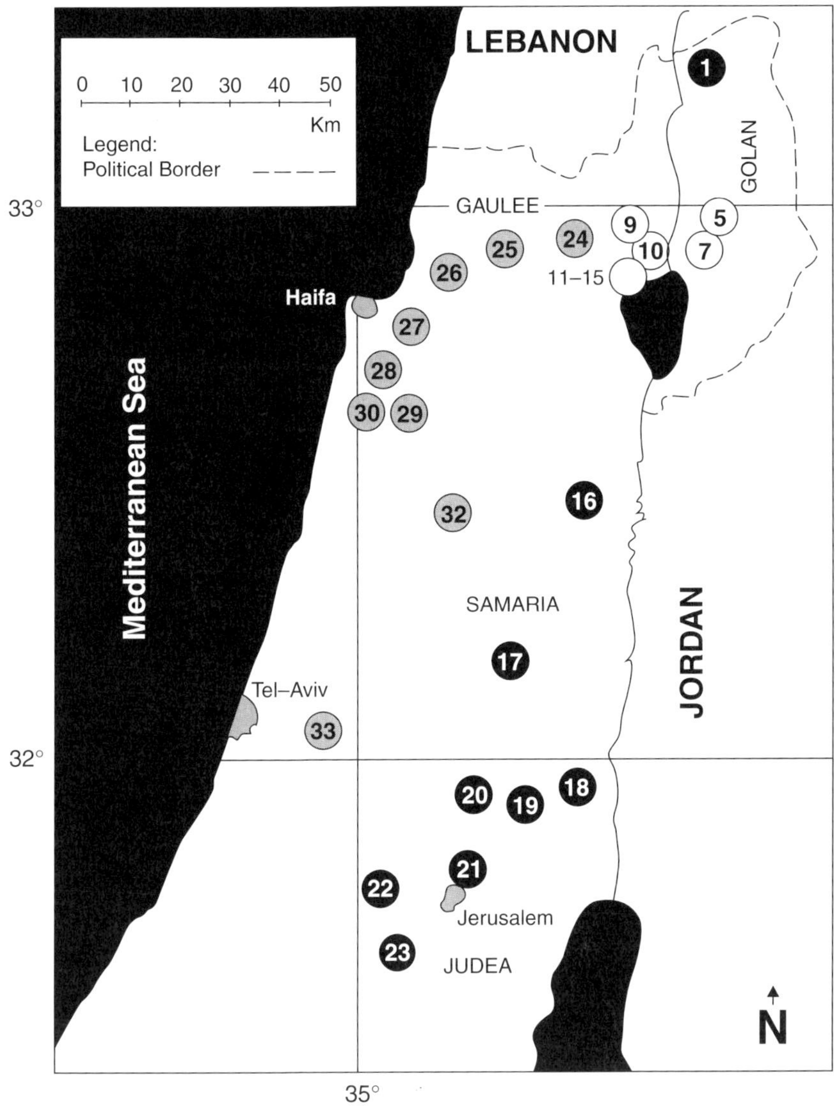

Fig. 6.10. Geographic distribution of 27 populations tested for photosynthetic capacity from three ecogeographic regions of wild emmer wheat, *T. dicoccoides*, in Israel. For names of numbered populations see Table 6.22. Central populations are indicated by *white circles*, mesic (western) populations by *gray circles*, and xeric (northern and eastern) populations by *black circles*. (Nevo et al. 1991a)

populations of *T. dicoccoides* having high photosynthetic capacity can be identified by climatic factors and isozyme markers. Out of 107 genotypes tested in discriminant analysis, 97 (91%) were correctly classified into their respective sets (Fig. 6.11). The best photosynthetic performance was that of

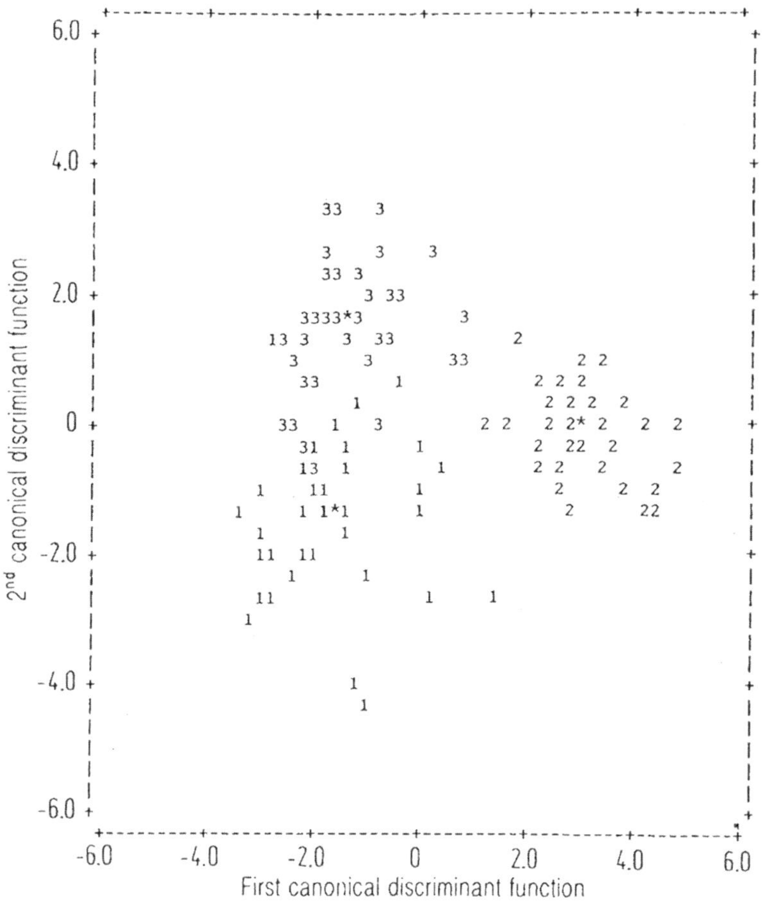

Fig. 6.11. Discriminant analysis between three ecogeographic regions of *T. dicoccoides* based on nine photosynthetic characteristics. *1* Central populations, *2* xeric margins, *3* mesic margins, * group centroid. (Nevo et al. 1991a)

the xeric-warm-marginal populations 16–23 (in black circles in Fig. 6.10, and group 2 in Fig. 6.11). These xeric-marginal populations were selected for high photosynthetic yield together with other traits, e.g., high levels of storage proteins (Nevo et al. 1986a) and high rDNA diversity (Flavell et al. 1986), to shorten the life cycle in the steppic environment to avoid the early hot summer. Identification by genetic mapping can facilitate the maximization of conservation, in situ or ex situ (Nevo 1998a), and the utilization of these photosynthetic genetic resources for improvement of hexaploid wheat (*T. aestivum*).

6.11 Diurnal Rhythms of mRNAs for the Chlorophyll a/b Binding Protein in Wild Emmer Wheat and Wild Barley in the Fertile Crescent

Biological rhythms in nature are entrained to a 24-hr (circadian) period by environmental cues (*Zeitgeber*). Circadian clocks are abundant in both plants and animals. In plants, morphological, cellular, biochemical, physiological, and photosynthetic oscillations have been observed (Britz et al. 1987); however, the molecular basis of these rhythms is still largely unknown. The mRNAs encoding the chlorophyll-a/b binding (CAB) proteins of the light-harvesting system were monitored in wild emmer wheat, *T. dicoccoides*, and wild barley, *H. spontaneum*, the progenitors, respectively, of all cultivated wheats and barley (Nevo et al. 1993b). Significantly different mRNA levels were detected at different time points during the 24-hr cycle, with generally low levels around sunrise, sunset, and midnight and maximum levels around noon (Fig. 6.12). These results indicate that there is diurnal control of CAB gene expression in these ancient species.

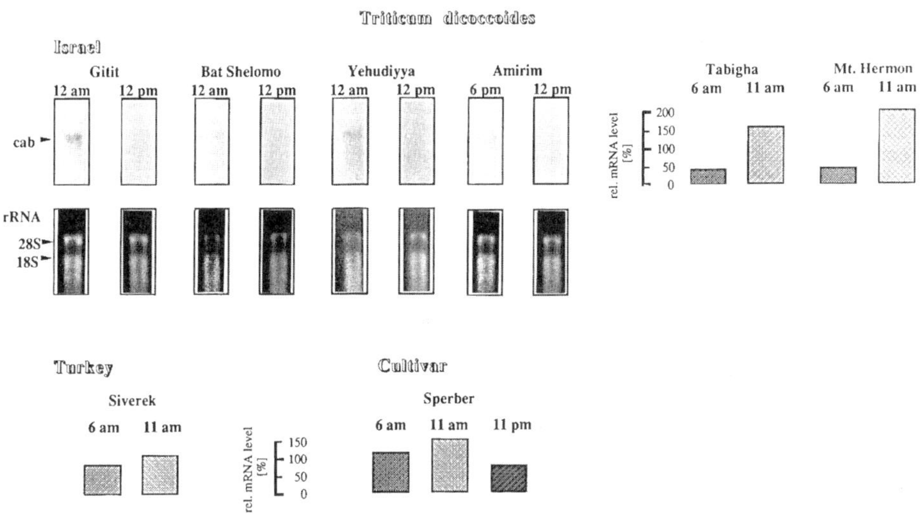

Fig. 6.12. Diurnal fluctuations of chlorophyll-a/b binding protein steady-state mRNA levels in wild wheat genotypes. Total RNA extracts from Gitit, Bat-Shelomo, Yehudiyya, and Amirim (Israel) genotypes were analyzed on formaldehyde-agarose gels. The autoradiograms of respective Northern blots are presented and the respective bands representing the *cab* transcripts are indicated. Quantitative analysis of *cab* probe hybridization is presented for the Tabigha, Mt. Hermon (Israel), and Siverek (Turkey) genotyeps. The cultivar *T. durum* cv. 'Sperber' was used as a control. For further details, see Fig. 1 in Nevo et al. (1993c)

Our study (Nevo et al. 1993c) suggested that a circadian clock is involved in *cab* gene expression in wild cereals, indicating that this regulatory mechanism represents an ancient process. Genetic polymorphism in diurnal *cab* gene expression is suggestive though not yet conclusive. Remarkably, however, such polymorphisms in photosynthetic performance were found both within and between populations of wild emmer wheat in Israel (Carver and Nevo 1990). These genetic polymorphisms are predictable by ecology and by isozyme markers (Nevo et al. 1991a), indicating adaptive photosynthetic patterns generated by natural selection.

6.12 Crop Improvement

6.12.1 Genetically Engineered Plants for Crop Improvement

Through recombinant DNA technology, molecular biology can now probe much deeper into the genome, giving detailed information on the structure and function of genes. It also provides plant breeders with numerous potential genetic markers (reviewed in Beckman and Soller 1986; Korol et al. 1994) for isolating genes of interest from genomic libraries. Dramatic progress has been made in the development of gene-transfer systems for higher plants (Austin et al. 1986; Plucknett et al. 1987; Schell 1987; Gasser and Fraley 1989; Murray 1991; Grieson 1991; Shewry 1992; Kung and Wu 1993; Prakash and Pierik 1993; Henry and Rolands 1994).

The ability to introduce foreign genes into plant cells and to regenerate viable, fertile plants has facilitated an explosion of our understanding of plant biology and provided an unparalleled opportunity to modify and improve crop plants. Genetic engineering of plants offers significant potential for crop improvement. Transgenic plants have already been produced in many species, including herbaceous dicots (e.g., tomato, potato, tobacco), monocots (e.g., asparagus, rice, corn, orchard grass, rye, and wheat) and woody dicots (e.g., poplar, walnut, apple) among others. In particular, virus-, herbicide- and insect-resistant plants have been produced, and other genes conferring traits of agronomic importance will be transferred in the future. A report on genetic modification (GM) technology and world agriculture was prepared in July 2000 by seven Academies of Sciences. The report discusses the need for GM technology in agriculture and provides examples that would benefit world agriculture. It deals with transgenic plants (TG) and human health and safety, TG and the environment, and the problem of intellectual property. Many millions of hectares of commercially produced transgenic crops, such as soybean, cotton, tobacco, potato, and maize, have been grown in several countries including the United States (28.7 million ha in 1999), Canada (4 million ha), China (0.3 million ha), and Argentina (6.7 million ha) (James 1999). The report of the seven academies deals with the extensive debate about the potential benefits and risks that may result from the use of such crops. The report offers scientific perspectives on the ongoing public debate regarding the potential role of GM technology in world agriculture. Predictions of biodiversity response to genetically herbicide-tolerant crops have been discussed by Watkinson et al. (2000).

Finally, the genomics era has sequenced complete viral, plastid, mitochondrial, bacterial, and some eukaryotic genomes (yeast, *Caenorhabditis elegans*, *Drosophila*, and humans). So far, genomics has only scratched the surface and has instead revealed how little is known about prokaryotic and eukaryotic species. In the next few years, by 2004, more than 100 projects for sequencing microbial genomes should be completed, providing the scientific community with information on more than 300,000 predicted genes of prokaryotes (Fraser et al. 2000), and more will be discovered about eukaryotic genes. A significant number of these genes will be novel and of unknown function. These novel genes represent exciting new opportunities for future research and potential sources of biological resources to be explored and exploited. The benefits of comparative genomics in understanding biochemical diversity, virulence and pathogenesis, and the evolution of species has been unequivocally demonstrated, and the usefulness of comparative techniques will improve as more genomes become available. One of the major challenges is to develop techniques for assessing the function of novel genes on a large scale and integrating information on how genes and proteins interact at the cellular level to create and maintain a living organism, i.e. to create a bridge between genomics and proteomics. It is not unreasonable to expect that, by expanding our understanding of microbial biology and biodiversity, great strides can be made in the diagnosis and treatment of infectious diseases, while discoveries in the microbial and fungal world could be applied to improving agricultural and industrial processes.

Most important for cereals, mapping of the rice genome (400 Mb) has been completed, and mapping of the sorghum genome (700 Mb) will be completed soon. Likewise, the expressed sequence tags (EST) will dramatize mapping by finding new functionally important genes and microarray novel technologies will bridge structural and functional genomics and revolutionize agriculture (see also I.T.M.I. 2000).

6.12.2 Potential and Actual Genetic Resources of *Triticum dicoccoides* and Future Wheat Breeding

Diverse studies in Israel and the Near East Fertile Crescent of wild emmer, *T. dicoccoides*, the progenitor of most tetraploid and hexaploid wheats, have revealed rich genetic resources applicable in wheat improvement due to their diverse single and multilocus adaptations to stressful abiotic (climatic, soil) and biotic (pathogens, parasites) environments. Most of the discovered resources are yet untapped, and it takes many years to explore, evaluate, and especially to utilize these genetic resources in improving wheat. The genetic resources have been described by several authors (Zohary 1970b; Feldman 1979; Lange and Jochemsen 1992; Grama et al. 1983; Nevo 1983a, 1989a, 1995, 2001b) and were previously reviewed. They include genotypic variation of agronomic traits such as (1) germination, biomass, earliness, nitrogen content, and yield (Nevo et al. 1984b; Grama et al. 1987a,b); (2) amino acid composition (Nevo and Beiles 1992); (3) protein content (Gerechter-Amitai and Grama 1977; Avivi 1979; Grama et al. 1983; Nevo et al. 1986a; Levy 1987; Levy and Feldman 1985, 1987); storage proteins (HMW glutenins) (Nevo and Payne 1987; Levy and Feldman 1988; Joppa and Cantrell 1990); (4) disease resistances, including powdery mildew (Moseman et al. 1984; Nevo et al. 1985; Reader and Miller 1991; Rong et al. 1998; Liu et al. 2001), leaf rust (Moseman et al. 1985; The et al. 1993), stripe rust (Nevo et al. 1986b; The et al. 1993; Sun et al. 1997; Fahima et al. 1998; Peng et al. 1999, 2000a–c, 2001), stem rust (Nevo et al. 1991b; The et al. 1993), and soil-borne wheat mosaic virus (Hunger et al. 1992); (5) high photosynthetic yield (Carver and Nevo 1990; Nevo et al. 1991a); (6) salt tolerance (Nevo et al. 1992a, 1993c); (7) herbicide resistance (Snape et al. 1991a,b; Nevo et al. 1992c, 1993a; Krugman et al. 1997); (8) amylases (Nevo et al. 1993a). This is only a preliminary list of the vast potential genetic resources existing in *T. dicoccoides* that remain to be exploited. For additional data, see Nevo (1983a, 1989a, 1995, 2001b) and elsewhere (e.g., Feldman 1979; Grama et al. 1983; Lange and Jochemsen 1992). The foregoing indicates that wild emmer harbors rich genetic resources for wheat improvement. QTL mapping and beneficial cryptic, agronomically important alleles, have now been extensively described in *T. dicoccoides* (Peng et al. 2000a,b, 2001). These will be elaborated on in the last part of this monograph.

The Near East in general, and Israel in particular (Nevo 1986a), are the centers of origin and diversity of wild emmer, where it developed wide genetic adaptations against multiple pathogens and diverse ecological stresses, contributing to its fitness in heterogeneous and stressful ecologies. Genetic diversity is transferable from the wild to the cultivated gene pool, so genes of wild emmer are of primary importance for future wheat improvement. The rich genetic resources and diversity of wild emmer, described here and previously, are geographically structured (Nevo et al. 1982; Nevo and Beiles 1989) and are predictable by ecology and molecular (allozyme and DNA) markers (Nevo 1987, 2001a). Extensive genome mapping and QTL mapping (Peng et al. 2000a,b, 2001; see below) provide rich genetic resources for wheat improvement. Consequently, exploration (optimizing sampling strategies) of in situ and ex situ conservation along with utilization programs (involving classical and modern technologies) should maximize the contribution of wild emmer to wheat improvement.

6.12.3
Utilization of *Triticum dicoccoides* in Breeding

Genes for improving wheat could be extracted from land races of wheat (e.g., Payne et al. 1984) and from wheat's wild relatives (both *Triticum* and *Aegilops* species). Among the potential donors, *T. dicoccoides* assumes a very important and unique position due to its direct ancestry of bread wheats and its rich and largely adaptive genetic diversity. This idea was first suggested by Aaronsohn (1913) and later elaborated on by many authors (for the historical perspective, see Feldman 1977). Its initial (1965) and further utilization have been described by Grama et al. (1983) and were reviewed earlier in this volume and by Peng et al. (2000a,b, 2001), and Nevo (2001b).

There are many, currently ongoing programs in Israel, Europe, and the United States utilizing genes of *T. dicoccoides* for wheat improvement, primarily genes coding for resistance to powdery mildew and yellow rust, high protein content, and improved baking quality. Clearly, these are long-term projects. Cultivar releases based on commercial lines of *T. dicoccoides* genes are likely to appear in the near future. With conventional methods, it takes 10–15 years between the cross and release of the variety. With *T. dicoccoides*, at least three crosses with bread wheat are necessary, and the most promising material at some stations has been crossed four times (Groenewegen and Van Silfhout 1988; Reader and Miller 1991). The likely positive impact of *T. dicoccoides* on the genetic improvement of wheat will certainly be appreciable in the future (Nevo 1983a, 1989a, 1995, 2001b; Peng 2000; Peng et al. 2000a–c, 2001).

The utilization of wild emmer in breeding began with the finding, in 1964, that Israeli populations harbor a valuable source of stripe-rust resistance (Gerechter-Amitai and Stubbs 1970). In 1967, it was discovered that grain protein content in several collections of wild emmer ranges from 20 to 24%, considerably higher than that found in cultivated wheat (Gerechter-Amitai and Grama 1977). Further extended research revealed protein values in wild emmer ranging from 13.9 to 28.9% (Grama et al. 1983). Similar results were reported by Avivi (1978, 1979), Nevo et al. (1986a), and Levy (1987). The breeding program at the Volcani Center of the Agricultural Research Organization (ARO) utilizes a selection of wild emmer, G-25, that is resistant to stripe rust and has a protein content of 20.5% and a kernel weight of 31.5 mg. Thus, it combines several desirable traits in a donor. The hexaploid wild emmer wheat derivatives developed at the ARO were grown under New Zealand conditions and tested for a number of milling and baking quality parameters and for gliadin and HMW glutenin protein composition. Superior qualities of protein content, dough strength, and baking quality were found to be related to the presence of specific protein components (Grama et al. 1987b). Likewise, four wild emmer derivatives grown in New Zealand proved superior to standard bread wheats in total content of grain nitrogen and flour protein after foliar urea sprays at heading (Grama et al. 1987a). The novel yellow-rust resistance genes, *YRH52* and *Yr15*, are under extensive study at the University of Haifa's Institute of Evolution (Sun et al. 1997; Fahima et al. 1998; Chagué et al. 1999; Peng et al. 1999, 2000a–c, 2001). *Yr15* is not currently used in commercial cultivars, but some lines possessing the gene are present in breeder populations in Australia (e.g., Avocet 5*6/*Yr15*, Fig. 6.13; Aroona 3*/*Yr15*; Spear *3/*Yr15*) and in the Netherlands (McIntosh et al. 1995).

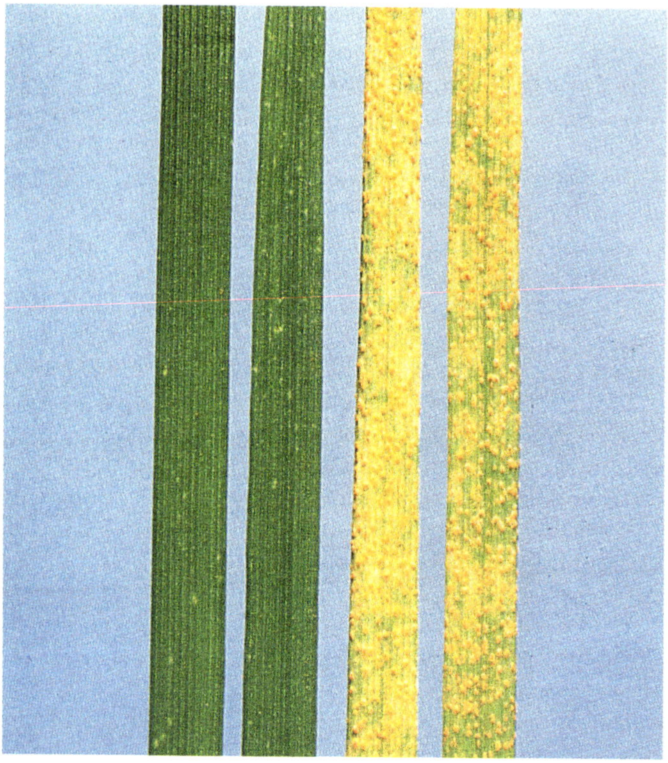

Fig. 6.13. Transfer of the stripe-rust resistance gene *Yr15* from *T. dicoccoides* into Australian cultivated bread wheat. Seedling leaves of (*left to right*): V763. 254 (resistant); resistant and susceptible F2 segregates from Avocet 5*6/V763. 254; and Avocet (susceptible); infected with strain 110 E143 A+. (McIntosh et al. 1995)

Kushnir and Halloran (1984) produced high kernel weight and high grain protein hexaploid wheat lines by crossing a high-kernel-weight and high-protein *T. dicoccoides* accession with Chinese Spring and its two homologous pairing mutants, ph_{1b} and ph_2. Those pentaploids were then crossed with two cultivars to produce stable, hexaploid high-kernel-weight and high-grain-protein lines. Extensive work on transferring genes for high protein content from *T. dicoccoides* to cultivated wheat is currently underway in several labs (e.g., M. Feldman, Weizmann Institute; L.R. Joppa, USDA, Fargo, ND; and J. Dubcovsky, University of California, Davis, CA).

Grain protein content (GPC) of wheat is important for improved nutritional value and is also one of the major factors affecting bread making and pasta quality (Finney et al. 1987; Dick and Youngs 1988). In spite of its importance, progress in breeding for high GPC has been slow and difficult. The first limitation is that genetic variation for protein content is small compared with variation due to differences in growing environments. The second limitation is that there is a strong negative correlation between GPC and grain yield; cultivars with high GPC tend to be low yielders. There are, however, exceptional genotypes that combine excellent yield potential and high GPC, probably by a more efficient relocation of nitrogen from senescing tissues to grain or by a more efficient uptake of nitrate and ammonia from the soil (Blackman and Payne 1987).

A promising source of high GPC was detected years ago in a survey of wild populations of tetraploid *Triticum turgidum* var. *dicoccoides* (accession FA15-3 from Israel; Avivi 1978). Substitution lines of the chromosomes of this *T. dicoccoides* accession in the cultivated durum cultivar Langdon (*Triticum turgidum* var. *durum*) showed that a gene for high protein content was present on chromosome 6B (Joppa and Cantrell 1990). Recombinant substitution lines from a cross between the LDN(DIC 6B) substitution line and Langdon were developed by Joppa et al. (1997). These lines were used to map a QTL for GPC on the proximal region of the short arm of chromosome 6B, which accounted for 66% of the variation in GPC present in that particular cross (Joppa et al. 1997).

Other researchers crossed the same *T. dicoccoides* 6B substitution line with high-yielding *durum* cultivars from North Dakota (Chee et al. 1998) and Canada (Kovacs et al. 1998) to introgress the high GPC gene. Results from the North Dakota recombinant lines showed a single-gene effect and no interaction between the increase in GPC content and the genetic background or growing environments. The effect of this gene was independent of protein quality, plant height, heading date, and yield (Chee et al. 1998). Results from the Canadian recombinant lines showed a positive effect on protein and quality (Kovacs et al. 1998). The incorporation of the high-GPC gene from *T. dicoccoides* into the Canadian lines resulted in increased protein content and a positive effect on pasta cooking quality. The high-GPC gene from *T. dicoccoides* increased protein content, sedimentation volume, wet gluten yield, and mixograph cooking quality (Kovacs et al. 1998). Kovacs et al. (1998) concluded that this gene would be a valuable resource to increase the protein level for durum wheat breeding programs.

The high GPC gene from *T. dicoccoides* has also been transferred to hexaploid wheat. Hexaploid cultivar Glupro was developed by Dr. R. Frohberg from a three-way cross between two bread-wheat cultivars and the same *T. dicoccoides* accession used to develop the substitution line LDN(DIC6B). Recent results have shown that the same *T. dicoccoides* chromosome segment is present in the high-GPC hexaploid variety Glupro (Mesfin et al. 1999; Khan et al. 2000). The hexaploid cultivar Glupro carries a distal segment of chromosome GBL and a proximal segment of 6BS from *T. dicoccoides* (Fig. 6.14; Khan et al. 2000). The positive effect of this segment on GPC in a bread-wheat genetic setting was recently demonstrated in double haploid lines (Humphreys et al. 1998) and recombinant inbred lines (Mesfin et al. 1999) developed from crosses between Glupro and high-yielding bread-wheat cultivars from Canada and the USA.

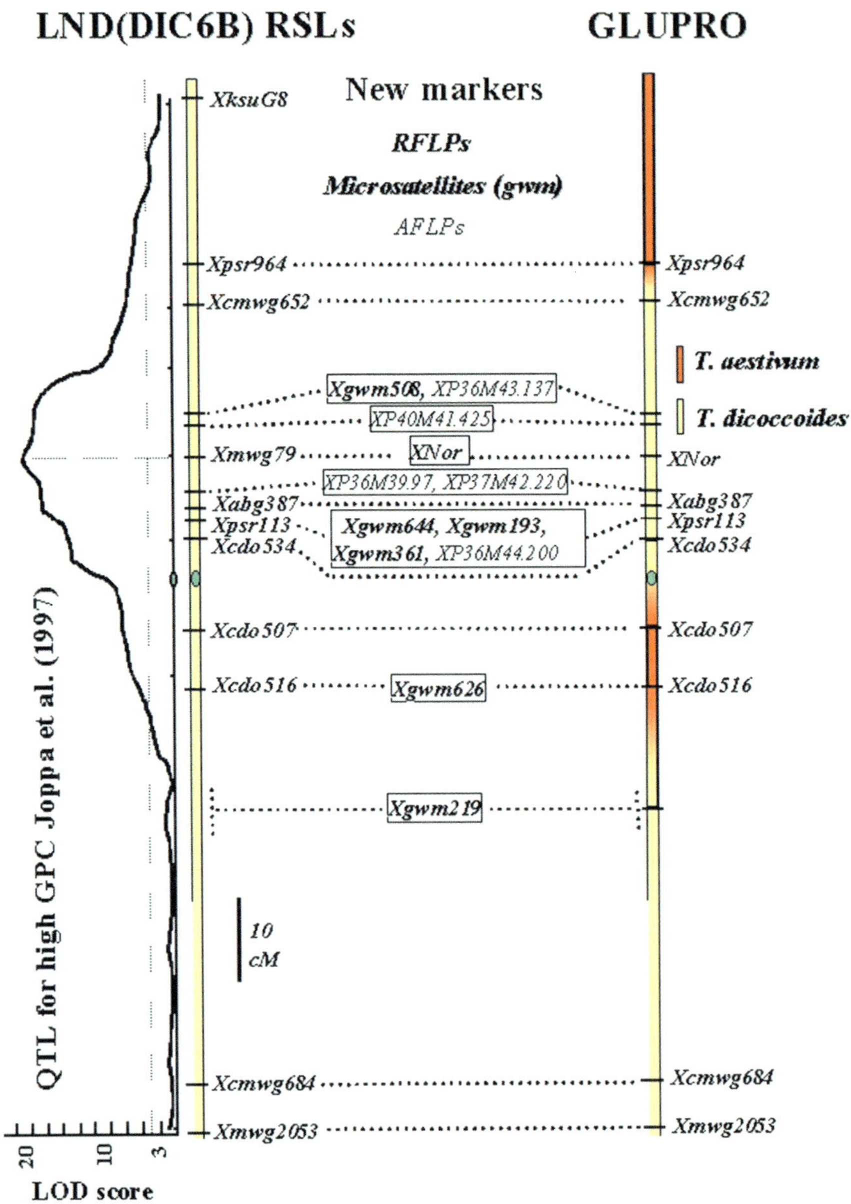

Fig. 6.14. *Left* Chromosome 6B map developed by Joppa et al. (1997) with quantitative trait loci (QTL) analysis for high grain protein content (GPC). *Center* Molecular markers added by Khan et al. (2000). *Right* Inferred distribution of *T. dicoccoides* and bread-wheat chromosome segments in the hexaploid cultivar Glupro based on the same molecular markers. RFLP markers indicated as present in *T. dicoccoides* 6B substitution line and Glupro are absent in Columbus and Len, the hexaploid progenitors of Glupro. Genetic distances are inferred from the map on the *left*. Distances between markers are proportional to genetic distances. Centromeres are indicated with a *circle*. (Khan et al. 2000)

A major gene for resistance to powdery mildew, caused by *Erysiphe graminis*, was introduced into the bread wheat *T. aestivum* from *T. dicoccoides* (Reader and Miller 1991). The donor was crossed to bread wheat and the pentaploid progeny was then self-pollinated. Plants having a near-stable hexaploid chromosome complement were selected in the F_3 progeny and top-crossing and back-crossing of these to a second wheat cultivar to improve the phenotype was carried out. Monosomic analysis of early back-cross lines showed the transferred gene to be located on chromosome 4A. The gene has been designated Pm16. Another powdery mildew resistance gene has been located on chromosome 2BS (Rong et al. 1998). Recently, a new powdery mildew resistance gene, *MlC20*, derived from wild-emmer, has been described (see end of section 6.9.1). The gene is located on chromosome 5BS and was physically mapped to the gene-rich regions of fragment length 0.41–0.43 using Chinese Spring deletion lines (Liu et al. 2001). The prospects of discovering additional novel genes are promising.

Part IV

Genome Organization and Genetic Mapping

Chapter 7

Genome Structure of *Triticum dicoccoides*

7.1 Molecular Genetic Maps

7.1.1 Introduction: Molecular Markers as a Tool for Genetic Mapping

Molecular linkage maps of many plant species have been obtained recently and utilized in quantitative trait analysis, gene tagging, genome organization, and evolutionary studies, as well as in improved selection activities (Tanksley and Orton 1983; Paterson et al. 1991; O'Brien 1993; Whitkus et al. 1994; Blanco et al. 1998). Molecular markers can provide a spectacular improvement in the efficiency and sophistication of plant breeding. It is now generally accepted that markers represent the most significant advance in breeding technology that has occurred in the last few decades and they are currently the most important application of molecular biology to plant breeding (Langridge and Chalmers 1998). Especially useful are microsatellite markers. Compared with RFLPs, microsatellites are PCR-based, easily handled, cheaper to use, suitable for automation, and highly reproducible, and therefore can be used on a larger scale. Molecular biology studies of bread wheat, *Triticum aestivum* (L.) Thell., have received much attention and several RFLP-based maps either for groups of homologous chromosomes (Chao et al. 1989; Devos et al. 1992, 1993; Nelson et al. 1995; Van Deynze et al. 1995; Gill et al. 1996a,b) or for all the chromosome groups (Liu and Tsunewaki 1991; Anderson et al. 1992; Gale et al. 1995; Mingeot and Jacquemin 1999) have been reported. By contrast, tetraploid durum wheat has received little research attention and an RFLP-based linkage map has been published only recently (Blanco et al. 1998). Genetic research on important agronomic and quality traits in *T. dicoccoides* has lagged far behind other wheat species and cereals in general. A detailed genetic map was urgently required as a basis for the effective utilization of the rich genetic potential of *T. dicoccoides* in the breeding programs of cultivated wheat, including bread and durum wheats, described elsewhere in this Volume and in Nevo (1983a, 1986a, 1989a, 1995, and 2001b).

Recently, Röder et al. (1998) have developed a set of hexaploid wheat microsatellite markers and constructed a molecular map consisting of 279 microsatellite loci amplified by 230 primer sets. The efficiency of these primer-sets in analysis of *T. dicoccoides* genomic DNA was previously demonstrated for the purposes of mapping (Chagué et al. 1999; Peng et al. 1999, 2001) and population genetics (Fahima et al. 1998; Li et al. 2000a–d, 2001a–d). Seventy-nine of these microsatellite markers were integrated into the above-mentioned RFLP linkage map in durum wheat (Korzun et al. 1999).

AFLP (amplified fragment length polymorphism) technology is based on the PCR amplification of selected restriction fragments of a total genomic DNA, and separation of labeled amplified products by denaturing polyacrylamide gel electrophoresis. A great advantage of the AFLP approach is that it allows simultaneous identification of a large number of amplification products (Van Eck et al. 1995; Vos et al. 1995). Compared with RFLP or other PCR-based marker systems, AFLP is fast, reliable, and cost-effective. It may be a good supplement to other marker systems (such as RFLP) in species such as wheat that present a low-level of polymorphism. Furthermore, it may be especially useful for high-density mapping in regions containing genes of interest (Ma and Lapitan 1998). In wheat, a complete AFLP linkage map has not yet been reported in the literature.

The main objectives of our mapping efforts (Peng et al. 2000b) were, therefore, to construct a DNA molecular genetic map of *T. dicoccoides* using microsatellite markers, complemented by AFLP and RAPD markers, to fill the gaps. This should contribute to an understanding of genome evolution in the wheat progenitor and to accelerating the utilization of this important genetic resource in wheat improvement programs. Our aim was also to use the generated information to characterize the marker-related anatomy of the *T. dicoccoides* genome and the segregation and recombination patterns upon crossing with its domesticated descendant *T. durum* (cultivar Langdon). This could contribute to tracking the evolution of wheat domestication.

7.1.2
Marker Polymorphism Between *Triticum durum* and *Triticum dicoccoides*

Hexaploid bread wheat shows relatively low levels of polymorphism for RFLP loci, most likely as a result of its narrow genetic base (Chao et al. 1989). Usually <10% of all RFLP loci are polymorphic in an intraspecific context (Röder et al. 1998). However, in the genetically distant cross of hexaploid bread wheat Opata 85 × W7984, 72% of the RFLP probes (Mingeot and Jacquemin 1999) and 80% of the microsatellite primers (Röder et al. 1998) exhibited polymorphism between the parents. In another relatively distant cross of hexaploid wheat, *T. aestivum* × *T. spelta*, 64% of the RFLP probes were polymorphic (Messmer et al. 1999). In a distant cross of tetraploid wheat, Messapia × MG4343, which is similar to our mapping population, 70.1% of RFLP probes (Blanco et al. 1998) and 84.4% of microsatellites (Korzun et al. 1999) detected polymorphism between the parents. Therefore, for distant crosses of both hexaploid and tetraploid wheats, about 70% of RFLP probes and 80% of microsatellite primer sets can detect polymorphism.

The level of polymorphism of microsatellite primer sets in the present mapping population was even higher (>90%) (Peng et al. 2000b). Furthermore, all of the 33 AFLP primer combinations screened detected polymorphism between the parents; 30.4% of overall AFLP loci were polymorphic, and 49.2% of the RAPD primers detected polymorphism. This clearly shows that highly significant genetic differentiation occurred between *T. durum* and *T. dicoccoides* during their evolutionary divergence caused by domestication, even though they share the same genomes and display no genetic or reproductive barriers. The two parental lines, *T. dicoccoides* accession Hermon H52 and *T. durum* cultivar Langdon, were also highly polymorphic in morphological and agronomic traits, i.e., plant height, growth duration, leaf shape, grain characteristic, spike fragility, stripe-rust resistance, etc. (Peng 2000b). The high level of DNA polymorphism between the two parental lines makes it possible to map these domestication-related and agronomically important traits (see Sect. 8.3), and, it is hoped, to map many candidate genes in wheat.

7.1.3 Construction of the Molecular Genetic Map

When 14 previously mapped microsatellite markers (Röder et al. 1998) were used as anchors of the 14 chromosomes of *T. dicoccoides* (one for each chromosome), 543 marker loci, the *YrH52* stripe-rust resistance gene, and the *Ws* gene conferring spike glaucousness were assigned to the 14 chromosomes. Among the 544 loci, 78.5% were assigned with logarithm odds (LOD) scores over 10, 20.2% with LODs ranging from 3.1 to 10, and only seven LODs (1.3%) ranging from 2.5 to 3.0. Therefore, the chromosomal assignment of the loci and the construction of the 14 linkage groups proved highly reliable (Lincoln et al. 1992).

Among the 315 assigned AFLP fragments, 20 bands (10 pairs) amplified by the same primer combinations were closely linked in repulsion phase and were converted into ten codominant AFLP markers. Thus, a total of 549 loci, comprising 545 PCR-based marker loci, two RFLP loci, and two genes, *YrH52* and *Ws*, were included in the mapping analysis. By means of the three- and multipoint analysis of MAPMAKER 3.0b (using Kosambi mapping function) at a minimum LOD of 3.0, two molecular genetic maps for each of the 14 chromosomes of wild emmer wheat, *T. dicoccoides*, were constructed (Fig. 7.1), each using codominant markers and dominant markers in coupling phase. The dominant markers of the Hermon (H) and Langdon (L) maps were based on PCR bands amplified from genomic segments of *T. dicoccoides* (the parental genotype derived from the Hermon population) and *T. durum* (Langdon), respectively. The H and L maps span a distance of 3,169 and 3,180 cM, respectively. The mean intermarker interval lengths of H and L maps were 9.9 and 10.7 cM, respectively (Table 7.1).

It is estimated that the genome sizes of tetraploid wheat (*T. dicoccoides* and *T. turgidum*) and hexaploid wheat (*T. aestivum*) are ~1.2×10^{10} and

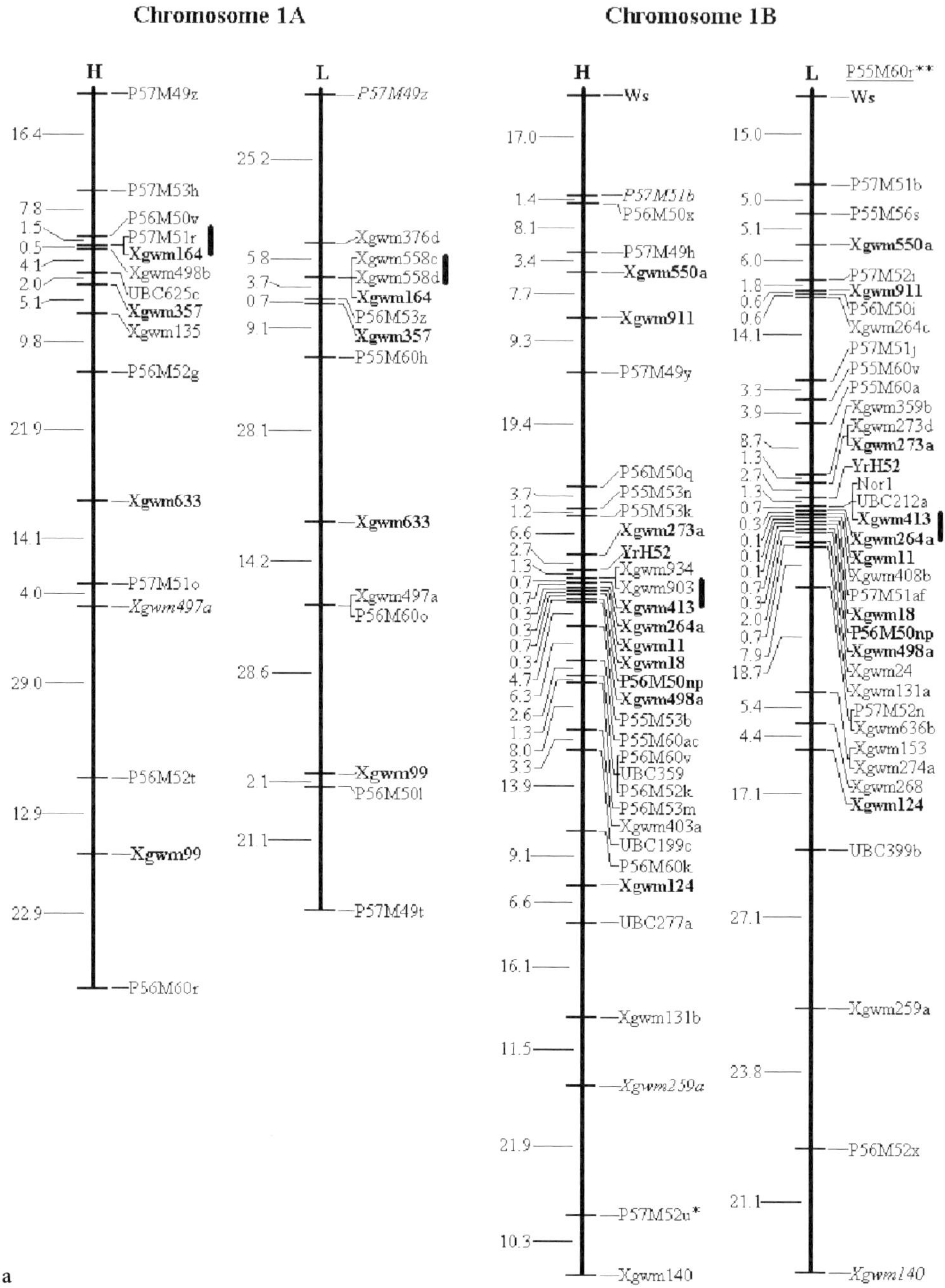

Fig. 7.1a–g. Molecular genetic maps of wild emmer wheat, *Triticum dicoccoides*; chromosome homologous groups from 1–7 (**a–g**, respectively). The centromeric location is cited from Röder et al. (1998) and it is approximately indicated by the *short vertical bar*. Short arms of chromosomes are at the *top*. Genetic distances on the *left* of the maps are given in centi-Morgans (cM) estimated with the Kosambi (1944) mapping function. The H map is essentially constructed using codominant markers and the *T. dicoccoides* H52-derived dominant markers. The L map is based on codominant markers and the *T. durum* Langdon-derived dominant markers. The *underlined* markers with relatively lower LOD scores (2.0–2.5) are placed in the most probable intervals. The markers in *italic* are from the opposite map and in repulsion phase. The markers in *boldface* are codominant. * and ** indicate that the markers significantly and highly significantly deviate from the expected segregation ratio (3:1 or 1:2:1), respectively. *Ws* on chromosome 1B is a gene for spike glaucousness. (Peng et al. 2000b)

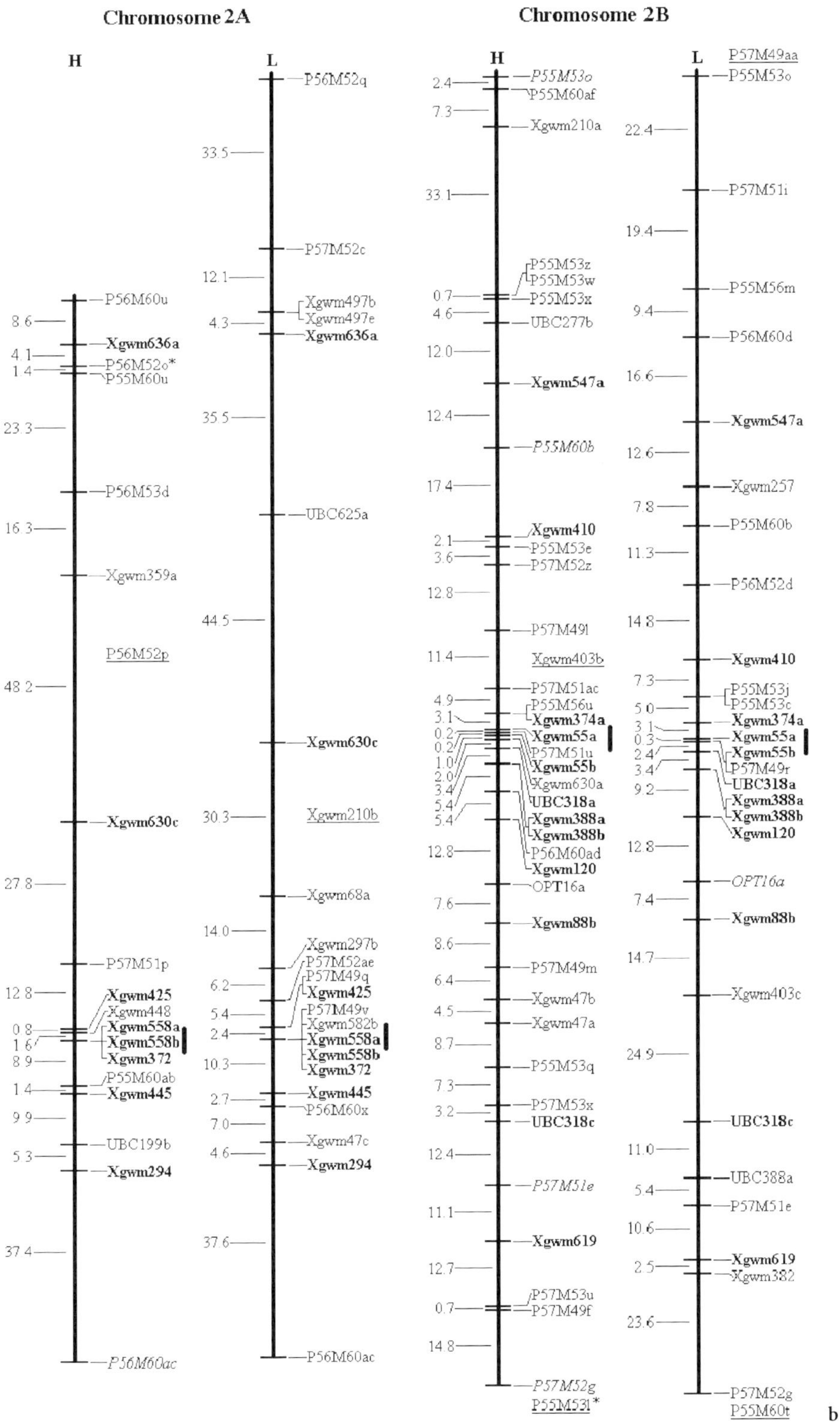

Fig. 7.1. *Continued* (Figs. 7.1c–g see pages 246–250)

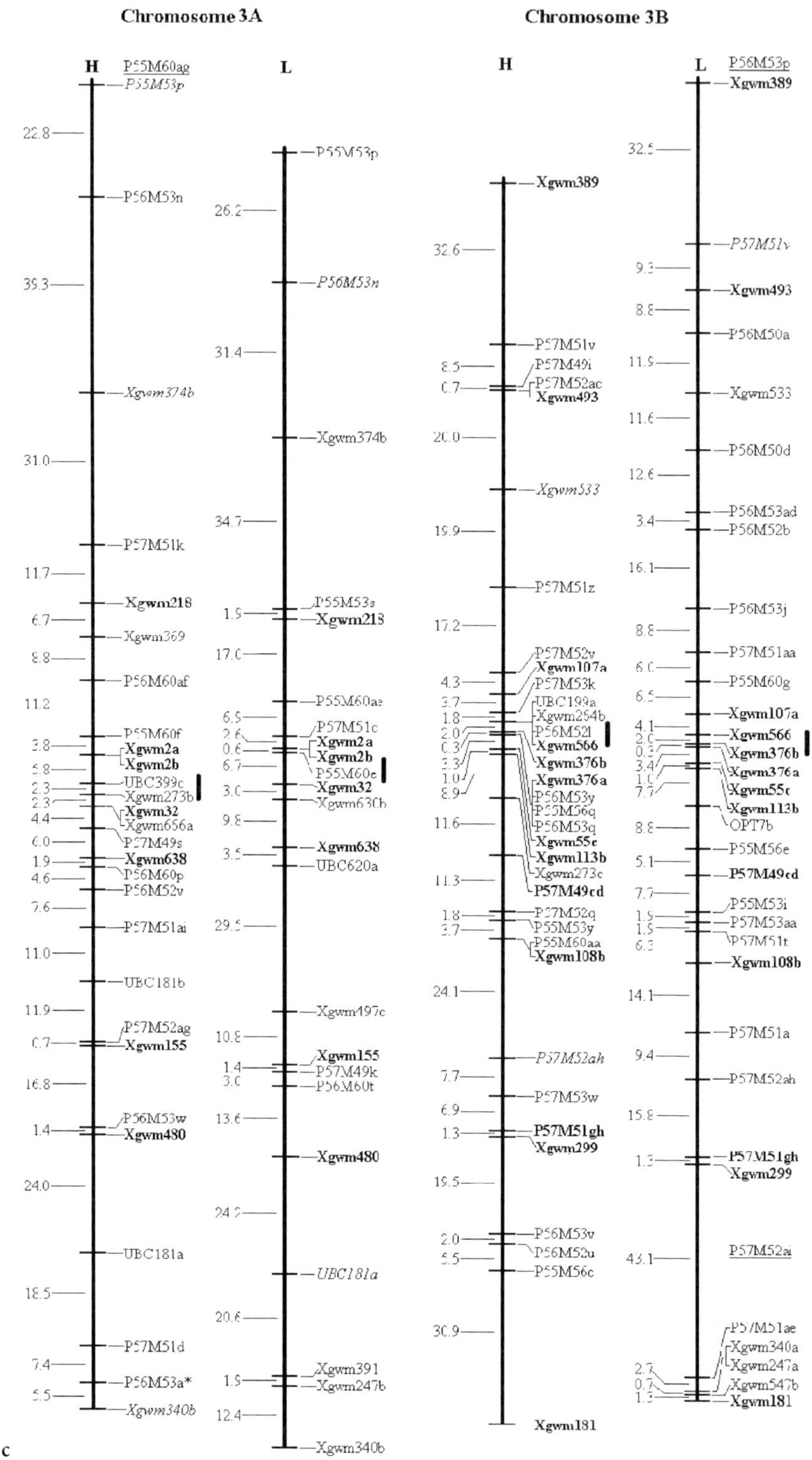

Fig. 7.1. *Continued*

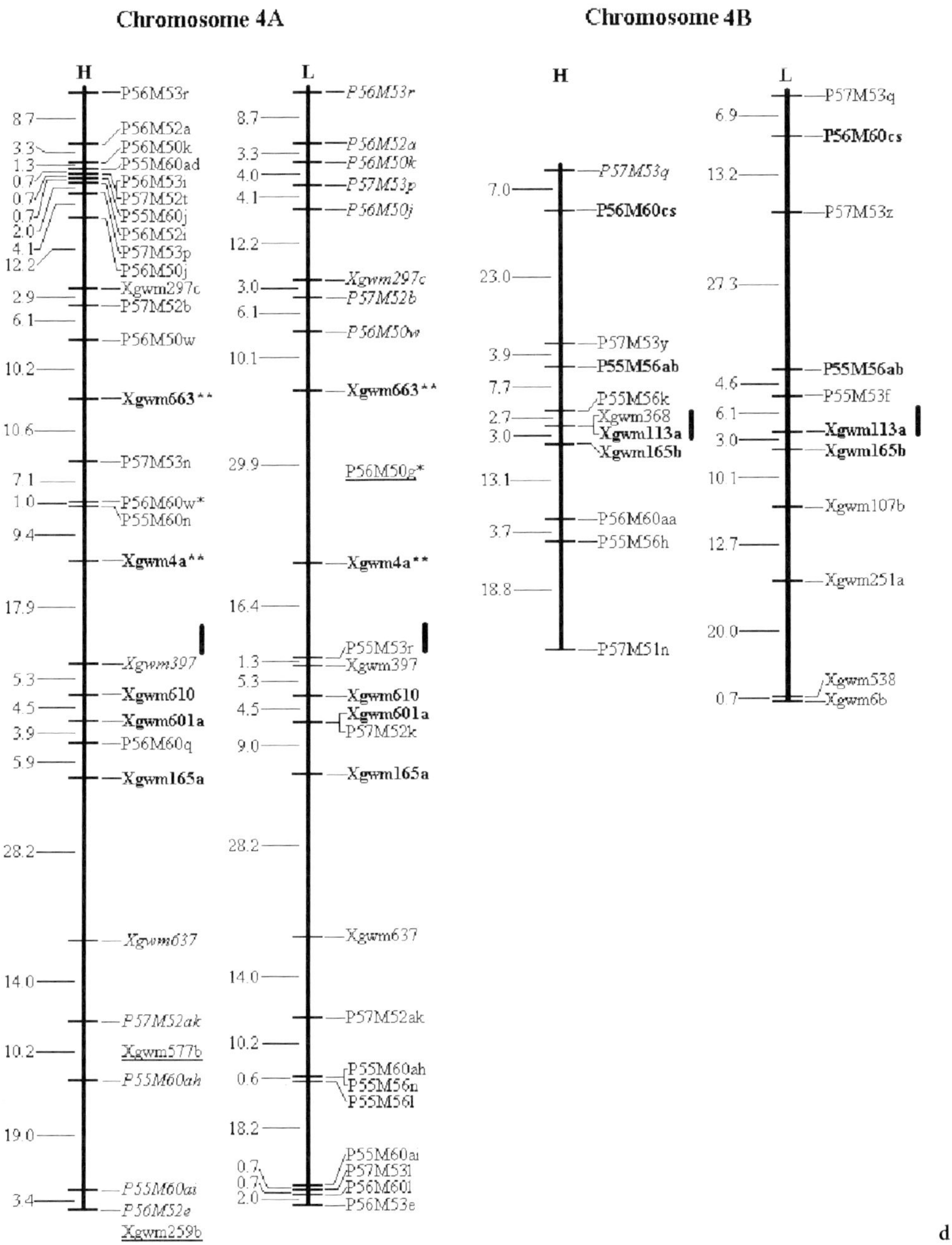

Fig. 7.1. *Continued*

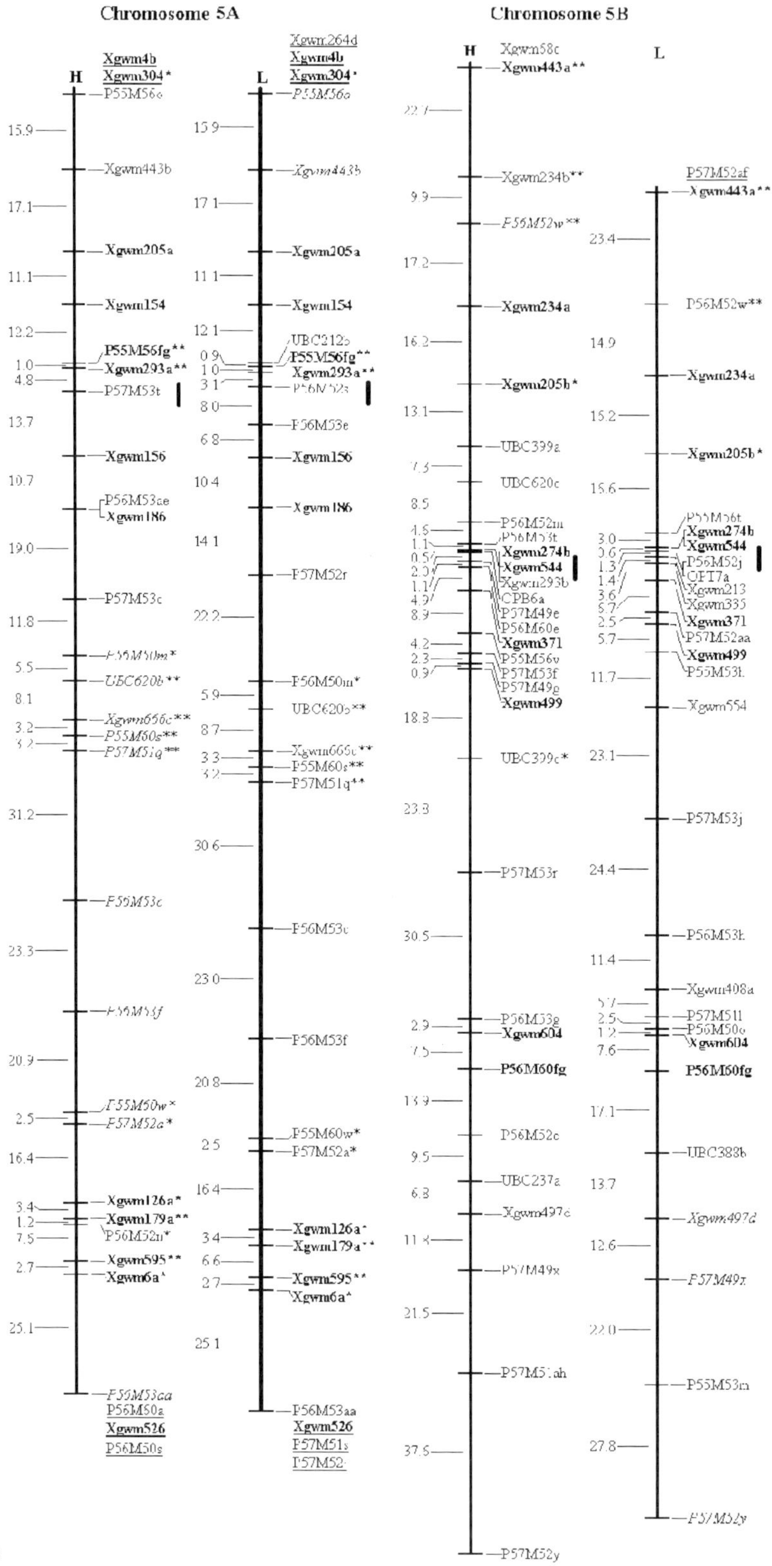

e

Fig. 7.1. *Continued*

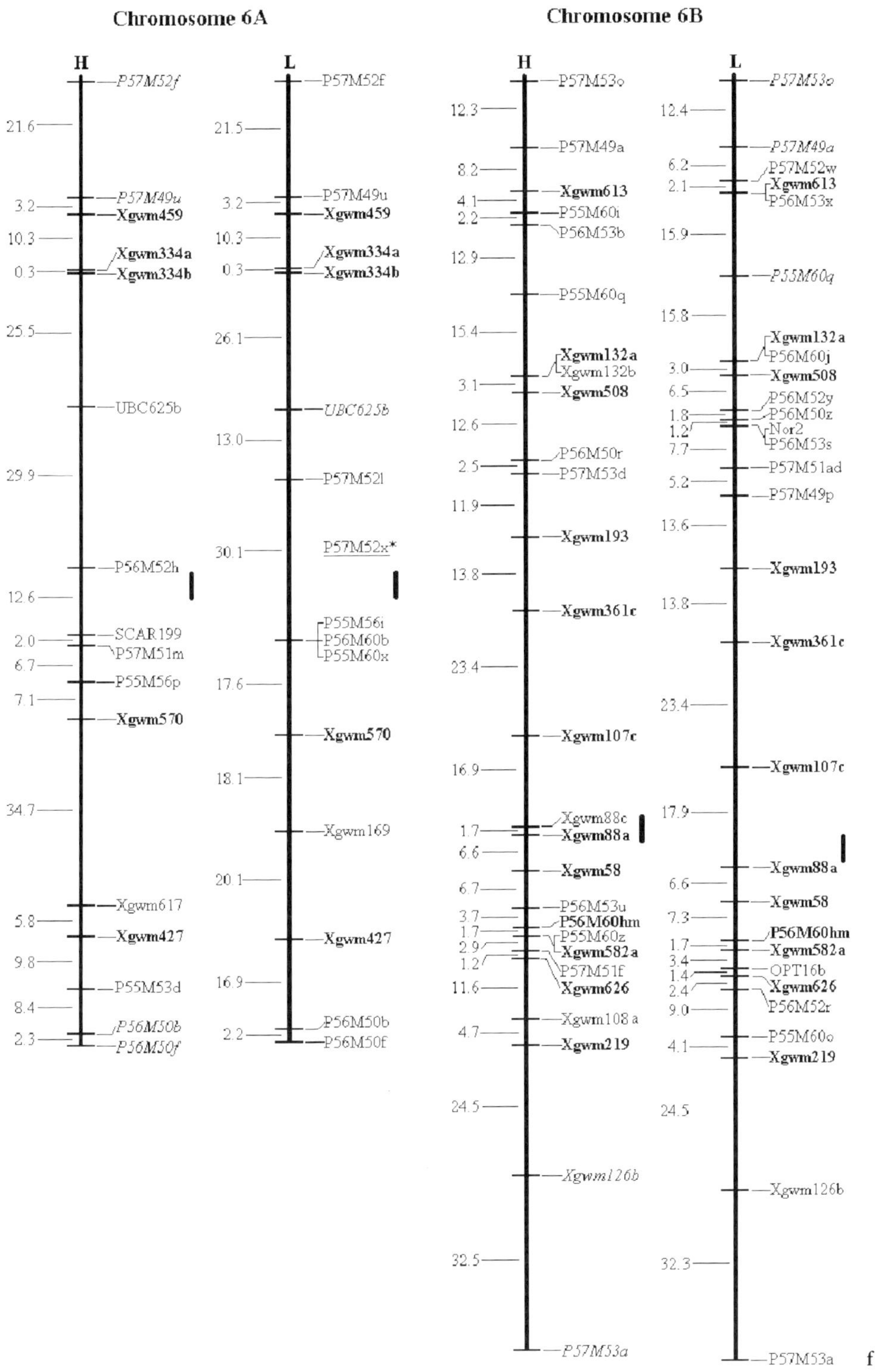

Fig. 7.1. *Continued*

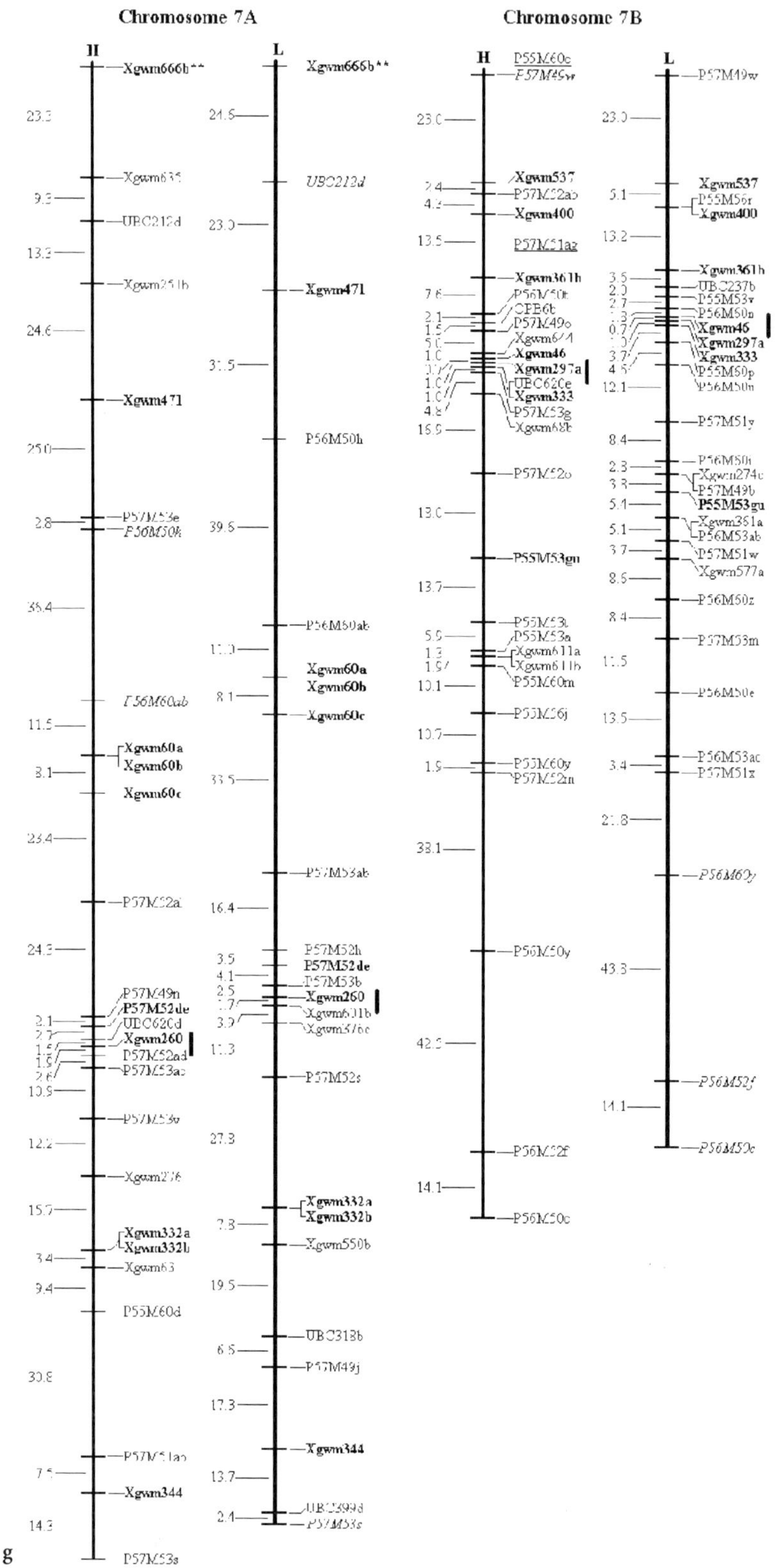

Fig. 7.1. *Continued*

Table 7.1. Distribution of molecular markers among genomes and chromosomes. (Peng et al. 2000b)

Chromosome	NM	ML		MIL		CV	
		H	L	H	L	H	L
A genome							
1A	24	152.0	138.6	10.86	13.86	82.78	79.87
2A	33	207.8	250.3	13.85	16.69	103.75	89.57
3A	40	267.4	261.7	10.70	12.46	91.40	89.25
4A	37	193.3	192.5	7.43	8.75	89.77	95.20
5A	38	271.6	274.9	11.32	11.00	74.20	76.73
6A	22	180.2	179.4	12.01	14.95	89.84	63.28
7A	37	317.0	309.8	13.21	14.75	76.76	77.76
Subtotal	231 (42.1%)	1589.3	1607.2				
WM				11.11	12.76	86.05	82.71
B genome							
1B	58	200.4	200.2	6.13	6.23	95.11	125.20
2B	53	256.2	257.9	7.54	10.76	87.00	64.50
3B	54	250.5	266.1	9.63	8.58	99.17	106.53
4B	17	82.9	104.6	9.21	10.46	80.89	78.30
5B	48	310.0	276.7	11.07	11.07	86.09	76.96
6B	39	237.1	239.2	9.88	9.57	83.70	86.83
7B	49	243.0	227.8	9.72	8.76	115.12	106.28
Subtotal	318 (57.9%)	1580.1	1572.5				
WM				8.88	9.12	93.27	95.01
Total	549	3169.4	3179.7				
GWM				9.88	10.66	90.03	89.79
Statistical test[a]							
Effect	χ^2			F value			
Genome	13.38**	0.01	0.05	5.29*	10.41**	0.88	1.34
HG	17.62**	3.52	3.84	0.55	0.32	0.47	0.74

NM Number of markers, *ML* map length (cM), *MIL* mean interval length (cM), CV coefficient of variation (%) of interval length, *WM* weighted mean, *GWM* grand weighted mean, *HG* homologous group. H and L correspond to, respectively, the H map, consisting of mostly dominant markers in which the dominant allele is derived from *T. dicoccoides* accession H52, and the L map, consisting of mostly dominant markers in which the dominant allele is derived from *T. durum* cultivar Ldn. (see Fig. 7.1)

[a] The LOG linear analysis was used for NM. ANOVA was used for ML, MIL, and CV. *Significant at the 5% level, **significant at the 1% level

1.7×10^{10} bp, respectively (Bennett et al. 1998). If the average chromosome length is assumed to be 200 cM (Messmer et al. 1999), the total map size of hexaploid wheat is estimated as 4,200 cM. The sizes of published maps for the populations Chinese Spring × *T. spelta* (Liu and Tsuneawaki 1991), Chinese Spring × Synthetic (Gale et al. 1995), Forno × Oberkulmer (Messmer et al. 1999), and W7984 × Opata 85 (Faris et al. 2000) in hexaploid wheat were 1,801,

2,575, 2469, and 3,700 cM, respectively. These maps covered 43, 61, 59, and 88% of the entire genome of hexaploid wheat, respectively.

The total map size of tetraploid wheat (*T. dicoccoides* and *T. turgidum*) could be estimated as 2,800 cM if the size of hexaploid wheat is 4,200 cM (Messmer et al. 1999). In tetraploid durum wheat, the total length of the published RFLP-based genetic map was 1,352 cM (Blanco et al. 1998) and was extended to 2,035 cM after integration of 79 microsatellite markers (Korzun et al. 1999). These two maps cover 48 and 73% of the entire genome of tetraploid wheat, respectively. In the present study, the total lengths for the H and L molecular maps in tetraploid *T. dicoccoides* were 3,169 and 3,180 cM, respectively (Table 7.1), covering, therefore, the entire genome of *T. dicoccoides*. The present molecular map (Fig. 7.1) thus has the largest relative coverage in wheat and covers the entire genome. Another advantage of our map is its relatively large population size (150 F_2 individuals, compared with 65–120 in other studies; Gale et al. 1995; Blanco et al. 1998; Röder et al. 1998). This is important for mapping quantitative trait loci (QTL) that confer the evolutionarily and agronomically important traits on this species.

In the present map, AFLP markers terminate the molecular maps at one or both ends of all chromosomes except 1B and 3B (Fig. 7.1). We assume that the high coverage of the present molecular maps resulted from the integration of AFLP markers derived from the restriction enzyme *PstI*. AFLP is more effective with wheat genomic DNA if *Pst*I rather than *Eco*RI is used to generate DNA fragments. This is probably because of the high G + C content of *Pst*I recognition sites relative to *Eco*RI, resulting in preferential targeting of low-copy/gene-rich regions of the genome. Consequently, use of *Pst*I rather than *Eco*RI allows better genome coverage and less clustering of marker loci (Langridge and Chalmers 1998). In maize, the total map length also significantly increased due to the addition of *Pst*I AFLP markers to the telomeric regions, where RFLP markers were poorly represented (Castiglioni et al. 1999).

7.1.4 Distribution of Molecular Markers Among the Genomes and Chromosomes

7.1.4.1 Clustering of Marker Loci

Marker distribution along the present molecular maps of tetraploid wheat was far from uniform, with clusters of tightly linked loci and regions with a low density of markers. Most of the 14 chromosomes had marker clusters in the centromeric regions of the genetic maps (Fig. 7.1). A statistical test proved the significance of the marker clustering on the B genome, but not on the A genome, despite clusters on a few A chromosomes (Fig. 7.2) Similar deviation from randomness is observed in each genome, and the overall test (A+B in Fig. 7.2) is also significant. This feature has been observed in most mapping

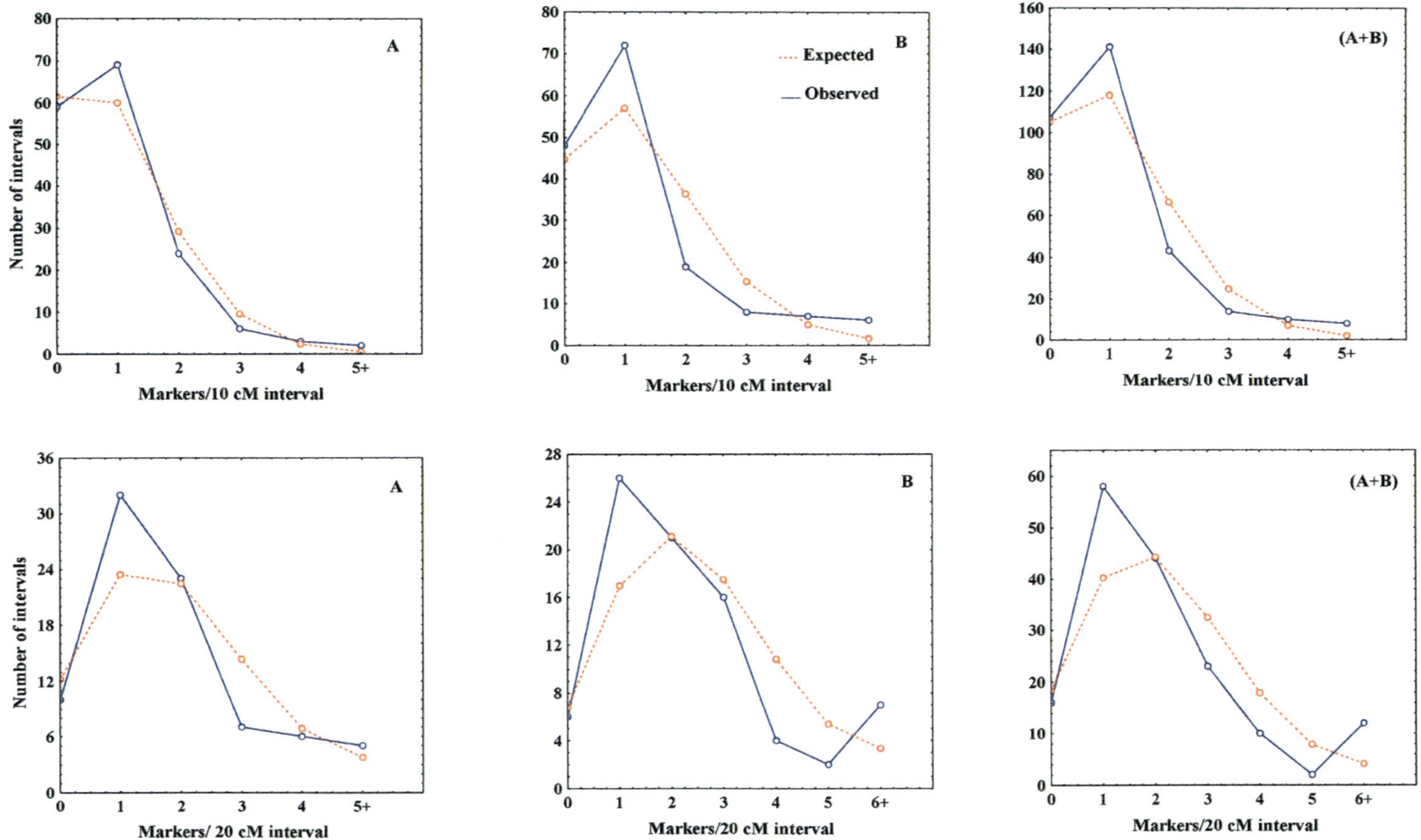

Fig. 7.2. Distribution of chromosome segments with respect to number of markers per interval. *Dashed line* indicates the expected (Poisson) distribution calculated by assuming no clustering; *solid line* indicates the observed distribution showing a deficit of moderately populated intervals and an excess of low populated (or empty) intervals and densely populated intervals. *A*, *B*, and (*A* + *B*) represent A, B, and the entire genome, respectively. (Peng et al. 2000b)

studies of wheat (Chao et al. 1989; Gale et al. 1995; Nelson et al. 1995; Blanco et al. 1998; Röder et al. 1998; Messmer et al. 1999) and in many other organisms (Korol et al. 1994). It was attributed to nonrandom crossover distribution along chromosomes, i.e., reduced recombination in the proximal and centromeric regions and hotspots of recombination in median or distal chromosome regions (Korol et al. 1994). The marker cluster involving Nor1 in the near-centromeric region of 1B in the present study is physically located on the distal part of 1BS (Gill et al. 1996b). Likewise, the cluster involving Xgwm371 and Xgwm499 in the near-centromeric region of 5B is physically located in the middle of 5BL (Faris et al. 2000). This discrepancy between physical and genetic distances is explained by the variation in the density of recombination events per physical unit length along chromosomes. In wheat, this intra-chromosomal variation may be more than 100-fold (Lukaszewski and Curtis 1993; Gill et al. 1996a).

7.1.4.2 Nonrandom Distribution of AFLP Markers

More than half (59.7%) of the AFLP markers in the present study were mapped to the B genome (Peng et al. 2000b). This difference between the A and B genome is highly significant ($p < 0.01$) and implies nonrandom distribution of AFLP markers among A and B genomes of tetraploid wheat. The same trend has also been observed for microsatellites (Röder et al. 1998; Li et al. 2000b, 2001d) and RFLP markers (Liu and Tsunewaki 1991). This may mainly reflect the higher amount of polymorphism in the B genome than in the A genome (Röder et al. 1998; Li et al. 2001c). AFLP markers derived from each primer combination are not evenly distributed among the chromosomes. However, the effects of homologous group and primer combinations were not significant (Peng et al. 2000b). This is consistent with the result in maize, showing the random distribution of *Pst*I-based AFLP markers among the ten chromosomes (Castiglioni et al. 1999). AFLP markers derived from *Pst*I preferentially target low-copy regions of the genome (Langridge and Chalmers 1998). Most AFLP markers correspond to unique positions on the genome (Vos et al. 1995). *Pst*I-based AFLP markers are preferentially located in the hypomethylated noncentromeric regions associated with genes in the maize genome (Castiglioni et al. 1999). Therefore, the nonrandom distribution of the *Pst*I-AFLP markers may mirror the genetic differentiation of structure and function among genomes and chromosomes between *T. dicoccoides* and *T. durum* during the domestication process.

The degree of genetic variation obviously varies among chromosomes and genomes of *T. dicoccoides* and *T. durum*. The differentiation between *T. dicoccoides* and *T. durum* is mainly attributed to the B genome, as indicated by AFLP markers. A higher differentiation has also been detected on the B genome than on the A genome between *T. aestivum* and *T. spelta* by use of RFLP markers (Liu and Tsunewaki 1991) and between *T. aestivum* and *T. durum* by use of microsatellites (Röder et al. 1998). For six of the seven homologous groups, the genetic differentiation between *T. dicoccoides* and *T. durum* for B chromosomes, as revealed by AFLP markers, is significantly higher than for the corresponding A chromosomes. Our study also indicates that the degree of genetic differentiation of chromosome 4A between *T. dicoccoides* and *T. durum* significantly exceeds that of chromosome 4B. This is probably due to the evolutionary translocation events involving chromosome 4A (Liu et al. 1992; Devos et al. 1995a; Gale et al. 1995).

7.1.5
Conserved Order of Microsatellite Loci and Structural Changes of Chromosomes

The chromosomal locations and map orders for most of the microsatellite loci in our present molecular map of *T. dicoccoides* (Fig. 7.1) are the same as those in *T. aestivum* (Röder et al. 1998) and in *T. durum* (Korzun et al. 1999). This result indicates that macrochromosomal organization is highly conserved among various species in the genus *Triticum*. It also further corroborates that wheat microsatellite markers are mainly wheat-genome-specific, and that the microsatellite markers developed in hexaploid wheat can be easily transferred to other species in the genus (Röder et al. 1998; Korzun et al. 1999).

However, some changes of chromosomal location and map order were also observed for a few microsatellite markers in the present map in comparison with the maps of Röder et al. (1998) and Korzun et al. (1999). The inconsistency of map order may be explained by the presence of additional loci and structural changes of chromosomes in the wheat genome (Korzun et al. 1999), and limitations caused by the population size (70 and 65 recombinant inbred lines in Röder et al. (1998) and Korzun et al. (1999), respectively, compared to 150 F_2 individuals in the present study). For example, the microsatellite locus Xgwm550 was mapped to chromosome 1BL by Korzun et al. (1999) but to 1BS by Röder et al. (1998) and in the present study (Fig. 7.1). Thus, the Xgwm550 locus in Korzun et al. (1999) is definitely different from that in Röder et al. (1998) and in the present study (Fig. 7.1). Xgwm498 was mapped to chromosome 1A and 1B by Korzun et al. (1999) and Röder et al. (1998), respectively. So, there should be two loci for this microsatellite and these two loci were mapped to chromosome 1A and 1B, respectively, in our present study (Fig. 7.1). In wheat, paracentric and pericentric inversions have been repeatedly observed on chromosome 4A (Liu et al. 1992; Devos et al. 1995a). Faris et al. (2000) observed an inversion within a small segment at the proximal region of the *T. dicoccoides* chromosome 5B map involving several closely linked markers. Thus, the changes of the map order of Xgwm397, Xgwn610, Xgwm601a, and Xgwm165a, genetically near the centromere of chromosome 4A, are possibly due to the inversion of this chromosome segment, but for this we should assume that the inversion is present in both *T. dicoccoides* and *T. durum*.

Joppa et al. (1995) found that a high proportion (70%) of *T. dicoccoides* genotypes had translocations, as was evident in the study of Kawahara and

Nevo (1996). Compared with the map of Röder et al. (1998), the changes of chromosomal location of Xgwm582b, Xgwm344, and Xgwm526 may be explained by translocations of chromosomal segments between 1B and 2 A, 7B and 7 A, and 2B and 5 A, respectively. Non-homologous translocations have been repeatedly reported in hexaploid wheat (Liu et al. 1992; Devos et al. 1993) and tetraploid wheat (Blanco et al. 1998). Adaptive translocation may be one of the forces maintaining the high genetic diversity of *T. dicoccoides* under the diverse natural environments in Israel (Joppa et al. 1995; Kawahara and Nevo 1996). Translocation frequencies (TFs) of various populations were correlated with environmental variables, primarily with water availability and humidity, and possibly also with soil type. In general, TF was higher in peripheral populations in ecologically heterogeneous frontiers of species distribution than in central populations located in the catchment area of the upper Jordan Valley (Joppa et al. 1995).

7.2 Segregation and Recombination upon Crossing *Triticum dicoccoides* with *Triticum durum*

7.2.1 Distorted Monogenic Ratios

A total of 573 marker loci including 203 microsatellites, 326 AFLPs, 39 RAPDs, two RFLPs, SCAR, and two genes (*YrH52* and *Ws*) were scored in the described study (Peng et al. 2000b). The χ^2 test was used to check whether the marker segregation in F_2 fitted the Mendelian model (1:2:1 for codominant and 3:1 for dominant markers). In total, 34 (5.9%) loci showed significant ($p < 0.05$) deviation from the expected ratio, which is close to the number one would expect to get by chance in a test including many loci, even if no real distortion exists at all. However, significant deviation at $p < 0.01$ was manifested by 19 (19/571 = 3.3%) out of these 34 loci, i.e., three times that expected by chance (deviating loci are marked * and ** in Fig. 7.1, for $p < 0.05$ and $p < 0.01$, respectively).

Besides the threefold excess of highly distorted loci over the level expected by chance, the genomic distribution of these loci indicates the reality of segregation distortion in our material. Indeed, of the 34 segregation-distorted markers, 31 were assigned to specific chromosomes. Out of these 31 markers, one each was assigned to chromosome 2A, 2B, 3A, 6A, and 7A, two were assigned to chromosome 1B, four (12.9%) to chromosome 4A, five (16.1%) to chromosome 5B, and 15 (48.4%) to chromosome 5A (Table 7.2 and Fig. 7.1). Actually, the segregation-distorted markers clustered in some specific regions on chromosome 5A (three clusters along the chromosome), and one cluster each was present on chromosomes 5BS and 4AS (Fig. 7.1). Distortion favored *T. durum* alleles for 23 segregation-distorted loci, including all the loci on chromosomes 5A and 5B, P57M52u on chromosome 1B, P56M50g on chromosome 4A, and P57M52x on chromosome 6A. Only for five loci, P55M60r

Table 7.2. Distribution of segregation-distorted markers and the corresponding allele frequencies. (Peng et al. 2000b)

Locus	Type	CHR	Percentage[a]	Number of genotypes					Allele frequency	
				HH	H-	HL	L-	LL	H	L
P55M60r	Dominant	1B	6.45	66			84		0.66	0.34
P57M52u	Dominant	1B			101			49	0.43	0.57
P56M52o	Dominant	2A	3.23		123			26	0.58	0.42
P55M531	Dominant	2B	3.23		124			25	0.59	0.41
P56M53a	Dominant	3A	3.23		123			23	0.60	0.40
Xgwm4a	Codominant	4A	12.90	29		96		24	0.44	0.40
Xgwm663	Codominant	4A		33		94		23	0.47	0.39
P56M50g	Dominant	4A		25			122		0.41	0.59
P56M60w	Dominant	4A			122			26	0.58	0.42
Xgwm6a	Codominant	5A	48.39	22		83		45	0.38	0.55
Xgwm126a	Codominant	5A		25		74		51	0.41	0.58
Xgwm179a	Codominant	5A		17		80		52	0.34	0.59
Xgwm293a	Codominant	5A		29		67		54	0.44	0.60
Xgwm304	Codominant	5A		23		78		49	0.39	0.57
Xgwm595	Codominant	5A		22		77		51	0.38	0.58
Xgwm666c	Dominant	5A		19			129		0.36	0.64
P55M56fg	Codominant	5A		27		63		55	0.43	0.62
P55M60s	Dominant	5A		21			129		0.37	0.63
P55M60w	Dominant	5A		9			141		0.24	0.76
P56M50m	Dominant	5A		23			124		0.40	0.60
P56M52n	Dominant	5A			98			51	0.41	0.59
P57M51q	Dominant	5A		23			127		0.39	0.61
P57M52a	Dominant	5A		10			140		0.26	0.74
UBC620b	Dominant	5A		21			129		0.37	0.63
Xgwm205b	Codominant	5B	16.13	24		77		49	0.40	0.57
Xgwm234b	Dominant	5B			94			56	0.39	0.61
Xgwm443a	Codominant	5B		17		83		49	0.34	0.57
P56M52w	Dominant	5B		19			130		0.36	0.64
UBC399e	Dominant	5B			100			49	0.43	0.57
P57M52x	Dominant	6A	3.23	26			124		0.42	0.58
Xgwm666b	Codominant	7A	3.23	29		100		19	0.44	0.36

CHR Chromosome, *H* allele from *T. dicoccoides* accession H52, *L* allele from *T. durum* cultivar Ldn., *HH* homozygous genotype for the alleles from the *T. dicoccoides* accession H52, *LL* homozygous genotype for the alleles from *T. durum* cultivar Ldn., *HL* heterozygous genotype from the two alleles from H52 and Ldn., *H-* = HH + HL, *L-* = LL + HL. The frequencies were estimated based on the ratio of homozygotic marker genotypes in the F_2 generation

[a] Percentage accounted for the total segregation-distorted markers

on chromosome 1B, P56M52o on 2A, P55M531 on 2B, P56M53a on 3A, and P56M60w on 4A, was the bias toward the *T. dicoccoides* alleles (Table 7.2). Thus, the distorted loci showed a significant ($\chi^2 = 11.57, p < 0.001$) bias toward a deficit of alleles of the pollen parent of the hybrid.

Distorted segregation of molecular markers has been observed in mapping populations derived from intra- and interspecific hybrids in many plants, including potato (Gebhart et al. 1989), corn (Gardiner et al. 1993), rice (Causse et al. 1994), common bean (Vallejos et al. 1992), and barley (Heun et al. 1991). In wheat, this phenomenon has also been repeatedly reported (Liu and Tsunewaki 1991; Devos et al. 1993; Nelson et al. 1995; Blanco et al. 1998; Messmer et al. 1999). In the present study, the portion of segregation-distorted marker loci (5.9%) was relatively low compared with the previous studies in wheat (Liu and Tsunewaki 1991; Blanco et al. 1998; Messmer et al. 1999). The possible causes for segregation deviation of molecular markers are chromosomal rearrangement (Tanksley 1984) and gametic or zygotic selection (Nakagahra 1986). Further indications of causes of deviation, such as presence of lethals, meiotic drive, and chromosomal rearrangements, could be obtained from the analysis of additional populations with segregation distortion (Blanco et al. 1998).

Whether the cause of distorted segregation is gametic competition or zygotic selection can be discerned by estimation of the frequencies of two alleles for a codominant locus. The obtained data indicate that the gametes carrying *T. durum* alleles have greater vigor and higher competition ability than those with *T. dicoccoides* alleles. Therefore, the cause for distorted segregation on chromosomes 5A and 5B would be gametic competition (Peng et al. 2000b). It is noteworthy that *T. dicoccoides* was a pollen parent of the F_1 hybrid, hence the cytoplasm was provided by *T. durum*. We could speculate that the indicated regions with presumably gametic selection carry loci involved in nuclear-cytoplasmic interaction. Faris et al. (1998) reported a segregation distortion locus within the homologous region of chromosome 5D in *Aegilops tauschii*. They also postulated a possible homeo-allele of the distortion factor on chromosome 5B causing skewed segregation of two markers (Faris et al. 2000). It seems that a common locus confers segregation distortion on group 5 chromosomes (5A, 5B, and 5D) in wheat through differentially affecting the vigor of gametes.

7.2.2
Quasi-Linkage: Nonrandom Segregation of Nonhomologous Chromosomes

With independent segregation of loci from nonhomologous chromosomes, the frequency of parental and nonparental combinations of corresponding alleles is 50%. In 1953, Michie and Wallace observed a departure from random segregation of markers on nonhomologous chromosomes in crosses between different strains of house mice (*Mus musculus*) (Michie 1953; Wallace 1953). Such a departure from independent assortment of unlinked genes was termed "quasi-linkage." This phenomenon was observed both in plants and animals, especially in interspecific hybrids (cotton, tomato, maize, *Coix*, mice, mule, dipteran insect *Sciara coprophila*, human) (for review see Sapre and Deshpande 1987; Korol et al. 1994).

High genome coverage by molecular markers makes our F_2 mapping population (*T. durum* × T. *dicoccoides*) especially suitable for testing quasi-linkage. Several pairs of nonhomologues manifested a highly significant deviation of recombination of their markers from the expected $r = 0.5$ level. These included either an excess or a deficit of recombinant genotypes, with the following highest deviating r values: $r(1A,2A) = 0.701$, $r(1A,6A) = 0.778$, $r(2A,2B) = 0.768$, $r(2A,7B) = 0.271$, $r(5A,7A) = 0.723$, and $r(3B,5B) = 0.726$.

However, the high pairwise significance of observed deviations may be an artifact caused by multiple comparisons. Indeed, if only one marker per chromosome is considered, a total number of possible nonhomologous pairs (and tests) for a genome with $N = 14$ is $N(N-1)/2 = 91$. The situation is even more complicated when each chromosome is represented by multiple markers. To cope with these complications in evaluating the genomewise significance of the observed numerous manifestations of quasi-linkage, we employed a computing-intensive permutation test. Its result indicates that quasi-linkage in our case could be declared "significant" ($p < 0.02$). Spurred by these results, we conducted a few other tests of quasi-linkage, with hexaploid wheat, maize, and *Arabidopsis* (data obtained from http://wheat.pw.usda.gov, http://ars-genome.cornell.edu/rice, http://www.agron.missouri.edu, and http://ukcrop.net/agr). It appeared that quasi-linkage might indeed be a significant phenomenon in wheat: for hexaploid wheat we obtained genomewise significance $p < 0.001$. The same test gave $p = 0.06$ for maize, $p = 0.05$ for rice and $p = 0.08$ for *Arabidopsis*.

Note that deviations of r values of linked loci to an unlinked one were positively correlated, as expected if the deviations from $r = 0.5$ are not related to misclassification of markers. In such a case, of special interest would be to conduct a test that will include not only the "representative markers" of nonhomologues showing the highest deviations, but also to take into account the average recombination rates characteristic of entire segments (Fig. 7.3). Application of permutation tests in such a formulation seems to provide much more reliable results (Table 7.3). The presented data show that the observed phenomenon indeed involves quite large segments of nonhomologous chromosomes. In some cases, even the mean(r) ± 3σ interval constructed by use of individual marker-marker recombination rates did not include the expected 50%. The permutation tests conducted for the marker segments and those that revealed 16 pairs of segments with significant ($p < 0.05$) deviation of mean(r) from 50% (Table 7.3) should be corrected for multiple comparisons. Indeed, out of 91 chromosome pairs, five pairs may exceed the significance level by chance. However, the probability of observing 16 such pairs by chance (H_0 hypothesis) is low ($<10^{-5}$, by binomial test). Moreover, eight of the foregoing 16 pairs were significant at $p < 0.01$ and three at $p < 0.001$; the probability of obtaining such results when H_0 is true is extremely low ($<10^{-7}$).

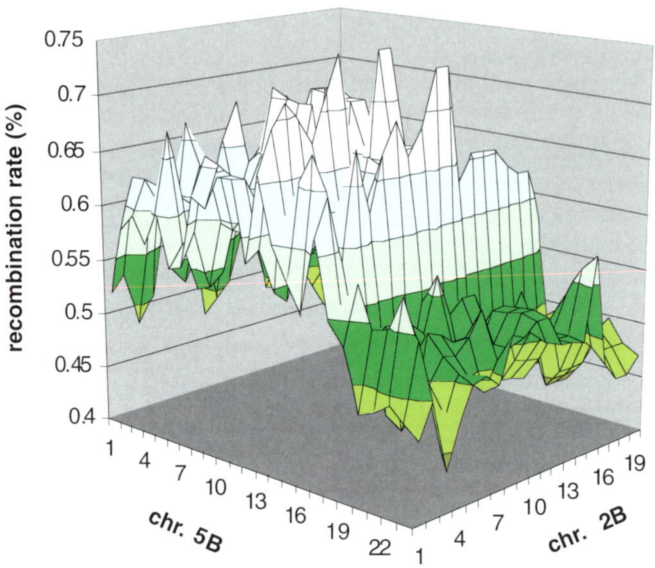

Fig. 7.3. Regional manifestation of quasi-linkage. Chromosome pair 2B and 5B was used as an example; it can be easily seen that deviation from the expected 50% recombination level is characteristic to entire regions of the nonhomologous chromosomes. (Peng et al. 2000b)

A departure from random segregation of markers on nonhomologous chromosomes, observed in crosses between different strains of house mouse 50 years ago, was explained by the tendency of segregating chromosomes of the same origin to migrate to the same pole during meiosis (the "affinity" hypothesis) (Michie 1953; Wallace 1953). Consequently, gametes with parental combinations of alleles at loci of nonhomologous chromosomes should appear with a higher frequency than the recombinant gametes. Reviewing the data in the literature, Wallace (1960a) postulated possible chromosome affinity in cotton. Even earlier, Malinowsky (1927) explained a number of cases of abnormal segregation in hybrids of lettuce, peas, tobacco, beans and wheat on the basis of the chromosome affinity hypothesis. He was probably the first to introduce the term "affinity." A pronounced effect of quasi-linkage was reported in an interspecific hybrid of *Coix* (Sapre and Deshpande 1987). There is evidence for nonrandom segregation of nonhomologues during the first meiotic division in humans (see Driscoll et al. 1979). Quasi-linkage has also been established in tomato (Wallace 1960b; Zhuchenko et al. 1977; Korol et al. 1989, 1994).

Beside nonrandom assortment (or affinity), quasi-linkage can also be explained by recourse to mechanisms of differential viability. Let us consider the MiMj/mimj heterozygote resulting from a cross between M_iM_j/M_iM_j and m_im_j/m_im_j. Suppose that parental combinations of whole chromosomes (M_iM_j and m_im_j) and recombinant combinations (M_im_j and m_iM_j) have differential

Table 7.3. Deviation from random segregation of marked segments of nonhomologous chromosomes. (Peng et al. 2000b)

Chromosome pair	Marked segments		L_1 and L_2 (cM)		Recombination rate (%)			
	m_1–m_2	m_3–m_4			Extreme	Mean (r)	σ	p
2A – 7A	Xgwm297b – Xgwm558b	P57M52h – Xgwm376c	14.0	15.7	66.9	59.90 ± 0.40	2.75	0.0157
3A – 4A	Xgwm497c – Xgwm480	Xgwm4a – P57M52k	28.8	27.5	30.6	39.68 ± 0.55	3.47	0.0039
2A – 1B	P57M49v – Xgwm294	Xgwm274a – Xgwm259a	24.6	54.9	36.9	42.45 ± 0.38	2.76	0.0245
2A – 2B	Xgwm497e – Xgwm630c	P56M52d – Xgwm55b	84.3	26.4	76.8	59.92 ± 1.08	5.72	0.0001
2A – 5B	Xgwm68a – P57M52ae	Xgwm205b – Xgwm544	20.2	19.6	29.3	41.49 ± 1.21	4.18	0.0180
3A – 2B	P57M49k – Xgwm247b	Xgwm410 – UBC318a	63.3	18.1	31.4	39.27 ± 0.41	2.60	0.0005
3A – 3B	P57M51c – UBC620a	P55M56e – P57M51t	26.2	16.6	79.8	60.57 ± 0.52	3.31	0.0070
3A – 7B	P55M53s – P55M60e	P57M49w – Xgwm333	29.0	53.1	31.1	42.36 ± 0.32	2.79	0.0222
4A – 6B	P56M52*e* – Xgwm637	Xgwm132a – P56M53s	46.4	12.5	65.6	60.40 ± 0.34	2.73	0.0081
5A – 4B	P55M60s – P57M52a	P55M53f – Xgwm251a	80.1	31.9	74.5	62.38 ± 0.66	4.85	0.0023
7A – 1B	P57M52h – P57M52s	P55M60v – UBC212a	27.0	17.9	38.1	41.76 ± 0.26	1.79	0.0208
7A – 7B	Xgwm332b – Xgwm344	P57M51y – P57M49b	51.2	11.2	33.3	40.39 ± 0.58	2.84	0.0130
2B – 5B	P57M51i – P55M53j	Xgwm274b – Xgwm554	99.2	36.5	74.7	62.08 ± 0.47	4.93	0.0002
2B – 7B	P55M56m – Xgwm55a	P57M51w – P56M50e	87.9	32.2	67.6	59.48 ± 0.53	3.90	0.0187
3B – 6B	P55M60g – Xgwm113b	Xgwm613 – P56M60j	17.3	18.0	35.1	39.58 ± 0.45	2.64	0.0012
4B – 7B	P55M56ab – Xgwm251a	P57M49b – Xgwm577a	36.5	18.0	34.7	41.80 ± 0.39	2.34	0.0273

Extreme value of recombination rate between individual markers as well as average and standard deviation over all pairs of markers from the defined segments (with $p < 0.05$) are presented. L_1 *and* L_2 denote the sizes of the segments. The significance of deviations of the mean (r) scores of segment pairs from 50% was tested based on 10,000 permutations for each pair selective values. Experimental evidence of this type has been obtained in *Drosophila* (Dobzhansky et al. 1965). Selective differences between parental and recombinant combinations may be reflected in the differential viability of zygotes, embryos, and adult individuals. The effect can also be expressed at the gamete stage in the form of differential fertilizing capacity of spermatozoa, differential rates of pollen tube growth, etc. (Korol et al. 1989, 1994).

The major problem with the old data on quasi-linkage was poor genome coverage of the morphological markers that could be followed up in one cross. This strongly limited the possibilities of discriminating among different explanatory hypotheses. Molecular markers solve this problem, but at the cost of another problem: the size of the available segregating populations is usually very small, restricting the detection power of the tests, so that only large deviations can be declared significant. Nonetheless, our data (Table 7.3), together with the few examples on molecular marker segregation in hexaploid wheat, rice, maize, and *Arabidopsis*, indicate that quasi-linkage may be a much more common phenomenon in plants (especially, cereals) than previously thought (Peng et al. 2000b). One practical aspect is related to a possible effect of quasi-linkage on the rate of false-positive detection in QTL mapping. The widespread approach of multilocus (composite) interval mapping (Jansen and Stam 1994; Zeng 1994) may be helpful in such cases.

7.2.3 Negative Crossover Interference

Positive crossover interference, i.e., a reduced frequency of adjacent double crossovers compared with that expected with the assumption of independence, is a characteristic of meiotic organisms, with only a very few exceptions (Egel-Mitani et al. 1982). Consequently, it is generally assumed that negative crossover interference, an excess of double exchanges, is mainly associated with intragenic recombination. Still, cases are known of higher-than-expected frequency of double crossovers in adjacent segments of small genetic length but large physical length. In *Drosophila melanogaster*, within a segment 4 cM long accounting for about 25% of the cytological length of chromosome 3 and spanning the centromere, a significant excess of multiple exchanges has been found (Sinclair 1975). Similar results have been obtained in other *Drosophila* studies with autosomes (Green 1975; Denell and Keppy 1979; Korol et al. 1994), but not with the X chromosome (Lake 1986). Significant negative crossover interference was found in barley (Søgaard 1977). Denell, and Keppy (1979) suggested that negative chromosome interference could be a characteristic of all regions exhibiting a very low density of recombination per unit physical length. This hypothesis fits the results of our study.

We found a significant excess of double exchanges in segments spanning or proximal to the centromere in nearly all chromosomes of both genomes, A and B (Table 7.4). The obtained estimates of coefficient of coincidence (c) indicate that negative crossover interference (i.e., $c > 1$) is characteristic of

Table 7.4. Islands of negative interference along the chromosomes. (Peng et al. 2000b)

Chromosome	Island[a]					
	1		2		3	
	c^b	χ^{2c}	c	χ^2	c	χ^2
1A	Xgwm164					
	9.84	19.82****				
1B	P57M51b		*YrH52*			
	>20	35.65****	>20	27.07****		
2A	Xgwm297b					
	3.21	12.75****				
2B	Xgwm120		Xgwm410			
	4.48	55.79****	5.36	21.12****		
3A	Xgwm218					
	2.44	35.31****				
3B	Xgwm566		Xgwm113b			
	>20	26.07****	3.43	8.27***		
4A	Xgwm397					
	3.74	25.12****				
4B	Xgwm107b					
	3.62	21.91****				
5A	Xgwm205a		XGwm186		Xgwm126a	
	3.86	20.1****	3.45	20.98****	4.35	5.65*
5B	Xgwm371		P57M49x		Xgwm205b	
	6.29	28.02****	3.66	20.39****	2.36	6.38*
6A	P56M52h					
	2.88	17.33****				
6B	P57M49a		Xgwm107c			
	9.99	15.90****	3.51	59.25****		
7A	Xgwm471		Xgwm60c			
	2.74	20.25****	4.26	19.08****		
7B	Xgwm361b		P56M60z			
	3.51	14.06****	6.57	13.61****		

[a] The central marker for two adjacent interval with maximum χ^2 is presented for each island of negative interference

[b] c is the coefficient of coincidence

[c] *Significant at 0.05, **significant at 0.01, ***significant at 0.005, ****significant at 0.001

the studied *T. durum* × *T. dicoccoides* hybrid. In fact, it is manifested in some regions of all chromosomes in genomes A and B, and for map versions H and L. Deviations from the "no interference" Haldane recombination scheme were highly significant. The maximum values of χ^2 (df = 1) (per chromosome) ranged, correspondingly, from 5.65 (with c = 4.35) for one of the islands of chromosome 5B to 59.25 (c = 3.51) for one of the islands on chromosome 6B. Clearly, the deviations from Kosambi interference in such cases should be even more significant. Each chromosome manifested a few islands of negative interference, from 1 to 3. As a rule, these islands were located in proximal regions, either spanning the centromere (chromosomes 1A, 3B, 4A, 6B, 7A, 7B), or excluding it. In the latter cases, the location of the islands could be to one (chromosomes 1B, 2A, 3A, 4B) or both (5A, 5B) sides of the centromere. These proximal segments comprise about 50–70% of the chromosome's cytological length but only 5–20% of the genetic length (Lukaszewski and Curtis 1993; Gill et al. 1996a). In some chromosomes, additional islands of negative interference were found in median or subterminal regions. An alternation of strong negative and full positive interference was characteristic of our data. When two or more islands were found per chromosomal arm, the location of extra islands was either median (3B) or terminal/subterminal (1B, 5A, 5B, and 7A).

The revealed trends were confirmed for the foregoing regions by use of larger intervals from one or both sides of the central point of any considered interval pair and by variation of the position of the central point. Correspondence between the results of the two map versions H and L served as an additional important test for the existence of a tract of negative interference. As in our previous results with chromosome 1B (Peng et al. 1999), a tendency for a higher level of negative crossover interference was found in regions proximal to or spanning the centromere. Likewise, the alternation of negative interference with segments with strong positive interference found earlier in chromosome 1B (Peng et al. 1999) appears to be a general phenomenon. In particular, for many such regions with positive interference, the estimated value of c was significantly smaller than predicted by the Kosambi mapping function.

Interest in the problem of coinciding crossovers is due to the current large-scale genome mapping efforts and the growing evidence that the length of the genetic maps of some plants tends to increase with the number of molecular markers employed. The simplest explanation is to assume that double crossovers can occur, at least in some organisms, at much smaller distances than generally believed. Our data support this view, corroborating other results (Gill et al. 1996b; Takahashi et al. 1997) and an apparent noncorrespondence between the total chiasma frequency and genome length of some species, including cereals (Nilsson et al. 1993).

Another simple explanation, that the size of the mapping population ($N = 150$) is too small to allow reliable conclusions about such kinds of patterns, cannot be considered plausible. How could one then explain the high resemblance of different chromosomes and excellent correspondence between the two map versions (H and L) as well as between dominant and codominant markers? Moreover, in light of the unusual fact of massive negative interference manifested by our data, we made some preliminary estimates of interference using recent mapping data available from public-domain Web sites. To our surprise, we discovered that negative interference seems not to be a rare phenomenon among other plant and animal species (to be published elsewhere).

The observed chromosomal distribution of islands of negative interference (either near-centromeric or median/subterminal) and the alternating positive/negative interference pattern make it possible to propose two models. (1) One is based on the hypothesis of Denell and Keppy (1979), that negative interference could be a characteristic of regions with low density of recombination per unit physical length. (2) The other is based on the recent findings in cereal genomics (Gill et al. 1996a,b; Faris et al. 2000; Kunzel et al. 2000) indicating the existence of gene-rich segments in wheat and barley chromosomes and higher recombination rates in these regions than in gene-poor segments. We can assume that the "positive-negative interference", doubted by Lukaszewski and Curtis (1993), may be a real phenomenon in wheat if double crossovers occur within the aforementioned islands and if recombination within an island reduces the chance of crossover in adjacent gene-poor segments. One important aspect of our results is that this recombination pattern was revealed mainly by means of microsatellite markers, which may not necessarily follow the island-like distribution of structural genes (half of the RFLP markers used by Gill et al. (1996b) were cDNA probes). If negative crossover interference is indeed a real phenomenon in wheat, then gene introgression via homologous recombination from relatively close, wild relatives to cultivated wheat can be considered even more optimistic than the estimates based on the association between gene distribution and recombination density in cereal genomes.

7.3 Coevolution of A and B Genomes in *Triticum dicoccoides*

7.3.1 Introduction

Polyploid formation is considered a unique evolutionary pathway exploited by many animals and plants. A few important crops, e.g., wheat, cotton, and tobacco, are allopolyploids. Evidently, allopolyploids exhibit the highest level of genetic complexity when two or more different genomes unify in one nucleus to create a new species (Wendel 2000). Polyploidization is an abundant process in plant evolution, hence present-day allopolyploids and their extant diploid progenitors provide a unique model for comparative investigations of both ancient and ongoing events that have allowed polyploidization and adaptive radiations. The recombination-domestication interface may have very interesting aspects when related to polyploids. These include the evolution of diploid-like chromosome behavior in meiotic pairing, recombination and segregation, the effect of various forms of recombination on the coevolution and sequence polymorphism of genomes sharing one nucleus, and the effect of selection on the foregoing processes. Detailed mapping and DNA sequencing efforts recently initiated on the foregoing crops, combined with comparative evidence on their diploid relatives and/or ancestors, may be considered an important way of highlighting the role of recombination in evolution and domestication of allopolyploids resulting in diploid-like behavior of meiotic chromosomes. To address this question, the interaction between meiosis-regulating genes (e.g., the Ph locus of 5B chromosome in wheat) and its target sequences should be studied.

Another related question is that of sequence coevolution of the partner genomes in the shared nucleus. Especially intriguing is the possible role of fast-evolving highly and moderately repetitive elements of the genome. At present, there is no reason to regard the process of allopolyploidization of *T. dicoccoides* as a mechanical combination of sequences from two genomes. But little is known about the interaction between sequences from different arrays in chromatin fractions (Wendel 2000). Repetitive DNA spreading from the A to D genome of allotetraploid cotton was demonstrated by in situ hybridization of 83 noncross-hybridized clones when some A-genome-dispersed repeats appear to have spread to D genome chromosomes (Hanson et al. 1998; Zhao et al. 1998). This phenomenon has great significance with respect to allopolyploid genome evolution, stabilization, and function.

The method of genomic in situ hybridization (GISH), with total genomic DNA as a probe in hybridization experiments, provides unique information about similarities between repetitive DNA from related species, as well as the physical location of conserved sequences on chromosomes (Fuchs et al. 1996; Heslop-Harrison and Schwarzacher 1996). This method can yield a generalized picture of the polyploidization phenomena. GISH was successfully used on mitotic and meiotic chromosomes of different cereal species to detect parental genomes in hybrids (Schwarzacher et al. 1989; Leitch et al. 1990; Benavente et al. 1996), allopollyploids (Mukai et al. 1993a; Chen and Armstrong 1994), and alien chromosomes or chromosomal segments in translocations and additions (Schwarzacher et al. 1992; Mukai et al. 1993b).

Compared with "classic" GISH, simultaneous GISH, with probe preannealing and overlapping, and the technique of molecular banding (Belyayev and Raskina 1998; Belyayev et al. 2001) permit the division of heterochromatin into several fractions and the analysis of differences/similarities between the content and chromosomal distribution of these fractions in the tested species. These types of data could shed light on the structure, function, and dynamics of repetitive DNA in the evolutionary process. One of the objectives of the studies conducted at the Institute of Evolution on the molecular cytogenetic level is to assess possible interactions between different repetitive sequence arrays in allotetraploid wheat. The experimental strategy includes GISH-based location of chromosomal positions of repetitive sequences that are common to *T. dicoccoides* and one of the proposed diploid progenitors (Belyayev and Raskina 1998; Belyayev et al. 2000). This approach provides new data both on the origin of polyploid wheat and on the process of allopolyploidy in plants. The experiments can be divided into two parts. First, chromosomes of *T. dicoccoides* are probed with DNA of the proposed progenitors of the A and B genomes to determine the chromosomal distribution of common repetitive sequences and to compare hybridization patterns of each diploid ancestor. Then chromosomes of *T. dicoccoides* are hybridized with DNA of species that show a high level of relatedness with B and A genomes corresponding to previous experiments to display the relationship between two genomes in the polyploid (Belyayev and Raskina 1998).

7.3.2 Probing of *Triticum dicoccoides* Genome with DNA from Its Putative Diploid Ancestors

To detect the chromosomal distribution of repetitive sequences common to *T. dicoccoides* and diploid progenitors of its A and B genomes, GISH was used (Belyayev and Raskina 1998). The hybridization mixture consisted of self-DNA of *T. dicoccoides*, detected in green, and DNA of one of the proposed ancestors, detected in red (probes overlapping). Thus, the green signal shows chromosome regions with preferential hybridization of *T. dicoccoides* DNA. The combined signal, visualized as red-orange fluorescence, shows regions of hybridization of both DNAs, i.e., the chromosomal distribution of common sequences. The intensity of the red-orange fluorescence shows the level of affinity between repetitive sequences in the tested genomes.

The results obtained from GISH experiments when DNA of species of the section *Sitopsis* were hybridized to the chromosomes of *T. dicoccoides* (Belyaev et al. 2000) are in agreement with the present understanding of the origin of the B genome. The bright-orange signal showed high homology of *Aegilops speltoides, Aegilops longissima*, and *Aegilops searsii* DNA to all chromosomes of *T. dicoccoides* in both euchromatic and heterochromatic regions, except in some of the pericentromeric heterochromatin regions (increased green fluorescence) (Fig. 7.4A–C). Close relatedness of sequences of those three species not only to the B genome, but also to the A genome can be explained by the presence of tribe-specific sequences usual for Triticinae (Anamathawat-Jonsson and Heslop-Harrison 1993) and by intrapolyploid sequences interaction (see below). DNA of *Aegilops sharonensis* and *Aegilops bicornis* showed less relatedness with DNA of *T. dicoccoides* (Fig. 7.4D,E).

The hybridization picture of probe-overlapping experiments with DNA of *T. urartu* and *T. monococcum* on the chromosomes of *T. dicoccoides* are different from those of S-genome species (Fig. 7.4F,G). In both cases, there is a distinct differentiation between the two genomes, and the B-genome chromosomes are easily detected by the presence of green bands (no hybridization signal from DNA of *T. urartu* and *T. monococcum*), which corresponds to C heterochromatin revealed on the same slide by Giemsa staining (Fig. 7.5) The presence of green bands on the chromosomes of the A genome in hybridization with DNA of *T. urartu* was also observed (discussed below).

Summarizing the above results, we can emphasize the following aspects with respect to the progenitors of A and B genomes: (1) In a comparison of hybridization of *Ae. speltoides, Ae. longissima*, and *Ae. searsii* DNA (proposed B-genome donors) with hybridization of *T. urartu* DNA (proposed A-genome

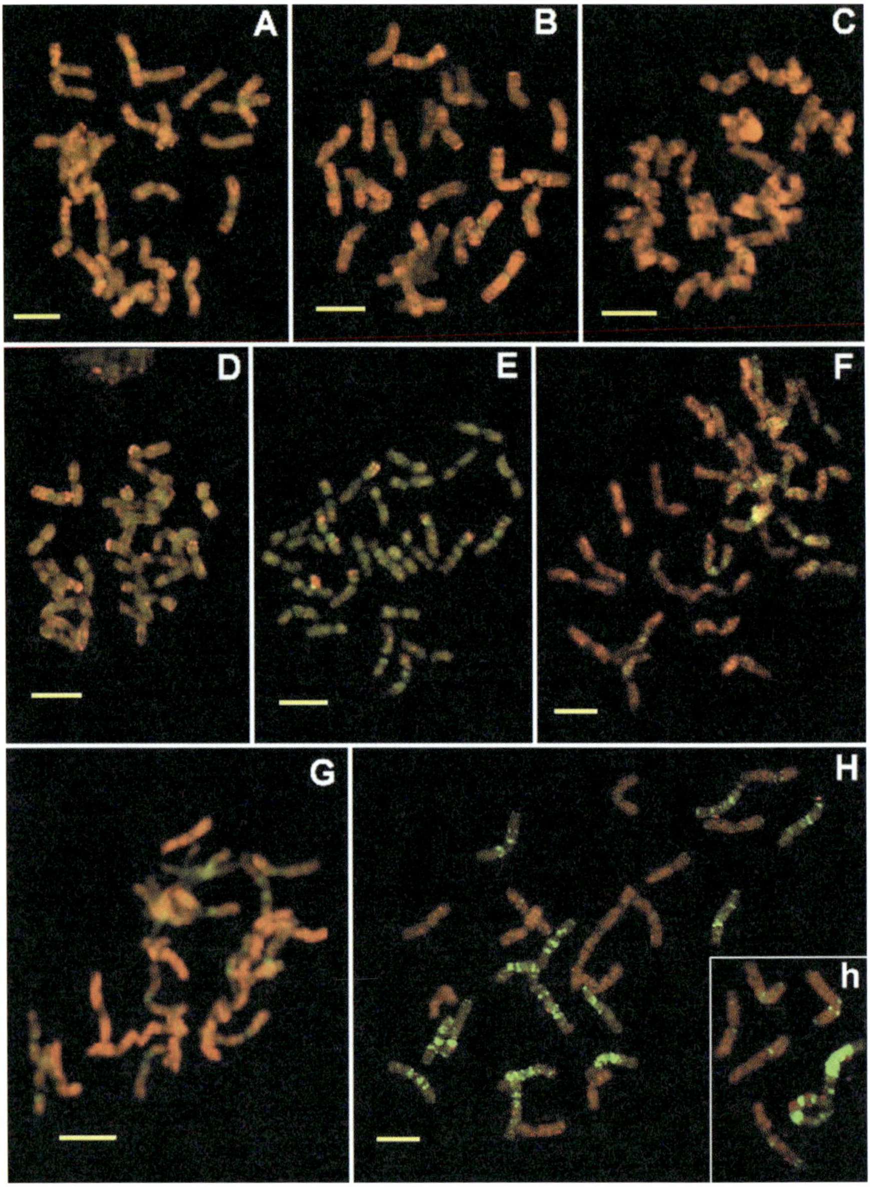

Fig. 7.4A–H. Simultaneous genomic in situ hybridization (GISH) with probe overlapping on metaphase chromosomes of *T. dicoccoides*. **A** Probe: DNA of *T. dicoccoides* (detected in *green*) plus DNA of *Aegilops speltoides* (detected in *red*). **B** Probe: DNA of *T. dicoccoides* (detected in *green*) plus DNA of *Aegilops longissima* (detected in *red*). **C** Probe: DNA of *T. dicoccoides* (detected in *green*) plus DNA of *Aegilops searsii* (detected in *red*). **D** Probe: DNA of *T. dicoccoides* (detected in *green*) plus DNA of *Aegilops sharonensis* (detected in *red*). **E** Probe: DNA of *T. dicoccoides* (detected in *green*) plus DNA of *Aegilops bicornis* (detected in *red*). **F** Probe: DNA of *T. dicoccoides* (detected in green) plus DNA of *T. urartu* (detected in *red*). **G** Probe: DNA of *T. dicoccoides* (detected in green) plus DNA of *T. monococcum* (detected in *red*). Simultaneous GISH with molecular banding on metaphase chromosomes of *T. dicoccoides*. **H** Probe: DNA of *Ae. speltoides* (detected in *green*) plus DNA of *T. urartu* (detected in *red*). *h* Fragment of another metaphase plate, same probe. *Bar* 10 µm. (Belyayev et al. 2000)

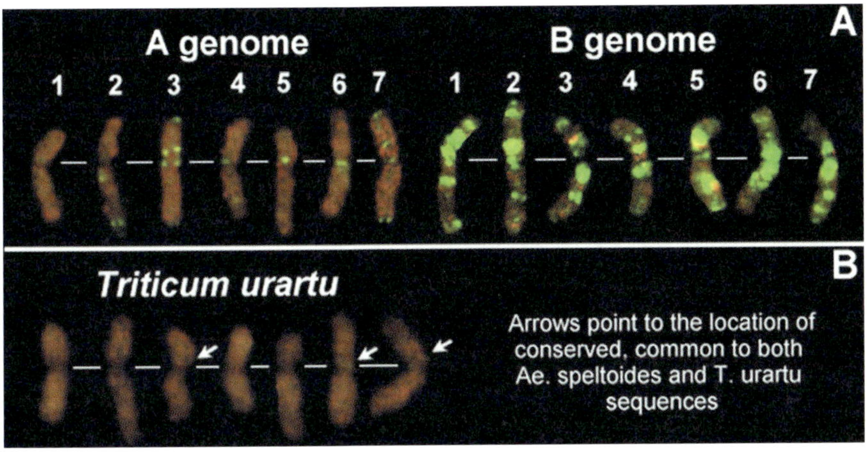

Fig. 7.5. **A** GISH-banding karyotype of *T. dicoccoides*. Probe: DNA of *Ae. speltoides* (detected in *green*) plus DNA of *T. urartu* (detected in *red*). **B** Karyotype of *T. urartu*, same probe. (Belyayev et al. 2000)

donor) on chromosomes of *T. dicoccoides*, we noticed that DNA of the three S-genome species revealed less contrast between A and B genomes; hence there are more common sequences between S-genome species and the A genome than between *T. urartu* and the B genome. (2) DNA of S-genome species does not hybridize to part of the pericentromeric regions of A-genome chromosomes (green bands). Likewise, DNA of *T. urartu* and *T. monococcum* does not hybridize to all pericentromeric regions of the B genome. Probably, sequences in those regions were involved in the interspecies divergence of these two groups of species at the diploid level.

7.3.3 Interaction of Two Genomes in the Polyploid

We used the technique of molecular banding to discriminate between A and B genomes of *T. dicoccoides* by simultaneous chromosome identification, and at the same time to display the relationship between the two genomes in the polyploid. The hybridization mixture consisted of two differently labeled DNAs, from *Ae. speltoides* (detected in green) and *T. urartu* (detected in red), that show a high level of relatedness with the B and A genomes corresponding to the previous experiments. The high temperature of hybridization allowed high-precision annealing of chromosome/probe sequences (Bostock and Sumner 1978), and at the same time it sharpened differences between the reassociation kinetics of euchromatin and heterochromatin, thereby revealing chromosome substructure. As a result, banded red chromosomes of the A genome and banded green chromosomes of the B genome were observed

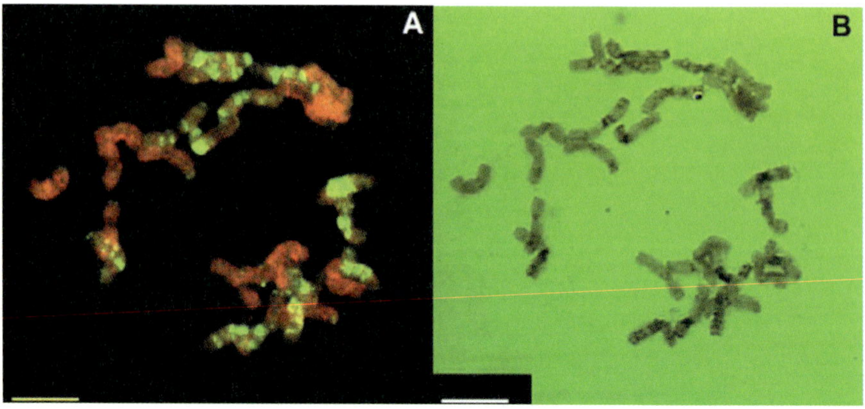

Fig. 7.6A, B. Successive Giemsa staining of the metaphase chromosomes after GISH. **A.** Simultaneous GISH with molecular banding on metaphase chromosomes of *T. dicoccoides*. Probe: DNA of *Ae. speltoides* (detected in *green*) plus DNA of *T. urartu* (detected in *red*). **B** The same plate stained with Giemsa. *Bar* 10 μm. (Belyayev et al. 2000)

(Fig. 7.4H,h). Molecular banding pattern corresponds to C heterochromatin (Friebe and Gill 1996; Belyayev and Raskina 1998) (Figs. 7.5A, 7.6) Hybridization revealed a combined red-orange signal from the two DNAs in euchromatic regions, but part of the A-genome chromosomes carried green heterochromatic blocks of B-genome type. This was true for all accessions of *T. dicoccoides* used in the present work. The control experiment, in which the same hybridization mix (DNA of *Ae. speltoides* plus DNA of *T. urartu*) was applied to chromosomes of three accessions of *T. urartu*, revealed only pericentromeric blocks on chromosomes 3, 6, and 7 – the location of conserved sequences common to both *Ae. speltoides* and *T. urartu* (Fig. 7.5B). Most likely, terminal heterochromatic regions of the A genome, recognized in many organisms as domains with fast-evolving species-specific sequences (Kipling 1995; Belyayev and Raskina 1998) during coevolution of two genomes in allopolyploids, have been enriched with sequences of B genome type.

Thus, interaction between the two genomes involves amplification of the B-genome satellite DNA (satDNA) at the "expense" of the A genome. This interpretation corroborates the experiment described previously, in which DNA of S-genome species that was hybridized to chromosomes of *T. dicoccoides* showed a high level of homology with the sequences of both B and A genomes (Fig. 7.4A–C). These results can be regarded as another example of directional interlocus homogenization (i.e., nonindependent molecular evolution) in polyploids (Elder and Turner 1995). The cause may be interlocus-concerted evolution that results in amplification and homogenization of repetitive DNA sequences (Elder and Turner 1995) and/or it may be colonization (Wendel 2000). The former involves sequence homology recognition between genomes followed by some sort of molecular gene conversion or crossing over, whereas the latter may simply be "jumping" of, for example, transposable elements. Such genomic changes may provide the physical basis for the several phenomena described in polyploid wheats, such as A-genome gene silencing (Harberd et al. 1987; White et al. 1994). Several studies have documented cases of interlocus homogenization of rDNA sequences following allopolyploid speciation in cotton species (Wendel et al. 1995; Wendel 2000) and in some hybridoform unisexual lizards (Hillis et al. 1991). Ribosomal DNA loci are the most evolutionarily stable structures that can be detected at the cytogenetic level; nevertheless, homogenization operates very strongly in these regions in all Triticeae genomes (Sallares and Brown 1999). Hence, it is not surprising that in the rapidly evolving repetitive fraction (Kipling 1995) processes of sequence homogeneity between homologous loci can also be observed.

GISH experiments evidently reflect general tendencies of intrapolyploid repetitive sequence interaction. In the old polyploid *T. dicoccoides,* we observed enrichment of part of A-genome heterochromatin clusters by satDNA of B-genome type. Thus, ancestral S-genome sequences have spread throughout the AB polyploid genome to a greater extent than ancestral A genome sequences. At present, the functions of heterochromatin are still not clear. But it can play a role in adaptive processes and in the process of species divergence and formation of interspecific barriers. Interlocus repetitive sequence directional homogenization acts in the opposite direction. It homogenizes arrays from formerly isolated taxa and can be regarded as one of the main mechanisms in speciation and stabilization of allopolyploid genomes.

We can now summarize the presented data on the coevolution of A and B genomes in allotetraploid wheat *T. dicoccoides* ($2n = 4x = 28$, genome AABB) obtained by GISH (Belyayev et al. 2000). Probing chromosomes of *T. dicoccoides* with DNA from the proposed A/B diploid genome ancestors points to a process of enrichment of the A genome with repetitive sequences of the B-genome type. Thus, ancestral S-genome sequences have spread throughout the AB polyploid genome to a greater extent than have ancestral A-genome sequences. The substitution of part of the A-genome heterochromatin clusters by satellite DNA of the B genome was detected by the molecular banding technique. The cause may be interlocus-concerted evolution and/or colonization. We propose that the detected high level of intergenomic invasion in the old polyploid might reflect general tendencies in speciation and stabilization of allopolyploid genome.

Chapter 8

Genetic Mapping of Agronomically Important Traits

8.1
Resistance to Diseases: High-Density Map of 1B Chromosome Region Harboring Stripe-Rust Resistance Genes *Yr15* and *YrH52*

8.1.1
Introduction

About 20% of the wheat harvest in the world is estimated to be lost each year due to diseases and pests. Stripe-rust (also called yellow rust), caused by *Puccinia striiformis* f. sp. *tritici* Westend, is considered the most or second-most important pathogen of wheat in six out of nine agroecological regions, as defined by Rajaram et al. (1988). Between 1970 and 1985, 16 moderate to major epidemics were recorded with estimated yield losses of 5–30% in Asia and Africa (Saari and Prescott 1985). The breeding of resistant varieties offers an effective approach to reduce or eliminate fungicide application and to minimize yield losses caused by this disease (Roelfes et al. 1992).

The best hope of future crop genetic improvement lies in exploiting the gene pools of the wild relatives of crop plants. The wild gene pools of wheat contain many economically important alleles, conferring resistance to diseases, etc., which might be transferred to cultivated wheat and utilized in its improvement (Feldman and Sears 1981; Nevo 1983a, 1989a, 1995, 2001b; Nevo et al. 1985, 1986a,b, 1991b). *Triticum dicoccoides* is indigenous to Israel and is a valuable source of stripe-rust resistance (Gerechter-Amitai and Stubbs 1970; Nevo et al. 1986b). Several novel stripe-rust resistance genes have been identified in wild emmer wheat (Gerechter-Amitai et al. 1989; Van Silfhout et al. 1989).

The tested *T. dicoccoides* accessions from Mt. Hermon, Israel, are highly resistant to stripe-rust at both the seedling and adult stages, and large numbers of the highly resistant accessions to stripe-rust in Israeli wild emmer wheat are from the Mt. Hermon population. Nevo et al. (1986b) tested 113 wild emmer accessions from 11 populations using rust race 2EO and found that all the accessions from the Mt. Hermon population were entirely immune to stripe-rust. The et al. (1993) tested 541 accessions from 23 Israeli wild emmer populations using rust race 110E143 A+ and found that all accessions from the Mt. Hermon population were highly resistant to stripe-rust, accounting for 74% of the total highly resistant accessions. Van Silfhout et al. (1989) tested about 850 wild emmer accessions collected from 31 locations using 28 stripe-rust isolates from 19 countries. They found that 19 (28%) of the 68 promising resistant accessions were derived from the Mt. Hermon area.

8.1.2 Why Mapping?

The development of molecular markers for disease and pest resistance genes in wheat can support resistance breeding in two ways. First, marker-assisted selection facilitates the selection process by detecting the presence of several resistance genes combined in one cultivar (gene pyramiding) and tracing the transfer of new sources of resistance from wild relatives. Second, molecular markers support the isolation of resistance genes through map-based cloning with subsequent transformation in recipient cultivars (Feuillet and Keller 1998). Moreover, recent studies of the genetic basis of disease resistance in many plant species revealed a common pattern of cluster organization of resistance gene families. Mapping can be considered as the first basic step in the genetic dissection of disease resistance. Last, but not least, is the importance of such markers in population-genetic analysis of the natural variation of resistance genes and host-parasite relationships (Fahima et al. 1998). Especially useful for such targets are microsatellites, PCR-based, easily handled, cheap, and highly reproducible markers that can be used on a large scale in plant-breeding programs (Langridge and Chalmers 1998). Röder et al. (1998) developed a set of hexaploid wheat microsatellite markers and constructed a molecular map. This set is also effective in mapping studies of tetraploid wheat (see Chap. 7 and Peng et al. 1999, 2000b). By means of this tool, two stripe-rust resistance genes derived from *T. dicoccoides* were genetically mapped at the Institute of Evolution. One of these, *Yr15*, which is derived from Rosh Pinna, Israel (Gerechter-Amitai et al. 1989), was earlier assigned to chromosome 1BS using cytogenetic analysis (McIntosh et al. 1996). A new stripe-rust resistance gene, tentatively designated *YrH52*, carried by accession Hermon (H52) from the unique Mt. Hermon population (Nevo et al. 1986b) of Israeli wild emmer wheat was also recently mapped on chromosome 1B (Peng et al. 1999). Therefore, our target was to conduct finer mapping of these genes. The development of a high-density genetic map is a prerequisite for gene isolation through map-based cloning strategies (Feuillet and Keller 1998). Marker-assisted selection of a gene of interest also greatly depends on the fineness of the map. Accordingly, the specific aims were: (1) to increase the marker density in the region harboring *YrH52* using additional molecular markers such as AFLP, microsatellite, and RAPD; (2) to improve the *Yr15* linkage map using *YrH52*-linked microsatellite markers, and (3) to discriminate preliminarily these two genes on chromosome 1BS.

8.1.3 Updated High-Density Molecular Map of *Yr15* and *YrH52* Regions and Preliminary Discrimination of *YrH52* and *Yr15* Genes

When the previously mapped microsatellite locus Xgwm131a (Röder et al. 1998) was used as the anchor of chromosome 1B, *YrH52* and 45 other linked markers were assigned to the same linkage group. The LOD scores of chromosomal assignment were higher than 10 for 41 marker loci and the gene *YrH52* (55.9) (Peng et al. 1999). Most of these loci (34) were located on a chromosomal region between P57M51j and UBC199c (54 cM in length) spanning the centromere and *YrH52* gene (Fig. 8.1). In a short interval, Xgwm359b–P55M53b (13.5 cM), containing *YrH52*, there were 19 marker loci and the mean interval length was 0.96 cM. As a result, *YrH52* was bracketed by markers Xgwm413 (*Nor1*, UBC212a) and Xgwm273a (Xgwm273d) with a map distance of 1.3 and 2.7 cM from either side, respectively.

Microsatellites linked to the *YrH52* gene were used to test the polymorphism between parental lines of *Yr15*-mapping populations (Sun et al. 1997). Eight of these lines were polymorphic and were used to genotype the population. In sum, 14 molecular markers, including the six detected in the previous studies (Sun et al. 1997; Chagué et al. 1999), appeared to be linked to *Yr15* with recombination frequencies of 0.03–0.26 and LOD scores of 3.0–19.6. By means of multiple-point analysis, an updated genetic map of *Yr15* region involving nine microsatellite loci, one RFLP locus, and four RAPD loci was constructed. The *Yr15* gene was bracketed by *Nor1* and UBC212a (Xgwm413) with map distances of 2.6 and 4.3 cM, respectively (Fig. 8.1).

On the basis of map distances from the common marker *Nor1*, and ecogeographic and pathological data, *YrH52* and *Yr15* were considered different genes on the same chromosome arm 1BS (Peng et al. 1999). However, it was shown that recombination frequency or map distance is highly variable among different genetic backgrounds or different hybrid plants derived from the same parents (Korol et al. 1994). The reliability of this assumption is thus challenged, and more markers are needed to verify the relationship between the two genes. In Peng et al. (2001a), ten common markers were found to be linked with both *YrH52* and *Yr15* genes, but their linkage or map distances to *YrH52* and *Yr15* were variable (Fig. 8.1) due to the difference in the genetic backgrounds of the two mapping populations (Sun et al. 1997; Peng et al. 1999). The three markers, *Nor1*, UBC212a, and Xgwm413, bracketing the *Yr15* gene (Fig. 8.1A) appeared to be closely linked to the *YrH52* gene and were mapped on the same position on chromosome 1BS; *Nor1*, UBC212a, and Xgwm413 obviously co-segregated in *YrH52*-mapping populations (Fig. 8.1B). Therefore, in *YrH52*-mapping populations, there was no crossover in

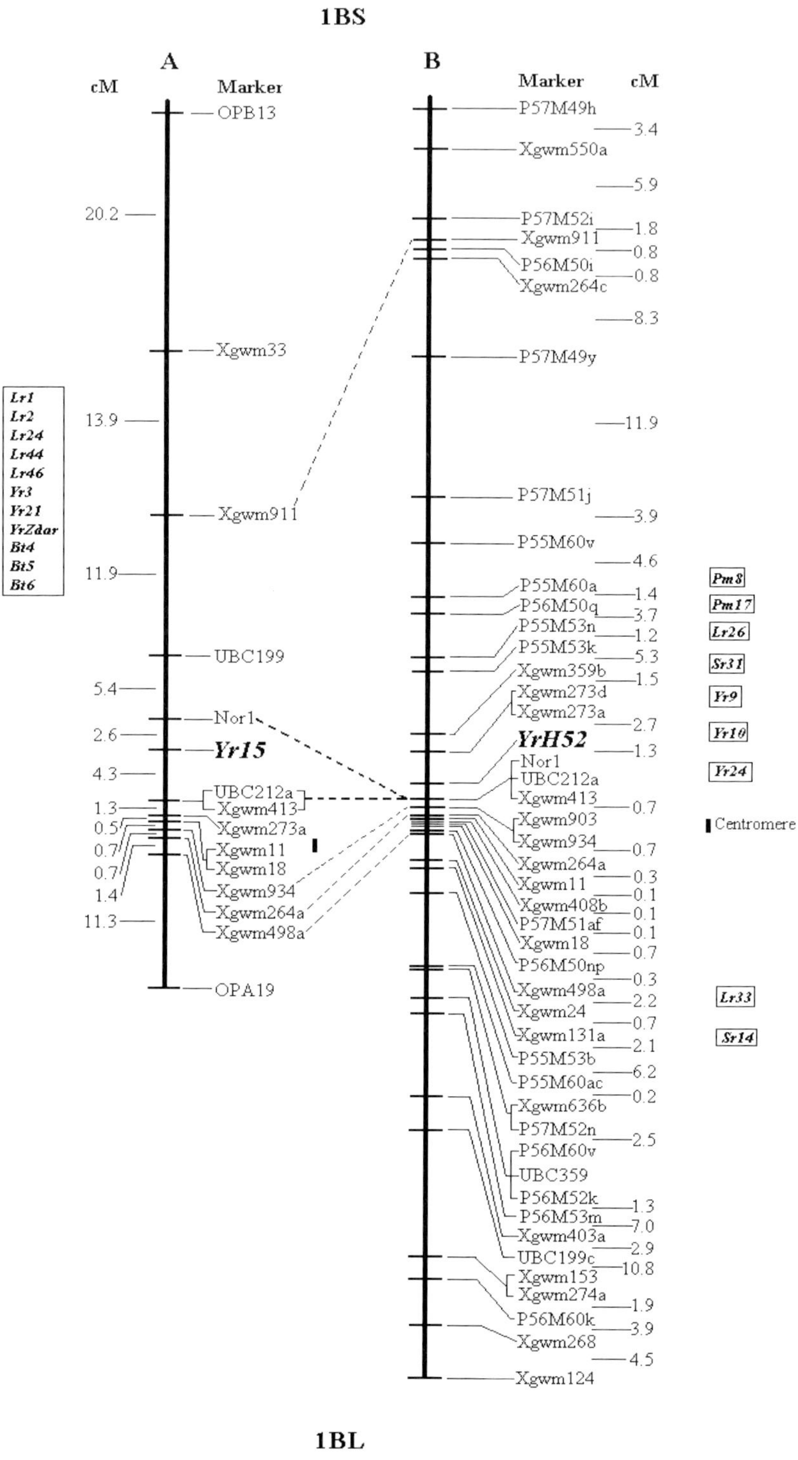

Fig. 8.1. High-density molecular map of stripe-rust resistance gene *YrH52* region (**B**) and linkage map of *Yr15* stripe-rust resistance gene (**A**) on chromosome 1B. The *vertical short bar* indicates the approximate location of the centromere according to Röder et al. (1998). (Peng et al. 2000c)

the chromosomal interval Xgwm413/UBC212a-*Nor1* carrying *Yr15*, and *YrH52* was located distally outside this interval. *YrH52* was thus assumed to be different from *Yr15*, and the RFLP marker *Nor1* may be located between these two genes. However, the population sizes (121 and 150 F_2 individuals) were theoretically small for the detection of rare recombination events between closely linked markers or genes, and a bigger mapping population is thus needed for the precise mapping and discrimination of these two genes.

Our results show that the *YrH52* gene and many of its linked markers were genetically clustered in a region spanning the centromere Peng et al. (2000c). This corroborates the known general pattern of wheat maps indicating that the recombination rate is higher in distal portions of the chromosomes (see Sect. 7.1). It has been revealed that markers genetically mapped near the centromere are physically located at a considerable distance from it (Delaney et al. 1995). The RFLP marker *Nor1*, which is closely linked to the stripe-rust resistance gene *YrH52* and other molecular marker loci, has been physically located in the gene/marker cluster in the distal region of chromosome 1BS (Gill et al. 1996b). Feuillet and Keller (1999) confirmed, at the microlevel, the presence of a gene-rich region at the distal end of chromosome group 1 in various grass genomes. It is thus reasonable to assume that the *YrH52* gene and its closely linked molecular marker loci (Xgwm359b, Xgwm273d, Xgwm273a, *Nor1*, UBC212a, Xgwm413, Xgwm903, Xgwm934, Xgwm264a, Xgwm11, Xgwm408b, P57M51af, and Xgwm18) belong to a gene/marker cluster that is located distally far from the centromere on chromosome 1BS. Other markers that are closely-linked to *YrH52*, like P56M50np, Xgwm498a, Xgwm24, and Xgwm131a, may reside physically far from *YrH52* because they are located on the long arm of chromosome 1B (Fig. 8.1B; Röder et al. 1998). In wheat, 1 cM is less than 1 Mb in some hot spots of recombination that contain a high density of marker loci (Delaney et al. 1995). *Nor* (nucleolar organizing region) is a hot spot of chromosome breakage (Tsujimoto et al. 1998). Therefore, the size of the chromosome 1BS fragment Xgwm413 (UBC212a, Nor1)–Xgwm273a (Xgwm273d) bracketing the *YrH52* gene can be estimated as 4.0 Mb or less. This situation is quite helpful but may not be sufficient for the map-based cloning of the stripe-rust resistance gene *YrH52*. Therefore, the finer mapping of this gene is under further study.

8.1.4 Resistance Gene Cluster on Chromosome 1BS

Classical genetic and molecular data show that genes for disease resistance are not randomly distributed over the genome of a species, but frequently occur in clusters on particular chromosomes (Islam et al. 1989). It has been observed repeatedly that disease resistance genes often are located in complex chromosome regions containing different pathogen race specificities (Hartl et al. 1995; Michelmore and Meyers 1998). Moreover, in wheat, many DNA markers and genes are clustered in a few regions within chromosomes. Most of the chromosome group *1* RFLP markers derived from cDNA clones (86%) and 14 agronomically important genes are present in five clusters that encompass only 10% of the physical length of the chromosomes (Gill et al. 1996b). In particular, among the 24 catalogued and allocated stripe-rust resistance genes, *Yr9*, *Yr10*, *Yr15*, and *Yr24* were located on chromosome 1BS (McIntosh et al. 1998; Fig. 8.1). Recently Ma et al. (2001) described another yellow rust resistance gene, *Yr26*, in this cluster close to YRH52. Our results (Peng et al. 2000c) show that the temporarily named stripe-rust resistance gene *YrH52* derived from the unique Mt. Hermon population of *T. dicoccoides* was also mapped on chromosome 1BS, corroborating the suggestion that stripe-rust resistance genes in wheat are not distributed randomly over the genome either. Furthermore, leaf-rust resistance gene *Lr26* (completely linked with *Yr9*), stem-rust resistance gene *Sr31* and powdery-mildew resistance genes *Pm8* and *Pm17* are also located on chromosome 1BS. Leaf-rust resistance gene *Lr33* and stem-rust resistance gene *Sr33* are located on 1BL. The following 11 disease resistance genes are also located on chromosome 1B, though their arm locations are not yet detailed: leaf-rust resistance genes *Lr1*, *Lr2*, *Lr24*, *Lr44*, and *Lr46*; bunt resistance genes *Bt4*, *Bt5*, and *Bt6*; stripe-rust resistance genes *Yr3*, *Yr21*, and *YrZdar* (McIntosh et al. 1998; Fig. 8.1). There are at least four independent RGA (resistance gene analogue) loci in the distal region of the short arm of group 1 chromosomes of wheat (Spielmeyer et al. 1998). Wheat RGAs are very similar to previously cloned RGAs from rice and barley. Phylogenetic analysis revealed that each wheat RGA class clustered more closely with RGAs from barley and rice than with other sequences from wheat (Goodwin and Hu 1998).

Accordingly, evidence on the genetics of wheat disease resistance fits the generalization that resistance genes tend to be clustered in the plant genome. Resistance genes are organized as clusters of genes that show varying levels of recombination between the component genes. In *Arabidopsis*, the size of the *MRC-J* locus containing nine resistance genes for five pathogens was about 15 cM (Kunkel 1996). The recombination rate at the *Mla* locus for 28 genes in barley was zero (Jorgenson 1994). The size of the *Dm1/3* locus containing nine resistance genes for two pathogens in lettuce was 20 cM (Kesseli et al. 1994). In tomato, the recombination ratio among three resistance genes on the *Cf4/9* locus was zero (Jones et al. 1993). Therefore, on chromosome 1BS of wheat, there may be a cluster of disease resistance genes, including the *T. dicoccoides*-derived *YrH52* and *Yr15* stripe-rust resistance genes.

Resistance genes were hypothesized to be functionally and evolutionarily related members of recognition systems (Michelmore and Meyers 1998). Richter and Ronald (2000) have presented several common themes that have shaped the evolution of plant disease resistance genes: the duplication of a progenitor resistance gene and subsequent divergence; unequal recombination and gene conversion in intergenic regions; intragenic recombination and gene conversion; adaptive selection on domains involved in recognitional specificity, such as leucine-rich repeats (LRR); and transposable elements. The short arm terminal region of chromosome 1B accumulates unique and variable sequences (Tsujimoto et al. 1998). Resistance gene clusters appear to evolve more rapidly than other regions of the genome (Richter and Ronald 2000). These may be the mechanisms that generate the cluster of resistance genes for stripe-rust, leaf-rust, stem-rust, and other diseases on chromosome 1BS of wheat.

8.2 Herbicide Resistance

Crop species can show differential responses to successful, widely used herbicides. Whereas some varieties are unaffected by application, others show symptoms of damage ranging from a slight reduction in vigor to complete plant death. The elucidation of the genetic control of such responses is important for developing strategies for breeding for herbicide resistance within crop species and also in understanding the modes of action of the herbicides and the evolution of resistance in weed species (Snape et al. 1991a).

To investigate this phenomenon in cereals, detailed studies of the control of responses of wheat to difenzoquat and to three phenylureas, chlorotoluron, metoxuron, and isoproturon, have been carried out. These studies have revealed that the primary difference in the response between resistant and susceptible varieties is due to single major genes, although the influence of other "modifier" genes has also been detected. It has also been shown that the responses to chlorotoluron and metoxuron are determined by the same gene. Studies of related wild grass species have indicated that polymorphisms for response also exist in such species. This suggests that the genes for differential responses evolved prior to the domestication of cultivated cereals and not in response to the development and use of the chemicals. The importance of these results for developing strategies for the incorporation of herbicide resistance in new varieties of crop species by conventional and nonconventional methods of plant breeding was discussed by Snape et al. (1991a).

Chlorotoluron (CT) is a selective phenylurea herbicide widely used for broad-leaved annual-grass weed control in cereals. These herbicides mediate their effect on weeds by interfering with electron transport in photosystem II (PSII). Variation in the response to CT was found in both *T. aestivum* L. and *T. dicoccoides* (Snape et al. 1991a,b). The genetic control of tolerance to CT in bread wheat is determined by a single major gene, *Su1*, located on the short arm of chromosome 6B (Snape et al. 1991a), but neither the mechanism of resistance nor any gene product have been identified so far. Moreover, the physiological mechanism of the varietal selectivity of wheat to CT is still not clear (Cabanne and Snape 1993). Studies of CT metabolism in tolerant varieties indicate that detoxification by ring-methyl oxidation represents the probable mechanism, while in susceptible varieties metabolism is generally slow and restricted to *N*-mono-dealkylation (Ryan and Owen 1983). Current opinions are that detoxification of the herbicide in bread wheat is correlated with the activity of cytochrome P450 mixed-function oxidases (Jones 1991); this has also been indicated in wild wheat (Levy 1995). The CT-tolerant phenotype in wild emmer wheat has been correlated with ecological factors and allozyme markers located on chromosome 6B (for example, *Adh, Est-4* and *Got*) (Nevo et al. 1992c). These findings imply that CT resistance has a similar mode of action in bread and wild wheat. However, the resistance alleles present in wild wheat might be different and could, for example, provide cross-resistance to several more herbicides (Snape et al. 1991a). Revealing the precise genetic control of herbicide response in cultivated and wild wheat is, therefore, essential for understanding the mechanism and evolution of CT resistance in wheat.

When differential responses to herbicides occur, there is generally a necessity for new varieties to incorporate resistance. Since the gene product is unknown, elucidation of the location of the genes will greatly simplify the process. Here we describe the comparative mapping of the *Sul* gene on chromosome 6B in bread and wild wheat.

In bread wheat, mapping was based on 58 F_4 plants of the cross between a genotype sensitive to CT, Chinese Spring (CS), and a resistant derivative, single chromosome substitution line, CS (Cappelle-Desprez 6B) [CS(CAP6B)] (developed by C. N. Law and A. J. Worland). The F_4 plants were derived by single-seed descent (SSD) so that each F_4 line represents a different F_2 plant. These single-chromosome recombinant lines were selected from F_3 families that were classified on the basis of their response to CT treatment (Snape et al. 1991a). The parental lines and ditelosomics of CS (DT6BL, DT6BS) (Sears 1954) were used to ascertain chromosomal arm locations of the fragments identified by hybridization with the DNA probes. In *T. dicoccoides,* mapping was based on 37 F_2 plants obtained from the cross between the CT-susceptible accession B-7 and the resistant accession B-35. Both originated from individual plants of a wild population collected at the Yehudiyya site, Israel (see Chap. 4). The response to CT was scored by estimating the decay of chlorophyll fluorescence induction kinetics. This was recorded following the exposure of detached leaves treated with CT to a single turnover flash light (Benyamini et al. 1991).

Nine RFLP probes spanning the centromere were chosen according to their location on the chromosome 6B linkage map (Devos and Gale 1993) and based on the provisional location of *Su1* close to the centromere (Snape et al. 199la): three known function probes, *α-Amy-1–3,* sub-fragment cDNA clone 2128 (described in Lazarus et al. 1985, kindly provided by D.C. Baulcombe); *Nor2,* wheat rDNA intergenic spacer (IGS)-3.0-kb *Taq*I-digested from the recombinant plasmid pTA71, wheat phosphoglycerate kinase (Longstaff et al. 1989); three anonymous cDNA probes-wheat leaf cDNAs *Xpsr106, Xpsr149, Xpsr88;* and three gDNA probes, *Xpsr312, Xpsr371,* and *Xpsr627* (kindly provided by M.D. Gale). Linkage analysis was carried out using Mapmaker 3.0b (Lander et al. 1987) for the F_2 mapping population and our own maximum-likelihood program for the SSD population. In the latter, frequencies between loci were calculated with minor modifications (see Krugman et al. 1994, 1997), in which the maximum likelihood estimate of *r* was obtained by numerical iteration. As described above, the progeny obtained after three generations of SSD was estimated for CT resistance, and equal numbers of resistant (R) and sensitive (S) genotypes were scored for their alleles at the molecular markers. Hence, the expected ratio of R:S as well as their actual proportion was not 1:1. Thus, the expected frequencies of marker classes were normalized within the two resistance groups by their totals in order to obtain the expectations corresponding to the experimental designs used.

The segregation ratio for CT resistance in the *T. dicoccoides* F_2 progeny was 14S:24R ($\chi^2 = 2.42$, $p > 0.5$, for 1:3 segregation), indicating that the difference between the resistance and susceptible responses is, as expected, due to a single major locus. The parents of each cross were screened for polymorphism using the different restriction enzyme/probe combinations. Polymorphism between the *T. aestivum* parents was detected for *α-Amy-l/Hin*dIII, *Nor2/Taq*I and *Xpsr371/EcoRl.* Polymorphism between the *T. dicoccoides* parents was detected by *Nor2/Taq*I, *Pgk2 (Xpsr141)/Hin*dIII and *Xpsr312/ Eco*RI. Chromosome and chromosomal arm locations of polymorphic bands in *T. aestivum* were determined using ditelosomic lines of CS (Krugman et al. 1997), whereas in *T. dicoccoides* the positions of the loci were determined by progeny tests in which the polymorphic probes were hybridized to DNA of the segregating populations. As expected, in both wild and bread wheat, each plant displayed either one (in homozygotes) or both (in heterozygotes) band patterns of the parental 6B polymorphic fragments (Table 8.1). The rDNA probe detected polymorphism in both crosses. In bread wheat, the *Nor* containing the 18S.26S ribosomal DNA clusters have been detected by in situ hybridization on chromosome arms 1AS, IBS *(Norl)*, 5DS *(Nor3)*, 6BS (*Nor*2), and 7DL, and on 1BS and 6BS in wild wheat (Appels and Dvorak 1982; Flavell et al. 1986). The fragment profile of CS consisted of a 3.1-kb fragment for *Norl* and 2.6-kb and 2.8-kb fragments for *Nor2.* Taken together, these results show that the CT resistance gene in *T. dicoccoides* is located on chromosome 6B, which may indicate that it is the same locus as *Su1* in *T. aestivum*. Hence, it will be referred to as *Su1*.

Using two-point analysis, we found *Su1'* and *Xpsr312* to be closely linked (5.6 cM, LOD = 6.13), while the other markers and CT resistance gene were unlinked (LOD < 2.5). The multipoint command "ripples" was used to establish the framework order of markers, showing that the order of loci was *Nor2-Xpsr312-Su1'-Pgk2,* with genetic distances between the genes of 24.8 cM, 5.3 cM, and 6.8 cM, respectively (Fig. 8.2). RFLP markers were identified as linked to the *Sul* gene in *T. aestivum: Nor2* on the short arm and *Xpsr371* and *a-Amy-1* on the long arm. In *T. dicoccoides Xpsr312* was linked to *Su1'*, and the order of the other markers along the 6B chromosome was determined to be *Nor2-Xpsr312-Su1'-Pgk2* (see Fig. 8.2). In a map of 6B of hexaploid wheat, *Xpsr312* is 2.6 cM from *Nor2,* and data from other crosses of *T. aestivum* (for example, doubled haploid lines from the cross of 'Highbury' × 'Sicco') indi-

Table 8.1. Segregation of RFLP markers and chlorotoluron response in *T. aestivum* F_4 single-seed descent (SSD) lines, and *T. dicoccoides* F_2 progeny. (Krugman et al. 1997)

Locus	Polymorphic fragment	Chlorotoluron		*n*	Genetic distance
		R	S		
T. aestivum **F_4 SSD lines**					**Recombination %**
Nor2	CAP 2.2 kb	24[a]	1	58	2.74 ± 1.65
	CAP/CS	5	1		
	CS2.6; 2.8 kb	1	26[a]		
a-Amy-1	CAP3.0; 6.5 kb	20[a]	4	57	9.84 ± 3.36
	CAP/CS	7	3		
	CS3.5; 4.0 kb	3	20[a]		
Xpsr371	CAP 7.0 kb	20[a]	0	50	5.15 ± 3.36
	CAP/CS	3	4		
	CS 6.5 kb	1	22[a]		cM
T. dicoccoides **F_2 progeny**					
Nor2	B35 2.8 kb	4[a]	2	37	24.8[c]
	B35/B7	19	3		
	B7 3.1 kb	1	8[a]		
Xpsr312	B35 6.0 kb	21[a]	1	33	5.6*
	B7 Null	0	11[a]		
Pgk2	B35 Null	10	0	32	6.8[c]
	B7 3.5 kb[b]	1	21[a]		

R, Resistant, *S*, susceptible, *LOD = 6.13, [c] LOD < 2.5

[a] Parental types

[b] dominant locus

cate that *Pgk2* is 10 cM from *a-Amy-1* (J.W. Snape, unpubl. results). This indicates that the linear order of the genes along the 6B chromosome is similar to that found in *T. dicoccoides*: *Nor2-Xpsr312-Su1'-Pgk2*. However, although the gene order is similar, the genetic distance of *Su1'-Nor2* in wild wheat is tenfold greater than in bread wheat. This difference in map distance between the two species may be a function not only of the chromosomal location of the markers, but, also of the genotypes studied, environmental conditions (Israel and England) and the size of the segregating populations (which was rather small in the wild wheat population of our study). Consequently, the distances represent approximations and not absolutes. Noteworthy, significant differences in recombination values between crosses were found earlier in wheat (Pogna et al. 1993). In both bread and wild wheat, a consistent noncorrespondence is observed between the physical and genetic locations of genes along the chromosomes. This discrepancy between the genetic map and physical map can vary between the species that were used in our crosses. Moreover, it is the regions with low ratios of genetic/physical distances that manifest the highest genetic variation and strongest reaction to recombination-inducing environmental conditions (Korol et al. 1994). Com-

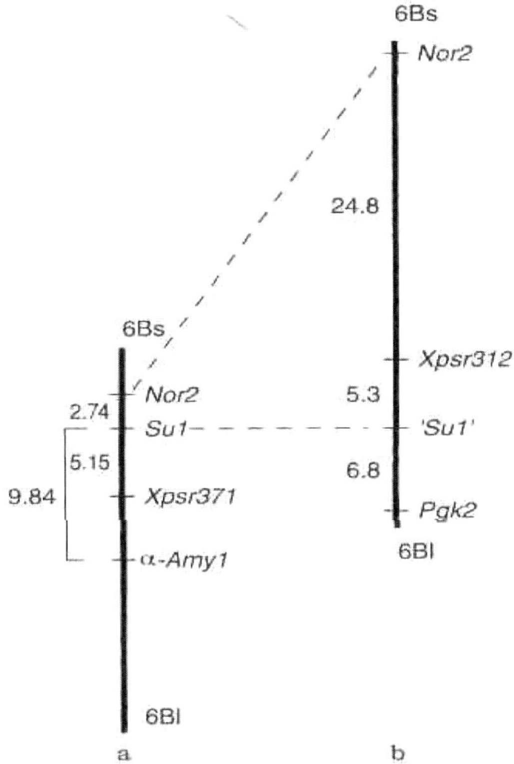

Fig. 8.2. Comparative linkage maps of 6B chromosome. **a** *Triticum aestivum*, based on F_4 of the cross CS × CS (CAP6B), **b** *Triticum dicoccoides*, based on F_2 of the cross B-7 × B-35. The genes and markers are given on the *right side* of the chromosome; map distances are given in centi-Morgans (cM) on the *left side*. (Krugman et al. 1997)

parative analysis of recombination in cultivated wheat and its wild progenitor deserves more attention, since it provides a system for examining the evolution of large genomes under domestication, especially in association with polyploidization.

Our results, together with indications that detoxification of the herbicide is correlated with the activity of cytochrome P450 mixed-function oxidases in both bread wheat (Gonneau et al. 1988; Jones and Caseley 1989; Jones 1991) and wild wheat (Levy 1995), imply that a similar genetic control of CT resistance exists in bread and wild wheat and that both species show response polymorphisms. Intriguingly, these differences must reflect genetic variation that already exists within the species and that the *Su1* gene for differential response to CT evolved prior to the domestication of cultivated wheat rather than in response to the development and use of chemicals.

8.3 Mapping QTLs for Agronomically Important Traits in *Triticum dicoccoides*

8.3.1 Introduction

Most crop traits are quantitatively inherited and are controlled by multiple genes. The advent of molecular markers has made it possible to genetically dissect this type of traits. By means of statistical analysis, the variation of a quantitative trait can be partitioned into the effect of individual genome regions, the quantitative trait loci (QTLs), linked to markers on a molecular-marker map (Paterson et al. 1988). Apart from localization of loci controlling quantitative traits, QTL mapping can also enable "cryptic" genetic variation to be uncovered, i.e., the identification of beneficial alleles that are otherwise hidden in a sea of deleterious alleles. This ability to detect cryptic variation can be important in exploration of wild germplasms (Soller and Beckmann 1988).

The analysis of the genetic basis of quantitative variation and the locations of specific QTLs used to be problematic in wheat. In the past, there were few reports of detailed analysis of individual genes, with perhaps a few exceptions, such as the work of Law (1966, 1967), who revealed the important role of genes on chromosome 7B. The major limitation was an insufficient number of morphological variants to allow the development of comprehensive genetic maps to provide landmarks for intrachromosomal locations of QTLs. The situation improved in the 1970s and 1980s with the development of electrophoretic systems to visualize differences in isozymes and grain storage proteins. But again, the levels of polymorphism and the numbers of available marker loci were too low to develop detailed maps. Only in the 1990s, due to the development of molecular marker systems, did this situation change, revolutionized by the availability of detailed genetic maps based on RFLP (restriction fragment length polymorphism) markers. This has resulted in a dramatic increase in the power of genetic analysis so that the totality of genetic variation for any trait in a cross can be described and attributed to specific locations in the genome (Snape 1998) allowing marker-assisted selection and even marker-based QTL cloning (Fridman et al. 2000). Till now, the QTL mapping in hexaploid wheat has been conducted primarily based on RFLP (restriction fragment length polymorphism) genetic maps at the level of either one chromosome (Araki et al. 1999; Kato et al. 1999) or the whole genome for a single trait or a few related traits. QTLs have been reported for resistance to preharvest sprouting (Anderson et al. 1993), kernel hardness (Sourdille et al. 1996), resistance to diseases (Faris et al. 1997, 1999; Nelson et al. 1998; Keller et al. 1999b; Waldron et al. 1999; Messmer et al. 2000), crossability (Tixier et al. 1998), resistance to lodging (Keller et al. 1999a), flour viscosity (Udall et al. 1999), grain protein content (Prasad et al. 1999), amylase content (Araki et al.

1999), tissue-culture response (Ben-Amer et al. 1997), and morphological and/or agronomic traits (Cadalen et al. 1998; Araki et al. 1999; Kato et al. 1999).

In tetraploid wheat, QTLs have been reported mainly for quality traits, i.e., protein content (Blanco et al. 1996; Joppa et al. 1997), sedimentation (Blanco et al. 1998), and free-threshing habit (Simonetti et al. 1999). Genetic research on important agronomic and quality traits in *T. dicoccoides* has lagged far behind that of other cereals and, at the present, detailed genetic information is urgently required as a basis for the effective utilization in the breeding programs of cultivated wheat, including bread and durum wheats. To date, there have been only isolated reports on QTLs controlling morphological and agronomic traits in *T. dicoccoides.*

In order to accelerate the utilization of *T. dicoccoides* in wheat improvement programs, we mapped QTLs controlling the major agronomic and morphological traits by means of molecular markers (Peng et al. 1999, 2000a–c, 2001). These markers mainly included microsatellites, AFLP, and random amplified polymorphic DNA (RAPD). For data analysis, we employed a new package, MultiQTL (Korol et al. 2001), based on QTL mapping theory developed at the Institute of Evolution (Korol et al. 1994, 1995, 1996b, 1998a,b, 2000, 2001; Ronin et al. 1996, 1999).

8.3.2
QTL Detection: The Experimental Design

Dr. L. R. Joppa (North Dakota Agricultural Experiment Station, USA) kindly provided an F1 hybrid between a highly stripe-rust resistant *T. dicoccoides* accession, Hermon H52 (H52), from Mt. Hermon, Israel, and a *T. durum* cultivar, Langdon (Ldn.) released in North Dakota. An F_2 mapping population consisting of 150 individuals was then developed in Haifa, Israel. In the greenhouse, the F_2 individuals were scored for hairy leaf sheath (Hs) and hairy auricle (Pa) and were bagged to ensure selfing for producing high-quality F_3 families. The field trial was conducted at the Neve Yaar Agricultural Experimental Station, Israel, during the 1997–1998 crop season. A completely randomized block design was adopted with one replication. A total of 162 single-row plots, i.e., 150 for the F_3 families and six for each of the two parental lines, was included in the trial. Eleven quantitative traits scored on F_3 progeny of F_2 plants genotyped for molecular markers (see Sect. 7.1; Peng et al. 2000b). The following were measured for about ten individual plants from each plot in the field or laboratory: plant height (HT), cm from the basal to the top of the tallest stem in one plant; heading date (HD), the days from sowing to heading; spike number/plant (SNP); spike weight/plant (SWP, g), including the grains, hulls and rachis; single spike weight (SSW, g); kernel number/plant (KNP); kernel number/spike (KNS); kernel number/spikelet (KNL); 100-grain weight (GWH, g); grain yield/plant (YLD, g); and spikelet number/spike (SLS). The glossiness of spike (Ws) and brittle rachis (Br) were also scored as qualitative traits for the individual plants.

Obvious segregation on the F_3 family means was observed for all of the 11 measured traits in the F_2 mapping population with coefficients of variation (CV) ranging from 2.7 to 50.9%, and all the traits showed normal or nearly normal distribution. Except for heading date (HD) (CV = 2.7%), the other ten traits showed high variability, with CV > 10%. Spike number/plant (SNP), spike weight/plant (SWP), single spike weight (SSW), kernel number/plant (KNP), kernel number/spike (KNS), grain yield/plant (YLD) and spikelet number/spike (SLS) were highly variable, with CV > 20%. Transgressive segregation was also observed for a majority of the traits except for GWH. Three traits, HD, SNP, and SLS, showed the strongest transgressive segregation (Fig. 8.3). To detect QTLs, scanning of the entire genome was conducted for each of the quantitative traits by the general interval mapping approach (Lander and Botstein 1989; Korol et al. 1996b). Before conducting the QTL analysis, the initial trait distributions were transformed to reduce the effect of deviation from normality on the mapping results. The transformation procedure of the MultiQTL package allows maintenance of the transformed trait the same range of individual trait values as in the initial one. Based on the relative magnitude of additive and dominance effects, a suitable genetic model (pure additive, full dominance of either *T. dicoccoides* or *T. durum* QTL alleles, or general effects) was chosen for each of the putative QTLs. Statistical significance of the putative QTLs was determined by permutation test (Churchill and Doerge 1994). All of these analyses were carried out using the MultiQTL package. The detected QTLs were symbolized basically following the nomenclature described by McIntosh et al. (1998).

8.3.3
Genomic Distribution of QTLs for Agronomic Traits

Simple interval analysis was employed to scan the genome for QTLs for each of the 11 traits. As an example, Fig. 8.4 shows the distribution of LOD scores in H-version map for QTLs of KNS and GWH on chromosome 1B, SWP on chromosome 2A, and HD on chromosome 5A. By applying the permutation test, 45 QTLs in total were declared at the significance level $p < 5\%$ for all of the 11 traits measured by using the H- or L-version map. Table 8.2 lists the estimates of important parameters. By pooling the results obtained from the H- and L-version maps, 48 QTLs were found for the 11 traits. Of these 48 QTLs, 42 (87.5%) were detected by both H- and L-version maps, and six others were detected only by either the H- or L-version map (Table 8.3). As to the significance of LOD scores for the putative (significant at 5%) effects revealed by 1,000 rounds of permutation test, 31 (68.9%) for map version H and 34 (71.6%) for map L were significant at the 1% level, whereas 20 (44.4%) and 21 (46.7%) were significant at the 0.1% level (Table 8.2), respectively. The number of QTLs for the specific traits ranged from two (HT) to seven (GWH) (Table 8.3).

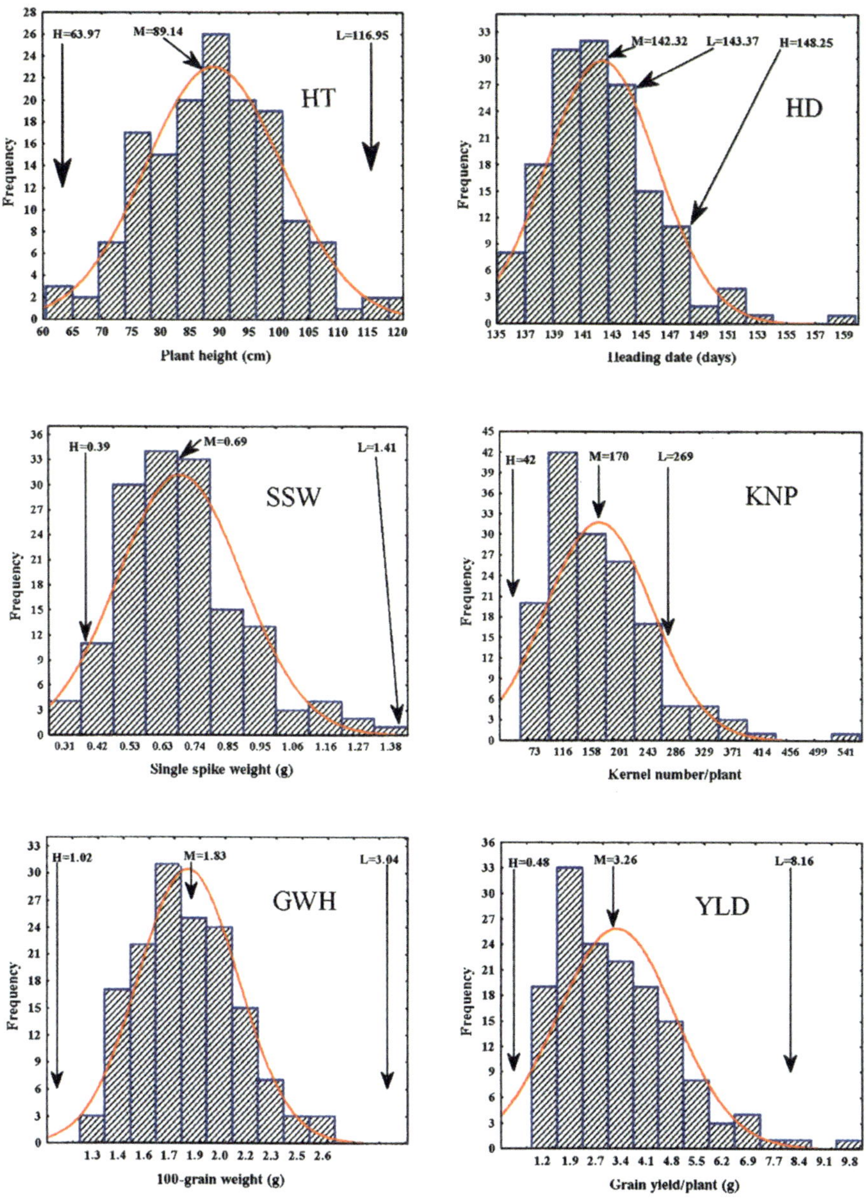

Fig. 8.3. Phenotypic distribution of agronomic traits in an F_2 mapping population of *T. durum* × *T. dicoccoides*. *HT* Plant height (cm), *HD* heading date, *SNP* spike number/plant, *SWP* spike weight/plant (g), *SSW* single spike weight (g), *KNP* kernel number/plant, *KNS* kernel number/spike, *KNL* kernel number/spikelet, *GWH* 100-grain weight (g), *YLD* grain yield/plant (g), *SLS* spikelet number/spike. (Peng 2000; Peng et al. 2001) (continued on page 288)

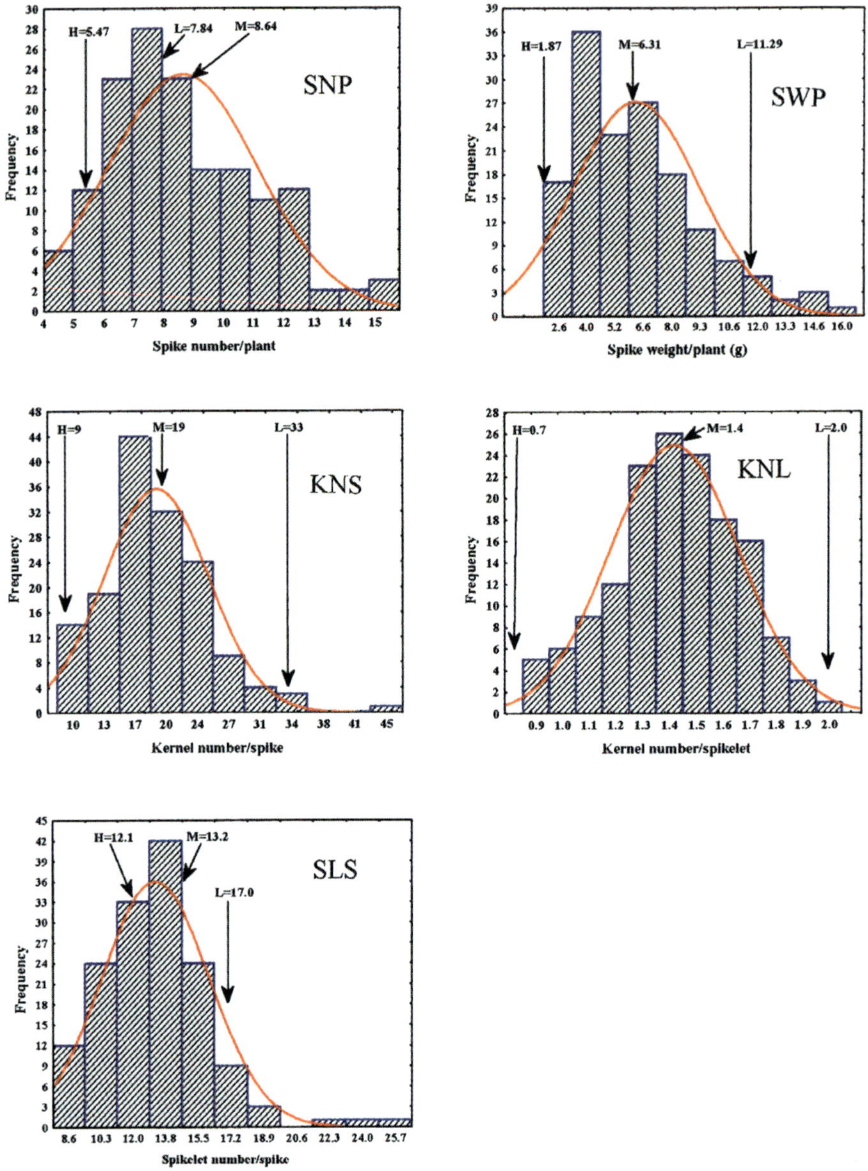

Fig. 8.3. *Continued*

As shown in Table 8.3, each version of the map, either H- or L-, detected 45 out of 48 significant chromosome-trait combinations, with 42 shared combinations. The difference between the two versions of the map for QTL number is thus small. This map-version difference for the individual traits on the levels of chromosome, genome, and the entire genome is also small (Table 8.3). For most of the traits (10/11), the number of the detected QTLs on the A genome

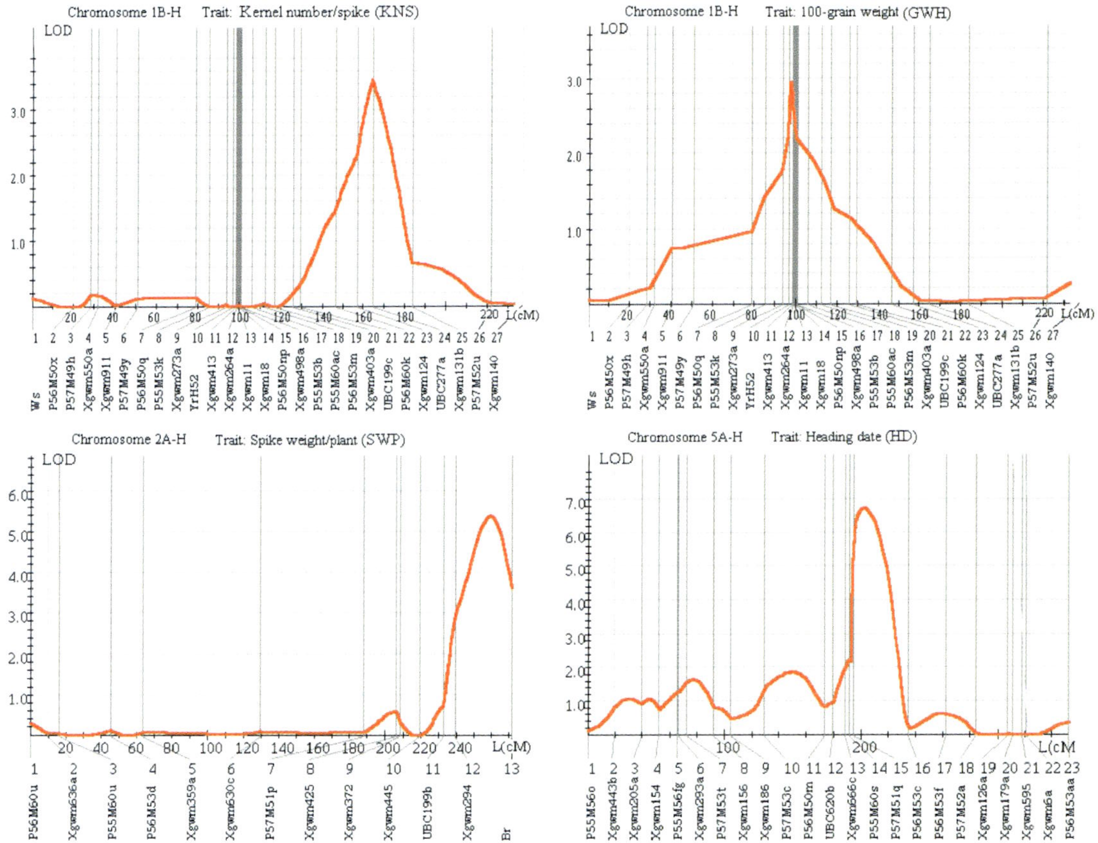

Fig. 8.4. LOD distribution pattern as a print-out from the computer program MultiQTL. *Horizontal scale* Molecular markers and the map distance. *Vertical scale* The LOD score. The peak of LOD distribution curve indicates the chromosome location of the corresponding QTL. (Peng et al. 2000b, 2001)

Table 8.2. Putative QTLs detected by single-QTL analysis. (Peng 2000; Peng et al. 2001)

Locus	H map				L map			
	LOD	PVE (%)	Effect	Mode of action	LOD	PVE (%)	Effect	Mode of action
Plant height (cm)								
QHt.hue-5A	5.13***	16.4	−14.60	R	5.25***	16.8	−14.70	R
QHt.hue-7A	2.12*	8.5	−10.07	A				
Total		**24.9**				**16.8**		
Heading date (days)								
QHd.hue-2A	2.30**	16.5	−4.72	D	2.34*	15.1	−4.44	D
QHd.hue-4B	2.02*	6.2	−2.72	R	2.00*	6.2	−2.71	R
QHd.hue-5A	4.75***	19.2	5.85	R	4.93***	19.2	5.92	R
QHd.hue-6B	2.64**	9.6	−3.86	A	2.23*	7.5	−3.37	A
Total		**51.5**				**48.0**		
Spike number per plant								
QSnp.hue-1B	2.90**	8.9	2.00	D	2.89**	8.5	1.95	D
QSnp.hue-2B	2.41*	7.8	1.85	D	1.97*	6.3	1.73	D
QSnp.hue-5A	2.16*	15.7	−2.98	A	2.22*	11.2	−2.35	A
QSnp.hue-7A					2.20*	18.2	−2.83	D
Total		**32.4**				**44.2**		
Spike weight per plant (g)								
QSwp.hue-1B	2.08*	6.8	2.26	D	3.60***	19.7	−4.12	A
QSwp.hue-2A	5.12***	20.3	−4.23	A	5.12***	20.3	−4.23	A
QSwp.hue-3A	2.59**	7.8	−2.40	D	2.42*	8.0	−2.46	D
QSwp.hue-5A	6.48***	36.1	−6.24	A	6.72***	36.5	−6.16	A
Total		**71.0**				**84.5**		
Single-spike weight (g)								
QSsw.hue-1B					2.75**	11.0	−0.19	R
QSsw.hue-2A	7.98***	29.5	−0.31	D	7.98***	29.5	−0.31	D
QSsw.hue-3A	3.07**	9.1	−0.17	D	3.23**	10.5	−0.18	D
QSsw.hue-3B	2.02*	6.4	−0.14	D				
QSsw.hue-5A	6.94***	32.0	−0.36	A	7.02***	32.3	−0.36	A
Total		**77.0**				**83.3**		
Kernel number per plant								
QKnp.hue-1B	1.99*	7.7	−76.68	A	4.46***	23.3	−128.34	A
QKnp.hue-2A	5.15***	20.2	−130.32	A	5.15***	20.2	−130.32	A
QKnp.hue-3A	2.37*	7.8	−75.22	D	2.37*	7.8	−75.22	D
QKnp.hue-5A	5.94***	33.3	−185.70	A	6.25***	33.6	−183.19	A
Total		**69.0**				**84.9**		
Kernel number per spike								
QKns.hue-1B	3.34***	10.0	−5.33	D	3.36**	16.9	−7.05	D
QKns.hue-2A	6.86***	25.9	−9.23	D	6.86***	25.9	−9.23	D
QKns.hue-3A	2.43*	7.3	−4.78	D	2.63*	8.2	−4.99	D
QKns.hue-5A	7.11***	19.6	−8.05	A	7.11***	19.6	−8.05	A
Total		**62.8**				**70.6**		

Table 8.2. *Continued*

Locus	H map				L map			
	LOD	PVE (%)	Effect	Mode of action	LOD	PVE (%)	Effect	Mode of action
Kernel number per spikelet								
QKnl.hue-1B	2.55*	8.0	−0.17	D	2.60**	13.5	−0.23	D
QKnl.hue-2A	3.79***	13.6	−0.24	D	3.79***	13.6	−0.24	D
QKnl.hue-3A	2.93**	9.6	−0.20	D	2.93**	9.6	−0.20	D
QKnl.hue-3B	2.13*	6.3	0.17	D	2.13*	6.3	0.17	D
QKnl.hue-5A	6.30***	17.6	−0.28	A	6.30***	17.6	−0.28	A
QKnl.hue-5B	2.83**	20.9	−0.31	A	2.69**	10.3	−0.22	A
Total		**76.0**				**70.9**		
100-grain weight (g)								
QGwh.hue-1B	2.78**	8.1	0.24	A	2.78**	8.1	0.24	A
QGwh.hue-2A	3.71***	15.8	−0.32	D	3.71***	15.8	−0.32	D
QGwh.hue-4A	4.86***	14.0	−0.33	R	4.79***	13.9	−0.33	R
QGwh.hue-5A	2.69**	9.0	−0.27	R	2.46*	8.2	−0.26	R
QGwh.hue-5B	4.01***	15.2	−0.29	D	2.91**	9.3	−0.23	D
QGwh.hue-7A	2.48*	7.8	0.23	D				
QGwh.hue-7B	2.15*	6.6	−0.22	A	3.32**	11.3	−0.29	A
Total		**76.5**				**66.6**		
Grain yield per plant (g)								
QYld.hue-1B					3.50***	15.3	−2.09	R
QYld.hue-2A	6.29***	23.9	−2.60	D	6.29***	23.9	−2.60	D
QYld.hue-3A	3.07**	10.1	−1.67	D	3.07**	10.1	−1.67	D
QYld.hue-5A	6.25***	35.6	−3.73	A	6.19***	35.9	−3.68	A
Total		**69.6**				**85.2**		
Spikelet number per spike								
QSls.hue-1B	2.04*	6.2	−2.05	D	2.54**	11.8	−3.16	A
QSls.hue-2A	5.65***	21.2	−4.34	A	5.65***	21.2	−4.34	A
QSls.hue-5A	5.17***	15.5	−3.66	A	4.98***	21.5	−4.56	A
QSls.hue-6B	2.40**	6.4	−2.47	R	2.36**	7.8	−2.39	R
Total		**49.3**				**62.3**		

PVE is the proportion of phenotypic variance (between F3 mean values) of F3 families explained by the detected QTLs, *D* and *R* dominance and recessiveness of trait increasing allele, respectively, *A* additive

was equal or exceeded that of the B genome. For H- and L-version maps, the number of QTLs on the A/B genome was 29/16 and 28/17, respectively. Overall, the number of QTLs on the A genome exceeded that on B genome (30/18) (Table 8.3). This contrasts with the polymorphism between the parental genotypes, which was employed to establish the current mapping population (Peng et al. 2000b; Sect. 7.1), and the proportion of polymorphic markers revealed in natural populations of *T. dicoccoides* (Li et al. 1999, 2000a–d, 2001a–d). One may speculate that the B genome plays a leading role in population

Table 8.3. Distribution of the detected QTLs among genomes and chromosomes. (Peng 2000; Peng et al. 2001)

Chromosome	Trait																							
	HT		HD		SNP		SWP		SSW		KNP		KNS		KNL		GWH		YLD		SLS		Total	
	H	L	H	L	H	L	H	L	H	L	H	L	H	L	H	L	H	L	H	L	H	L	H	L
Genome A																								
1A																								
2A			+	+			+	+	+	+	+	+	+	+	+	+	+	+	+	+	+	+	9	9
3A							+	+	+	+	+	+	+	+	+	+			+	+			6	6
4A																	+	+					1	1
5A	+	+	+	+	+	+	+	+	+	+	+	+	+	+	+	+	+	+	+	+	+	+	11	11
6A																							0	0
7A	+					+											+						2	1
Sub-total	2	1	2	2	1	2	3	3	3	3	3	3	3	3	3	3	4	3	3	3	2	2	29	28
Genome B																								
1B					+	+	+	+		+	+	+	+	+	+	+	+	+		+	+	+	7	9
2B					+	+																	1	1
3B									+						+	+							2	1
4B			+	+																			1	1
5B															+	+	+	+	+	+			2	2
6B			+	+																	+	+	2	2
7B																	+	+					1	1
Sub-total	0	0	2	2	2	2	1	1	1	1	1	1	1	1	3	3	3	3	0	1	2	2	16	17
Total	2	1	4	4	3	4	4	4	4	4	4	4	4	4	6	6	7	6	3	4	4	4	45	45

HT, Plant height (cm), *HD*, heading date (days), *SNP*, spike number, *SWP*, spike weight/plant (g), *SSW*, single-spike weight (g), *KNP*, kernel number/plant, *KNS*, kernel number/spike, *KNL*, kernel number/spikelet, *GWH*, 100-grain weight (g), *YLD*, grain yield/plant (g), *SLS*, spikelet number/spike adaptation to environmental conditions, whereas in domestication evolution genetic variants at loci of the A genome were more important. As revealed by the two versions of the map, the number of QTLs located on chromosome 1B, 2A, 3A and 5A was 33, which accounted for more than 68% of the QTLs detected. There were a few (0–1 each) QTLs on chromosomes 1A, 4A, 6A, 2B, 4B, and 7B (Table 8.3). On specific chromosomes, QTLs clustered in a few regions (especially in chromosomes 1B, 2A, 3A, 5A, and 5B). Among the 24 major QTLs (significant at $p < 0.1\%$), the majority (18) were located on chromosomes 2A and 5A (Table 8.2). These results may reflect either linkage or pleiotropy of the corresponding QTLs. It should be noted that the foregoing results are based on single-QTL mapping models. Our preliminary analysis showed that for some chromosomes the linked-QTL model better fits the obtained data. Therefore, analysis with linked QTLs will increase the number of detected QTLs and slightly change the described pattern. Likewise, a more detailed analysis based on a multiple-trait mapping approach (Korol et al. 1995, 1996a,b, 1998a,b, 2001; Ronin et al. 1996, 1999) is underway, which may allow detecting minor QTLs that appeared as nonsignificant or "nearly significant" ($p < 10\%$) with single-trait analysis.

8.3.4 Summary of the Revealed QTL Effects

Plant height (HT): Two QTLs were found for this trait. One major QTL, *QHt.hue-5A*, with LOD scores of 5.13 ($p < 0.001$) was located on the long arm of chromosome 5A and co-segregated with an AFLP marker, P57M51q; it can explain 16.4–16.8% of the phenotypic variance. Another minor QTL was detected on chromosome 7A (map version H only) with LOD score 2.12 ($p < 0.05$). These two QTLs together can explain 16.8–24.9% (depending on the employed map version) of the total phenotypic variance of this trait (Table 8.2). The alleles derived from *T. dicoccoides* for the QTLs on chromosome 5A and 7A obviously reduce the plant height of wheat.

Heading date (HD): Four QTLs were detected for this trait. One major QTL, *QHd.hue-5A*, with LOD score > 4.7 ($p < 0.001$) was located to the long arm of chromosome 5A and closely linked with P57M51q and major QTL *QHt.hue-5A* for plant height. It can explain about 19% of the phenotypic variation. All four QTLs could explain 48–51% of the total phenotypic variation (Table 8.2). The growth of *T. dicoccoides* accession H52 is usually very slow, so the heading date is obviously later than that of *T. durum* cultivar Langdon. The QTL analysis revealed that the *T. dicoccoides* alleles for the major QTL on chromosome 5A could be responsible for the late heading, but the QTLs on chromosomes 2A, 4B, and 6B could accelerate plant development and thus shorten the heading dates.

Spike number/plant (SNP): Four QTLs, located on chromosomes 1B, 2B, 5A, and 7A, were detected for this trait. These QTLs could together explain

32–44% of the total phenotypic variation (Table 8.2). The *T. dicoccoides* alleles for *QSnp.hue-1B* and *QSnp.hue-2B* could increase the number of spikes, but for the *QSnp.hue-5A* and *QSnp.hue-7A* they had the opposite effect.

Spike weight/plant (SWP): Four QTLs were uncovered for this trait and were mapped on chromosomes 1B, 2A, 3A, and 5A (Table 8.2). These QTLs could together explain 71–84% of the total phenotypic variation. *QSwp.hue-2A* and *QSwp.hue-5A* were major QTLs, with LOD scores of 5.12–6.72 ($p < 0.001$) and PVE (proportion of explained variance) values of 20.3–36.5%. The *T. dicoccoides* alleles for these QTLs lowered the trait value, except for *QSwp.hue-1B* which acted positively.

Single spike weight (SSW): Five QTLs were detected for this trait. These QTLs were mapped onto chromosomes 1B, 2A, 3A, 3B, 5A (Table 8.2). Together, these five QTLs could explain up to 77–83% of the total phenotypic variation. *QSsw.hue-2A* and *QSsw.hue-5A* on 2AL and 5AL are major QTLs with LOD scores of 6.94–7.98 ($p < 0.001$) and PVE values of 29–32%. The *T. dicoccoides* alleles for all of these QTLs negatively affected SSW (Table 8.2).

Kernel number/plant (KNP): Four QTLs were found for KNP and were located on chromosome 1B, 2A, 3A, and 5A (Table 8.2). These QTLs could explain 61–85% (depending on the employed map version, H or L) of the total phenotypic variation. *QKnp.hue-2A* and *QKnp.hue-5A* on 2AL and 5AS are major QTLs, with LOD scores of 5.15 and 5.94 ($p < 0.001$) and PVE values of 20.2 and 33.3%, respectively. The *T. dicoccoides* alleles for all of these QTLs reduce the trait value.

Kernel number/spike (KNS): Four QTLs were revealed for KNS and were mapped on chromosomes 1B, 2A, 3A, and 5A (Table 8.2). These QTLs could explain 63–71% of the total phenotypic variation. *QKns.hue-2A* and *QKns.hue-5A* on 2AL and 5AS are major QTLs, with LOD scores of 6.9–7.1 ($p \ll 0.001$), and PVE values of 25.9 and 19.6, respectively. *QKns.hue-3A.1*, in the region genetically near the centromere of chromosome 3A, may also be a major QTL, with a LOD score of 3.45 ($p < 0.005$) and PVE value of 35.2%, as revealed by the L-version map. The *QKns.hue-1B* also has a highly significant LOD score ($p < 0.01$). The *T. dicoccoides* alleles for all the QTLs lowered the mean value of KNS (Table 8.2).

Kernel number/spikelet (KNL): Six QTLs were detected for this trait. They were distributed on chromosomes 1B, 2A, 3A, 3B, 5A, and 5B (Table 8.2). These QTLs could explain 70–76% of the total phenotypic variation. *QKnl.hue-2A* and *QKnl.hue-5A* on 2AL and 5AS are major QTLs, with LOD scores of 3.79 ($p < 0.001$) to 6.30 ($p < 0.001$), and PVE values of 13.6–17.6%. The *T. dicoccoides* alleles for most of the QTLs had a negative effect on the trait, except for *QKnl.hue.3B* which had a positive effect (Table 8.2).

100-grain weight (GWH): The number of QTLs for GWH reached seven, the maximum among the 11 traits. These QTLs were distributed on chromosomes 1B, 2A, 4A, 5A, 5B, 7A, and 7B (Table 8.2) and together could explain

66–76% of the total phenotypic variation. *QGwh.hue-2A*, *QGwh.hue-4A*, and *QGwh.hue-5B*, on chromosomes 2A, 4A, and 5B, were major QTLs, with LOD scores of 3.83–4.86 (each significant at $p < 0.001$) and PVE values of 9.6–15.8%. Although the *T. dicoccoides* alleles for most of the QTLs (5/7) lowered the trait value, those for *QGwh.hue-1B* and *QGwh.hue-7A* could increase GWH.

Grain yield (YLD): Four chromosomes, 1B, 2A, 3A, and 5A, were identified for YLD (Table 8.2). These QTLs could explain 70–85% of the total phenotypic variation of YLD. *Qyld.hue-2A* and *Qyld.hue-5A*, on chromosome arms 2AL and 5AL, were major QTLs, with LOD > 6.1 ($p \ll 0.001$) and PVE values 23–35% of the phenotypic variance. For all of the YLD QTLs, the contribution of *T. dicoccoides* alleles was negative compared with those of *T. durum*.

Spikelet number/spike (SLS): Four QTLs were revealed for this trait. They were mapped on chromosomes 1B, 2A, 5A, and 6B (Table 8.2). Together these QTLs could explain 49–62% of the entire phenotypic variation. Two major QTLs, *QSls.hue-2A* and *QSls.hue-5A*, on chromosome 2A and 5A, contributed from 15.5 to 21.5% of the phenotypic variation. The contribution of *T. dicoccoides* alleles at all of the detected QTLs was negative compared with those of *T. durum*.

Our QTL mapping results may be very helpful for QTL introgression and cloning. We also found that beneficial alleles existed in wild emmer wheat for some of the traits, i.e., HD, SNP, KNL, and GWH (Table 8.2), which will be useful for wheat improvement. Therefore, our results can accelerate wheat genetic analysis and be especially helpful for marker-assisted introgression of beneficial QTL alleles from *T. dicoccoides* to cultivated wheats.

Wild emmer wheat has proven to be a valuable germplasm source of wheat improvement based on phenotypic performance and population-genetic analysis (Nevo 1995, 2001b). Soller and Beckmann (1988) pointed out that QTL mapping can facilitate the recovery of "cryptic" genetic variation, i.e., the identification of beneficial alleles that are otherwise hidden in a sea of undesirable alleles. In the present study, one parental line, *T. dicoccoides* accession Hermon H52, possessed stripe-rust resistance, short stature, and high tillering capacity, but also many undesirable features for agronomic traits such as of slow growth (late heading), small grains, less spikes, small spikes, low spike weight, and low yield. Among 48 QTLs for the 11 traits, the *T. dicoccoides* alleles for 7 (14.6%) were beneficial (Table 8.2).

Chapter 9

Molecular Evolution and Ecological Stress in Wild Emmer Wheat at Regional and Local Scales: Natural Selection in Action

9.1 Overview

The major problem we investigated, *regionally* and *locally*, in wild emmer wheat and other organisms, was how much of the genetic diversity in macro- and microgeographic scales, at single and multilocus structures of both protein and DNA levels, is parallel, adaptive, processed by natural selection, and contributes to differences in fitness. Our results suggest that, at all geographic levels (Near East Fertile Crescent, Israel, microsites within Israel), coding (allozymes) and largely noncoding (RAPDs, SSRs) genomic diversity is subject to natural selection. We discovered that the multilocus organization of wild emmer wheat *regionally* and *locally*: (1) is *nonrandom*, (2) displays parallel genetic patterns, (3) is heavily structured, and (4) frequently displays similar trends to wild barley, *Hordeum spontaneum*, and other, unrelated plants and animal taxa (Nevo 1998a). The genetic diversity and divergence of wild emmer are positively correlated with abiotic and biotic ecological diversity and stress, displaying genetic, genome, chromosome, and motif patterns that largely correlate with environmental heterogeneity and stress. These results are inconsistent with the predictions of the neutral (Kimura 1983) and nearly neutral (reviewed in Ohta and Gillespie 1996) theories of molecular evolution. Theoretically, both spatial and temporal ecological variation are of prime importance in originating and maintaining genetic diversity in wild emmer, particularly by diversifying and balancing selection. Aridity stress, combining the interaction of water availability and temperature stresses, seems to be the major selective driving force across the genome at both micro- and macrogeographic scales. The following sections present a brief summary of the evidence followed by a theoretical analysis of these conclusions.

9.2 Evidence

Triticum dicoccoides in Israel and Turkey was tested for proteins encoded by 42 gene loci in 1,815 plants, representing 37 populations, 33 from Israel and four from Turkey. The plants were sampled in 33 localities from 1979 to 1987 and showed the following patterns (Nevo and Beiles 1989): (1) Regional and local polymorphism was evident. (2) Altogether, there were 119 alleles, 114 in Israel. (3) Genetic divergence was primarily regional and local, not clinal. (4) Most variant alleles were common (>10%) and not widespread, but rather *localized* or *sporadic*, displaying an *archipelago* type of population genetics and ecology structure. (5) Allozymic variation comprised 40% within populations and 60% between populations. (6) Genetic divergence was sharp over short distances. (7) Regional, central, marginal, and microscale *soil* and *climatic* populations could be well-discriminated. (8) Linkage disequilibria and multilocus organization were substantial and positively correlated with humidity. (9) Allozyme diversity, overall and at single loci, was significantly correlated with and partly predictable by climatic and edaphic factors. (10) Migration did not seem to be a prime factor in population differentiation. (11) Wild emmer harbored large amounts of adaptive genetic diversity and resources that are exploitable in genetic mapping and for the purpose of wheat improvement. (12) *Regional* (Israel and Turkey) analyses of *coding* (allozymes) and partly or largely *noncoding* (RAPD and SSR) *genomic* regions (Fahima et al. 1999, 2001) gave basically *similar* results regarding adaptation.

However, the levels of diversity at both the regional and local scales were: SSR > RAPD > allozymes. Classification and predictability by ecology suggest that natural selection, presumably through aridity stress at all scales, is responsible for the adaptive evolution of the *coding* and *noncoding* genome. Basically, similar results were obtained in microscale analysis at three microsites in northern Israel, Yehudiyya, Tabigha, and Ammiad, that sharply differed ecologically in their microclimate (sun vs shade, Yehudiyya), soil (terra rossa and basalt, Tabigha), and topography-lithology (Ammiad) (Li et al. 1999, 2000a–d, 2001a–d). Results of the micro- and macroscale analyses showed that the level of diversity was SSR > RAPD > allozymes and that it was always higher in all three systems in drier, more stressful habitats. Likewise, genetic diversity was classifiable and predictable ecologically. In all analyses, *regional* and *local*, *protein* and *DNA*, *climatic*, *edaphic*, and *topographic*, natural selection, through *aridity stress*, appears to be a major evolutionary driving force, possibly originating and certainly maintaining genetic diversity of both the *coding* and *noncoding* genomes and resulting in microscale adaptive structures.

The foregoing macro- and microscale evidence across the *coding* and *noncoding* genomic regions displays impressive correlates with abiotic (climatic and edaphic) and biotic (pathogens, vegetation, habitats) environmental stresses. In general, higher levels of genetic polymorphisms, at both the protein (allozyme) and genic and intergenic DNA levels (RFLP, RAPD, SSR, chromosomal, linkage disequilibria and genome organization), are correlated with higher environmental heterogeneity and/or stress. Based on the genetic patterns and thousands of permutation tests, it can be concluded that the patterns observed are *nonrandom* and derive from climatic natural selection governed by aridity stress. In the following sections, we present our theoretical analysis.

9.3 Theory

9.3.1 Evolutionary Forces and Adaptive Complexes

How can the foregoing results be explained theoretically? What is the relative importance of mutation, migration, genetic drift, and natural selection in wild emmer wheat with respect to *macro-* and *micro*geographic genetic diversity, divergence, and architecture at the single locus and multilocus level, for both protein and DNA?

Evidence derived from the genetic structure and differentiation of populations across the Near East Fertile Crescent, from Israel and Turkey, at both regional and local scales and involving several thousands of genotypes, many populations, and numerous genes and repeats across the genome, indicate *nonrandom* structures on a massive scale. The *coding* and *noncoding* genomes both seem to be subject to local and regional natural selection. Genetic diversity is correlated with and predictable by the combination of a few ecological variables that are primarily associated with *aridity stress*, i.e., water availability and temperature. Regional and local ecological differences are largely independent of population size (except for very small and isolated populations) and strongly implicate natural selection in population and genotype differentiation. Various forms of selection, primarily through *diversifying*, *balancing*, and *cyclic* selection regimes, appear to be heavily involved, singly or in combination, in determining genetic structure and differentiation of populations at the various life cycle stages of wild emmer. Only critical future studies can identify these specific single or combinatorial selection pressure(s) operating in each particular case. Presumably, in ecologically contrasting niches, *diversifying selection* operates between two (Levene 1953) or more (Karlin 1979) niches. Heterosis cannot be an explanatory model in inbreeding wild emmer. Other evolutionary forces, including mutation, migration, and genetic drift, may interact with natural selection, either directly or indirectly, but they appear to be secondary in importance. Natural selection seems to override migration and genetic drift in establishing population-genetic structure and multilocus organization, both *locally* and *regionally*.

9.3.2 Maintenance of Genetic Diversity in Wild Emmer Wheat

Explaining the maintenance of genetic diversity in natural populations has been a central problem of evolutionary genetics since the discovery of abundant genetic polymorphisms in nature (e.g., Lewontin 1974; Nevo 1978, 1989a; Mitton 1994). The problem climaxed following the discovery of enormous DNA polymorphisms across phylogeny and the discussion on the levels of quantifying genetic diversity (Barton and Turelli 1991). The following discussion focuses mainly on models of multilocus systems subjected to *spatiotemporal*, *stabilizing*, *diversifying* selection regimes in fluctuating and variable environments.

9.3.2.1
Diversifying and Balancing Natural Selection

Theoretically, spatial and temporal variations of selection ("diversifying selection") could maintain genetic polymorphisms, although the conditions of their applicability are narrowly limited (Levene 1953; Haldane and Jayakar 1963; Karlin and Levikson 1974; Lande 1975; Felsenstein 1976; Hedrick et al. 1976; Ewens 1979; Hoekstra 1978; Maynard-Smith and Hoekstra 1980; Karlin and Levikson 1974; Karlin and Avni 1981; Karlin 1982; Hedrick 1986). Spatial variation appears to be more effective than temporal variation, though their coupled action could reinforce the maintenance of polymorphism (Ewing 1979; Karlin 1979, 1981, 1982). Most results related to selection variation in time analyze the one-locus case. However, the maintenance of polymorphism may be reinforced in the case of two loci (Gavriletz and Hasting 1994, 1995; Korol et al. 1994, 1996a; Kirzhner et al. 1995; Zivotovsky and Feldman 1996) or multilocus structures (Hedrick et al. 1976; Gillespie 1978; Karlin 1979, 1981, 1982; Hedrick 1986; Nevo 1978, 1989a,b; Kirzhner et al. 1996).

Our *regional* and *local* studies of wild emmer wheat suggest that climatic factors, primarily *aridity stress*, either directly as water availability factors (rainfall, evaporation, humidity) or indirectly as different soil types, select for *niche-specific* or *niche-unique* allozyme, RAPD, or SSR genotypes, or alleles. The *regional* pattern across Israel and Turkey (Nevo et al. 1982; Nevo and Beiles 1989; Fahima et al. 1998, 2001) is mirrored by local patterns at the three microsites at Yehudiyya, Tabigha, and Ammiad (Nevo et al. 1988a,b, 1991c; Li et al. 1999, 2000a–d, 2001a–d), reinforcing the case for selection at all geographic scales. Generally, and very remarkably, the diversity of the largely *noncoding* genome, as demonstrated by SSR and RAPD, allowed differentiation between environments, habitats, and niches in a manner parallel to that demonstrated by *coded* allozymes. Significant differentiation at a distance of just a few meters (*sun* vs *shade* in Yehudiyya; *terra rossa* vs *basalt* in Tabigha; and *Karst*, *Ridge*, *Slope*, and *Valley* in Ammiad) or across a mesic-xeric geographic gradient of several hundred kilometers in Israel, reflects the intimate relationship at the molecular-genetic level between wild emmer wheat and its immediate *local* or *regional* environments.

Neither migration, which homogenizes diversity, nor genetic drift, which leads to fixation and decrease of ecologically largely irrelevant diversity, explains our data. Autocorrelation analysis excludes migration as a dominant factor. The large-scale genomic data, involving both *coding* and *noncoding* genomic regions, of wild emmer wheat are explicable only by direct *balancing* and *diversifying* selection in different environments. This ecological differentiation is *robust*, in that thousands of permutation tests never achieve the significance of the original data, thus ruling out a stochastic effect. The *specificity* and *uniqueness* of genotypes or alleles at both the *micro*- and *macroscale* clearly display *diversifying* selection in action.

Generally, and most remarkably, similar environmental correlates with the combined water availability and temperature factors were found for *coding* and *noncoding* regions of the genome. However, these correlates, while similar and in the same direction, were not identities. Sometimes, high diversity values were displayed by allozymes, but the SSR diversity values were low (Li et al. 2000a–c). This incongruity of the *coding* and *noncoding* genomic regions suggests that the general similarity is not the result of hitchhiking due to the highly inbreeding mating system of wheat; instead there may be some independence of the structural and regulatory genomic elements. This conclusion needs critical testing by transplant experiments, QTL analysis, and other methods.

SSR repeat structure testifies to balancing selection of size constraints (Li et al. 2000a–c). These results cannot be explained by biased mutation (Garza et al. 1995; Dermitzakis et al. 1998) or size-dependent mutability, unless this bias is driven by ecological stress (Falush and Iwasa 1999). Natural selection appears to act on both the short and long SSR arrays. This implies that repeated arrays at some SSR loci are functional and are maintained by selection within certain minimum and/or maximum thresholds. Theoretically, it has been repeatedly demonstrated that natural selection can control the length of SSR repeats (Harding et al. 1992; Stephan and Cho 1994; Samadi et al. 1998). Some evidence supports this theory. In the promoter of the URA3 reporter gene in yeast, the threshold number of $(CTG)_n$ for URA3 expression is below 25 units (Miret et al. 1998). Thus, natural selection acts as an upper-limit truncating mechanism. The distinct structural differences between euchromatic and heterochromatic tandem arrays in most eukaryotes are, at least partially, due to balancing selection (Stephan and Cho 1994; and additional examples in Li et al. 2001d).

9.4 Effect of Ecological Selection on Single SSR Locus

9.4.1 Mutational Mechanisms of SSRs

The genomic abundance of SSRs and their various functions are associated with its mutation rate, which is very high (10^{-2}–10^{-6} events per locus per generation) compared with the rates of point mutations which vary among species (reviewed in Hancock 1999). The variation is predominantly manifested as changes in the number of SSR repeats. Two mutational mechanisms can be invoked to explain such high rates of mutation. The first involves a single DNA double-helix slippage strand mispairing (slippage) during DNA replication (Tachida and Iizuka 1992). The second involves recombination between DNA strands (Harding et al. 1992). The efficiencies of the two mechanisms may also depend on the effects of environmental factors (Li et al. 2000a–d, 2001a–d). Many changes in the repeat number at SSR loci are caused by slip-strand mispairing errors during DNA replication (reviewed in Eisen 1999). Some of these errors are corrected by exonucleolytic proofreading and mismatch DNA repair, but many escape repair and become mutations (e.g., Moore et al. 1999). It has been proved that microsatellite instability is associated with DNA mismatch repair genes. When these genes mutate or become defective, SSR instability consequently increases (e.g., Boyer et al. 1995; Sia et al. 1997; Clark et al. 1999; Flores and Engels 1999). Recombination could potentially change the SSR length by unequal crossing over or by gene conversion (Hancock 1999). Most repeat-length changes arise through complex recombination events including intra-allelic duplications/deletions and inter-allelic transfers of repeats (Buard et al. 1998). Unequal exchanges in combination with random genetic drift and selection can have a strong effect on the accumulation of tandem-repetitive sequences in the genome (Charlesworth et al. 1994). In particular, accumulation of highly repeated sequences is expected to occur in regions with very low recombination rates and weak selective constraints on array length (Charlesworth et al. 1986).

9.4.1.1 Interaction of Replication Slippage and Recombination

Recombination could potentially change the SSR length by unequal crossing over or by gene conversion (Hancock 1999; Brohele and Ellegren 1999; Jakupciak and Wells 2000; Richard and Pâques 2000). Most of such changes arise through complex recombination events including intra-allelic duplications/deletions and inter-allelic transfers of repeats (Buard et al. 1998). Structural analysis of mutant sequences derived from six alleles in human sperm revealed that it is the rate of intra-allelic rearrangement that increases with array size, and that intra-allelic duplication events tend to cluster within homogeneous segments of alleles. Both phenomena resemble features of trinucleotide repeat instability. Unequal exchanges in combination with random genetic drift and selection can have a strong effect on the accumulation of tandem-repetitive sequences in the genome (Charlesworth et al. 1994).

Recent studies point to the important role of non-reciprocal recombination (gene conversion) in destabilization of tandem repeats numbers (for both micro- and minisatellites) (Richard and Pâques 2000; Jakupciak and Wells 2000). Depending on the motif, the non-reciprocal exchange may yield unidirectional (say, only contraction) or bidirectional changes. The effect may be associated with either meiosis or mitosis, albeit with different rates.

In our study of microsatellite diversity in wild emmer wheat, we found a very strong effect of interaction between mean repeat length and SSR locus distance from the centromere on the number of alleles and variance in repeat size at SSR loci (Li et al. 200c; 2001a). We suggested that this effect may reflect possible influence of replication slippage during recombination-dependent DNA repair. Indeed, strand exchange between two homologous chromosomes should create a four-stranded configuration, called Holliday structure associated with mismatched (heteroduplex) DNA regions. These regions undergo replication-dependent correction, hence a slippage mechanism may also work in recombination tracts involving SSR arrays (Brohele and Ellegren 1999; Gendrel et al. 2000; Li et al. 2000c, 2001a). In yeast, such "repair-slippage" events associated with gene conversion and leading to SSR expansion or contraction may occur 800-fold, compared to "replication-slippage" (Richard and Pâques 2000). Recombinational repair during gene conversion was hypothesized to play a major role in trinucleotide expansion for a series of human neurological disorders (Jakupciak & Wells 2000; Richard & Pâques 2000).

9.4.1.2
Functional Perspectives of SSRs

Although SSRs are generally considered DNA markers, the functional significance of at least some SSRs has been proved by various critical experiments in many biological phenomena (reviewed by Kashi et al. 1997; Kashi and Soller 1999; Li et al. 2001d). These functions include the regulation of gene activity (transcription, translation), chromatin organization, genome size, recombination, DNA replication, and the cell cycle. SSRs showed remarkably similar distribution patterns in wheat and rye, indicating the special role of SSRs in chromosome organization of the tribe Triticeae (Cuadrado and Schwarzacher 1998). SSR DNA sequences are capable of forming a wide variety of unusual DNA structures with simple and complex loop-folding patterns (e.g., Pearson and Sinden 1996; Zheng et al. 1996; Karlin et al. 1998; Catasti et al. 1999). The formation of such stable structures offers a mechanism of unwinding that is advantageous during transcription and which provides unique protein-recognition motifs (Catasti et al. 1999). The centromeric region of chromosomes in many species comprises numerous tandem repeats that affect the organization of the centromere (e.g., Centola and Carbon 1994; Schmidt and Heslop-Harrison 1996; Brandes et al. 1997; Cambareri et al. 1998; Areshchenkova and Ganal 1999). Numerous SSR and minisatellite DNAs have been proposed as hotspots for recombination (Jeffreys et al. 1998; Templeton et al. 2000). Some support for this idea has been provided by experiments with SV40 (Wahls and Moore 1990a), yeast (Treco and Arnheim 1986), human cells (Aharoni et al. 1993; Templeton et al. 2000), mammalian cells (Wahls and Moore 1990b), and bacteria (Murphy and Stringer 1986). Dinucleotide repeat sequences are preferential sites for recombination because of their high affinity for recombination enzymes (Biet et al. 1999). In addition, SSRs may affect DNA replication (Field and Wills 1996; Caligo et al. 1999). For example, SSRs may serve as an arrest site for DNA replication (Baran et al. 1987) and they can affect enzymes controlling the cell cycle. Some genes controlling the cell cycle carry short repeated sequences, which are important in cell fidelity and growth control (e.g., Hawn et al. 1995; Bertoni et al. 1999; Codegoni et al. 1999; Monckton et al. 1999; Johannsdottir et al. 2000). SSRs in promoters can act as regulatory elements that enhance or suppress transcription. Modification of SSR tracts can markedly change transcriptional activity of the promoter (e.g. McKeon et al. 1984; Biggin and Tjian 1988; Johnson et al. 1988; Hoffman et al. 1990; Punt et al. 1990; Lafyatis et al. 1991; Stallings et al. 1991; Sandaltzopoulos et al. 1995; Chen and Roxby 1997; Aranda et al. 1998; Li et al. 1998; Meloni et al. 1998; Shah et al. 1998; Kenyon and Craig 1999). SSRs can affect gene translation. There are also many examples of SSRs increasing or decreasing the rate of translation (e.g., Ivanov et al. 1992; Sandberg and Schalling 1997; Henaut et al. 1998; Martin-Farmer and Janssen 1999; Timchenko et al. 1999). Some SSRs, found in upstream activation sequences, serve as binding sites for a variety of regulatory proteins (e.g., Lue et al. 1989; Aharoni et al. 1993; Epplen et al. 1993). SSR repeat number also affects protein binding (e.g., Solomon et al. 1986; Winter and Varshavsky 1989). Analysis of human genetic disorders provides rich evidence of an important functional role for SSRs (e.g., Rubinsztein et al. 1995; De Fonzo et al. 1998; Awad et al. 1999; Fujisawa et al. 1999; Rubinsztein 1999; Warpeha et al. 1999; White et al. 1999; Leung et al. 2000).

All of these examples demonstrate the functional importance of SSRs within both *coding* and *noncoding* sequences of the genome. As such, these repetitive elements are subject to evolutionary forces that shape the structure of the genome and which enable the populations to adapt to their changing environments. Here we suggest that different forms of natural selection appear to shape allele size, frequency, and distribution in natural populations of *T. dicoccoides*. However, more critical experiments are needed to determine the function of the *T. dicoccoides* SSRs and their involvement in the adaptation of wild emmer wheat populations to different ecological conditions.

Viewed as functional elements of the genome (Kashi et al. 1997; Kashi and Soller 1999), the special characteristics of SSRs as mutational hotspots, for example in humans (Aharoni et al. 1993; Templeton et al. 2000), have led to the proposal that SSRs may be a major source of genetic diversity and evolutionary adaptation (Kashi et al. 1997; Kashi and Soller 1999). Some SSRs, at least functional SSRs, seem to be selective targets per se (Innan et al. 1997; Kashi et al. 1997; King et al. 1997; Kashi and Soller 1999; King and Soller 1999; Li et al. 2000a–d). These functional SSRs may respond to environmental changes and adjust gene expression in relation to timing and context (King et al. 1997; King and Soller 1999). In other words, ecological factors may be evolutionary forces selecting certain SSRs for organismal fitness (Angers et al. 1999; King and Soller 1999). There is only circumstantial evidence for this assumption, however, and further experiments are needed. Our research on *local* and *regional* SSR divergence patterns showed that microclimatic and edaphic factors affect SSR diversity and divergence in wild emmer wheat (Li et al. 2000a–d, 2001a–d; Fahima et al. 2001), as reviewed below.

9.4.1.3 Microclimatic Selection

In the Yehudiyya microsite *sun*-vs-*shade* niches, which are separated by a few meters, 21 *niche-specific* and five *niche-unique* alleles appeared at a significantly higher frequency in a single niche, either *sun* or *shade* (Li et al. 2000a). Permutation test suggests that explaining the observed *niche-specific* or *niche-unique* patterns of allele distribution by random events is improbable ($p < 0.0002$). The observed niche-*unique* alleles distributed in neighboring trees may reflect microclimatic *diversifying selection* because *T. dicoccoides* seeds can migrate for about 1 m per generation (Golenberg 1987). Without microclimatic *diversifying selection*, which overrides the homogenizing effects of migration, one would expect that these *niche-unique* alleles would be spread over neighboring areas, in either *shady* or *sunny* niches, but this was not the case. Therefore, we believe that the revealed patterns reflect an adaptation to microclimatic *specificity* of the two, very close (meters apart) microclimatic niches. The lower temperature and radiation intensity in the *shady* niche, and drought stress and high radiation in the *sunny* niche, may favor certain *niche-specific* or *niche-unique* (Li et al. 2000a–d, 2001a–d) alleles of *T. dicoccoides*, as demonstrated earlier for nuclear allozyme alleles (Nevo et al. 1988b). Transplant experiments could assess the significance of these *unique* alleles and their contribution to fitness. In wild barley at "Evolution Canyon" such experiments showed a reduction in the fitness component in alternative niches (Lavie et al. 1993).

9.4.1.4 Edaphic Selection

Significant, differential, edaphic SSR diversity and divergence were revealed at both microgeographic and regional scales (Li et al. 2000b). After exclusion of rare alleles, observed in <15 individuals, eight *basalt-specific* and seven *terra-rossa-specific* alleles were found in the two edaphic groups at *micro-* (100 m) (Li et al. 2000a) and *macro-* (10,000 m) geographic scales (Li et al. 2000b). One *terra-rossa-unique* and one *basalt-unique* allele were found among these *soil-specific* alleles, notwithstanding the gene flow by seed dispersal caused by animals (ants, rodents), wind, and casual outcrossing (~1%, Golenberg 1987). For example, the *terra-rossa-unique* allele (repeat size = 21) at locus GWM18 was found in the terra rossa subpopulation of the 50-m transect at the Tabigha microsite (12 copies) and in the Ammiad population (42 copies) 4 km away (Fig. 5.7). This pattern cannot be explained by isolation by distance, i.e., a positive correlation between geographic and genetic distance (Wright 1943). Permutation tests showed that out of 10^6 randomly shuffled sets, in no case did the number of *soil-specific alleles* exceed four (in real data: 15 *soil-specific alleles*). Thus, the significance of the actual observed allele *soil specificity* is at least $<10^{-6}$. Moreover, in no case did random permutations produce a *soil-unique* allele. This result suggests with high confidence the *non-randomness* of the revealed soil-specific population differentiation. Moreover, such allele *soil-uniqueness* may indicate that these two alleles are not only markers linked to the selected loci, but may be the selected targets themselves (or sites within selected genes). The edaphic-ecological selection seems to act upon at least these two SSR loci (Li et al. 2000b). Similar *soil-specific* and *-unique* alleles at the *local* and *regional* scales were also found for RAPDs and allozymes.

9.4.1.5
Topographic and Microgeographic Drought Effects

In the Ammiad microsite, *habitat-specific* and *-unique* alleles were observed, particularly in the Karst subpopulation (Li et al. 2000c). SSR diversity tended to be higher in drier habitats than in the relatively wet habitats, as is also true for allozymes at the regional scale (and 21 other diverse species; Nevo and Beiles 1988), as well as for regional SSR in wild emmer wheat (Fahima et al. 2001). These results suggest that natural selection, presumably through *aridity stress*, the major regional stress factor, acts at least upon a part of the largely *noncoding* SSR diversity and divergence, as it does on allozymes (Nevo 1998b).

9.4.1.6
Natural Selection and Linkage Disequilibria of SSR Alleles

Linkage disequilibrium (LD) in wild emmer wheat provides a very useful genomic measure to unravel evolutionary dynamics at both the regional and local scales. In the three populations of wild emmer wheat, we also found much more significant LD than expected by chance (10–36% vs 5%). Moreover, a large number of significant linkage disequilibria (LDs) were niche-*specific* or niche-*unique* among geographically very close micro-environments, as was also shown for single alleles. In the Yehudiyya microsite (*shady* vs *sunny* niches), 86–94% of the LDs were shade-*specific* at significance levels of $p < 0.05$, 0.01, and 0.001. Permutation tests showed a high level of significant LDs ($p < 0.0002$) for shade specificity. At the Tabigha microsite, 83% of the LDs were *terra-rossa-specific* and 65% were *basalt-specific*. At the Ammiad microsite, 94% and 96% of LDs were *habitat-specific* between the two extremely opposite habitats, Valley and Karst. Permutation tests indicate a very low chance of such differences being due to a sampling effect ($p < 0.0002$). Among these neighboring subpopulations at the three microsites, *niche-unique* LDs were also found. Some authors have argued that such associations are evidence of selection (Allard et al. 1972), while others have pointed out that they could arise by chance from hitchhiking (Hedrick 1980) or founder effects (Avery and Hill 1979). The studied populations are located in the center of origin and diversity of wild emmer wheat and involve tens of thousands of individuals. This feature rules out a founder effect on the observed LDs. The linkage effect of loci on chromosomes was not significant either, suggesting little effect of linkage hitchhiking on the observed LDs.

Selfing, as a cause of hitchhiking, exerts a stronger influence on LD than linkage does, since it affects all of the genomic regions simultaneously. But selfing is repeatedly interrupted by relatively high outbreeding bursts during humid spells, up to 9% (and possibly more by outcrossing) (see evidence in wild barley, Brown et al. 1978). Hence, the genomic LD selfing hypothesis may be repeatedly contradicted. These outbreeding spells may provide the opportunity for the generation of novel clones to be selected by environmental dynamic stresses in space and time. Even without periodic bursts, which may reinforce genetic exchange, the outcrossing rate in humid environments is several times higher than in xeric environments, possibly permitting continuous genetic exchange followed by clonal microniche severe selection. This hypothesis awaits critical theoretical and empirical testing.

The evidence described above was derived from our three microgeographic tests and leads us to believe that natural selection, in its diverse forms, primarily *diversifying selection* following the Levene (1953) model, plays an important role in SSR, RAPD, and allozyme microecological divergence, as reflected by the *niche-specific* and *niche-unique* alleles of these genetic systems. No evolutionary force other than natural selection seems able to generate the genetic divergence found in our microscale climatic, edaphic and topographic ecological contrasts. Transplant experiments and their differential fitness effects as well as genetic mapping could further verify our conclusion.

9.4.2 Ecological Selection and Clustering of Allele Size Frequency Distribution

Our studies revealed *regional* and *local* geographic clustering patterns of allele-size frequency distribution in *T. dicoccoides* located in contrasting edaphic habitats (Fig. 5.10a, b) For example, an allele cluster $(GA)_{36-38}$ at locus GWM186 was found in the terra rossa subpopulation, but not in the neighboring (meters away) basalt subpopulation at the Tabigha microsite (Fig. 5.10a). Across the three microsites, the terra rossa and basalt soil groups obviously differed in the number of alleles and allele frequencies in the two clusters at locus GWM219. At locus GWM99, allele distribution seems to be limited at $(AC)_{17}$ in the terra rossa group, but expanded to 21–24 repeats and forming a small cluster of alleles at $(AC)_{21-24}$ in the basalt group. A small *terra-*

rossa-specific allele cluster was revealed at loci GWM415 and GWM18 (Table 5.19a and Fig. 5.10a, b). These findings may indicate an association between formation of allele clusters and edaphic ecology, supporting the selection hypothesis (Epplen et al. 1993; Garza et al. 1995; Dermitzakis et al. 1998) but refuting the biased mutation model (Garza et al. 1995; Bell and Jurka 1997; Dermitzakis et al. 1998). While, transplant experiments could test this hypothesis, it is nonetheless reassuring that allozymes, RAPDs, and SSRs display microclimatic, edaphic, and topographic *niche specificity* and *uniqueness*. This parallel pattern in *coding* and *noncoding* genomic regions reinforces the selection interpretation of the results.

9.4.3 Ecological Effect on SSR Mutational Mechanisms

DNA replication slippage (Tachida and Iizuka 1992; Amos and Rubinstein 1996; Dermitzakis et al. 1998) and recombination (Harding et al. 1992) are thought to be two major mutational mechanisms. Likewise, it has been shown that environmental conditions may play an important role in determining the extent of DNA mismatch-repair and mutability (e.g., oxidative stress, Brentnall et al. 1996; stress of chronic inflammation, Jackson et al. 1998) as well as recombination rates (Korol et al. 1994; Korol 1999).

In the present study of *T. dicoccoides* across the three microsites, there was a significant interaction effect of average repeat number (ARN) × soil type (*terra rossa* vs *basalt*) on dinucleotide SSR diversity (Li et al. 2000a,b). ARN may be considered an indirect estimate of replication slippage, owing to the known positive correlation between ARN and replication slippage (see Innan et al. 1997). These results may also indicate that edaphic stresses accelerate replication error or decrease the mismatch repair capability of *T. dicoccoides*, thereby increasing SSR diversity. This source of mutation, accompanied by *diversifying* selection, could explain why the subpopulation on the drier terra rossa soil showed higher SSR diversity than that on the relatively wetter basalt soil. Similarly, in the Yehudiyya microsite study of wild emmer in neighboring *shady* vs *sunny* microniches, a significant effect of the interaction niche × ARN was found for variance in repeat sizes and number of alleles (Table 5.14). These results suggest that microecologic stresses may potentially affect either replication slippage or mismatch DNA repair, and when complemented by *diversifying* selection result in edaphic divergent patterns of SSR diversity.

It was found that the interaction soil × locus distance from the centromere significantly affected the number of alleles at an SSR locus, suggesting that stresses of different soil types may affect SSR diversity by influencing recombination events (Li et al. 2000b). The recombination rate can increase as a result of adaptation to stressful conditions and decrease under optimal stable conditions (Hoffman and Parsons 1991; Korol et al. 1994; Korol 2001), as demonstrated in the microscale model of "Evolution Canyon" in the soil fungus *Sordaria fimicola* and in male recombination in *Drosophila melanogaster* (Nevo 1997).

9.5 Natural Selection and RAPD-DNA Diversity

In our microsite research, we used RAPD markers to detect genetic diversity within natural populations of *T. dicoccoides* from the Ammiad and Yehudiyya microsites. At the Yehudiyya microsite, significant genetic differentiations were found at single-, two- and multilocus structures between the neighboring *shady* and *sunny* niches across 12 trees (Li et al. 1999). These DNA polymorphisms appear to be associated with microclimatic stresses (high radiation and high temperature in the *sunny* niche vs. low radiation, lower temperature, and high competition for water in the shady niche). The effects of niches, trees, and the niche × tree interaction on allele frequencies were analyzed. The results confirmed that *sunny* vs *shady* microsite diversity existed around the 12 sampled trees (Li et al. 1999). We found that the proportion of two-allele/two-locus associations was significantly higher (13% and 9% for the *shade* and *sun* subpopulations, respectively) than the 5% expected by chance. In addition, *niche-specific* LDs were also observed. The pronounced niche effect on the *niche-specific* LDs suggests that microclimatic *diversifying selection* directs the two-locus associations. The *niche specificity* of LDs in each of the two microniches, beside the LDs shared by both niches, suggests *niche-specific* and *-unique* natural selection effects on two-locus associations. Multilocus structures also seem to result mainly from natural selection, and not by chance from population subdivision or founder effects. Gene diversity of RAPDs is higher in populations in the presumably more temporally variable *sun* niche than in those in the *shade*. These results are largely parallel to the previous allozymic results (Nevo et al. 1988b) and SSR patterns (Li et al. 2000d) at single-locus and multilocus levels. Like the allozymic and SSR results, the RAPD-DNA results suggest that *microclimatic diversifying* selection overrides migration and appears to play an important role in DNA differentiation patterns (Li et al. 1999). In the Ammiad microsite containing North, Valley, Ridge, and Karst microhabitats, *habitat-specific* RAPD alleles were found, and the RAPD diversity (*He*), as well as allozymic *He* and SSR *He*, were *negatively* and significantly correlated with soil water-content in the early growing seasons of wild emmer wheat. These results suggest that climatic and habitat selections through aridity stress act upon both protein-*coding* and RAPD largely *noncoding* DNA regions. Transplant experiments could test this conclusion.

Other studies also showed a correlation between environment and RAPD diversity. In wild barley, *Hordeum spontaneum,* a similar allozyme and RAPD pattern was also found at the "Evolution Canyon" microsite (Nevo 1997, 2001c; Owuor et al. 1997; Nevo et al. 1998), with opposing north-facing (NFS) and south-facing (SFS) slopes, higher genetic diversity was found in the more stressful SFS than in the less stressful NFS. RAPD polymorphism generally increased up-slope with stress. Even mobile elements, e.g., the BARE-1 retrotransposon of wild barley, displayed an increase in copy number with the height and dryness of the slope in "Evolution Canyon" (Kalendar et al. 2000).

Owuor et al. (1997) also found that RAPD polymorphism appeared to parallel allozyme diversity at the single-locus level and regarding LDs, displaying positive correlation with stress (Nevo et al. 1997). In 21 populations of *H. spontaneum* sampled in Israel, Turkey, and Iran, high RAPD diversity, like allozyme diversity, was significantly associated with more stressful environments, either with hot or cold steppes or deserts. The positive correlation indirectly suggests that natural selection is a significant contributing factor to both RAPD and allozyme diversities, both of which are correlated with environmental stress (Nevo et al. 1998). Intriguingly, parallel correlates between molecular and ecological diversities are displayed by wild barley and wild emmer wheat, again supporting a *selective* rather than *nonselective* model.

In *Spalax ehrenbergi* Nehring, 1898, a mole rat superspecies in Israel (now split into four species, Nevo 1991), the level of RAPD polymorphism, e.g., DNA fingerprints, extranuclear mitochondrial DNA (mtDNA), allozyme, and minisatellite DNA, increased southward, toward the Negev Desert, and eastward, towards the Syrian Desert, with aridity stress. This trend was observed not only for the number of polymorphic loci, but also for the frequency distribution of different bands (Nevo et al. 1996). Natural selection in outbreeder *Spalax* shapes the *coding* and *noncoding* sequences, similar to the shaping in inbreeder wild cereals. This demonstrates that it is not the mating system but ecological stress that shapes these genomes, usually selecting for higher genetic diversity in more stressful and heterogeneous environments in diverse taxa across life at both the *macro*- and *micro*scale (Nevo 1998b).

The occurrence of environmentally induced heritable changes (temperature, imbalance of nutrients) in certain flax varieties has been shown to be accompanied by changes in the genomic DNA (reviewed in Cullis et al. 1999). In varieties with extreme changes, termed genotrophs, hybridizing restriction digests and blotting have shown that 253 polymorphic bands produced by 320 RAPD primers were derived from all parts of the genome, namely the highly repetitive, middle-repetitive, and low-copy-number sequences (Cullis et al. 1999). The pattern observed with the RAPD markers is consistent with earlier data on nuclear DNA estimations (i.e., 5S rRNA genes, Schneeberger and Cullis 1991) between and within different genotrophs. The results are in agreement with earlier studies (Cullis 1990) showing that a large fraction of the genome is susceptible to variation when environmentally induced heritable changes occur. Within this fraction of the genome, the variations that occur in response to environmental perturbations in susceptible varieties of flax appear to be distributed throughout the genome. However, the process does not appear to be the result of random variation (Cullis et al. 1999) and suggests *selection*.

The foregoing and other studies (Dawson et al. 1993; Nevo et al. 1995, 1996, 1998; Owuor et al. 1997; Owuor and Nevo 2001) repeatedly showed a correlation between RAPD diversity and environmental heterogeneity and stress, and a parallel pattern between RAPD and other molecular markers. These results suggest that natural selection, through environmental stress, is a major cause of RAPD diversity, affecting not only *coding* but also *noncoding* genomic regions, probably often without the assumed hitchhiking effect (see below).

9.6 Natural Selection and Allozymic Diversity

In the analysis comparing allozymes, RAPDs, and SSRs in Ammiad (Li et al. 2000d), spatiotemporal diversity at 35 allozyme loci was analyzed in 1,207 individuals of wild emmer wheat over a period of 6 years at the Ammiad microsite in northern Israel. This study used new ecological local data in the correlation analysis and two additional sample sets (1988 and 1993) in addition to the previously published allozymic data (1984–1987) (Nevo et al. 1991c). Significant spatial and temporal variations of allele frequencies and levels of genetic diversity were detected in the North, Valley, and Karst subpopulations. Significant associations were observed among allele frequencies and gene diversities at different loci, indicating that many allele frequencies changed coordinately *over time*. The variation in soil-water content and rainfall distribution in the growing season *significantly* affected allele frequencies, numbers of alleles, and gene diversity at a number of loci. The *locus-specific* spatiotemporal divergence and strong allelic correlation or locus correlation, as well as the functional importance of allozymes, suggest that natural selection, presumably through water stress, is an important force adaptively directing spatiotemporal allozyme diversity and divergence in wild emmer wheat at the Ammiad microsite. Similar results were deduced in the earlier comprehensive allozyme study at Ammiad (Nevo et al. 1991c).

Previous studies presented extensive evidence emphasizing that the number of amino acid substitutions responsible for adaptation is generally quite small, as are the effects on fitness (for review, see Gillespie 1991). For example, lactate dehydrogenase (LDH) showed adaptations to different temperatures and pressures in ectotherms (reviewed in Somero 1983; Hochachka and Somero 1984); and in killifish (cited in Mitton 1997). Similar microadaptation to temperature, pH, and other stresses was also found for other proteins, such as phosphoglucose isomerase, hemoglobin, osmoregulatory proteins, insulin, and lysozyme c (reviewed in Gillespie 1991; Mitton 1997; Nevo 1998b). Clearly, allozyme diversity is largely functional, as suggested by parallel *local*, *regional* and *global* population-genetics studies, biochemical and physiological differential studies of genotypes, and convergence in molecular evolution studies.

The allozymic genetic diversity found in nature – at *global*, *regional*, and *local* scales – is *nonrandom*, heavily structured, frequently displays parallel trends in unrelated taxa, is positively correlated with abiotic and biotic ecological diversity and environmental stress (for review, see Nevo et al. 1984a; Nevo 1998a, b), and is often negatively correlated with population size. These results are inconsistent with the predictions of the neutral theory of molecular evolution and instead appear to be primarily driven by natural selection. Laboratory experiments using marine organisms show that inorganic and organic pollutants cause fast, differential genotypic mortality, indicating that the environment selects allozymic genotypes. Therefore, natural selection can drive allozymic diversity and divergence (Nevo 1986b). The grand parallelism across all geographic scales and in diverse taxa across phylogeny appears to be robust evidence for an interface and association between environmental stress and heterogeneity and genome evolution.

9.7 Random Genetic Drift and Neutral Theory of Molecular Evolution

The constancy of gene frequencies in the absence of selection, mutation, and migration cannot be explained to be absolute in populations of limited size. Wright (1931) was the first to point out the importance of random variation of gene frequency in understanding the evolution of a Mendelian system. In his evolutionary model, the frequency of a given gene in a population may be modified by a number of conditions, including recurrent mutation to and from it, migration, selection of various sorts, and, last but not least, mere chance variation (i.e., random sampling of gametes from generation to generation in a finite population; Wright 1931). Such random chance variation is called *random genetic drift*. Later, the neutral theory (Kimura 1968) or nearly neutral theory (Ohta 1976, 1992) asserted that the majority of evolutionary changes at the molecular level are caused *not* by Darwinian selection but by random drift of selectively neutral or nearly neutral mutants. According to both theories, evolutionary changes and polymorphisms are mainly due to neutral mutations and random drift.

As more protein sequence data became available, it became clear that the rate of protein evolution differs greatly and that protein evolution cannot be fully explained by these two neutral theories (see Ohta and Gillespie 1996 for review). In the present study, as well as in previous allozyme studies (reviewed in Nevo 1998b), significant associations were revealed between protein (allozyme) polymorphisms and ecological factors (e.g., aridity index) and temporal climatic changes at Ammiad and other microsites. Random genetic drift cannot explain the *ecological divergence*, driven by aridity stress, at these microsites. Also, the three microsite studies of wild emmer in northern Israel are located at the center of the genetic diversity and origin of wild emmer wheat, with each microsite located within a large population containing thousands of individuals of wild emmer. In such large populations, effects of random genetic drift and stochastic events should not be large, or may even be negligible. In the light of the present research of the molecular evolution of wild emmer, it seems that natural selection is a significant driving force, shaping the structure of the three populations at the microgeographic sites and regionally across them, as well as the regional genetic patterning across the Near East Fertile Crescent.

In the early 1980s, according to DNA sequence data, substitutions in *noncoding* DNA and silent substitutions were considered neutral or nearly neutral (see Ohta 1997 for review). Later, however, studies on DNA polymorphisms, provided evidence that the neutral and nearly neutral theories are not applicable to all data (Ohta and Gillespie 1996). The microgeographic studies in wild emmer wheat also showed that these two theories may not be suitable for the observed allozyme, RAPD, and SSR DNA diversity and divergence. Random genetic drift as a major force cannot result in the *niche-specific* and *niche-unique* alleles and LDs at allozymes, SSR, and RAPD loci revealed among neighboring subpopulations of *T. dicoccoides* at the Ammiad, Tabigha, and Yehudiyya microsites. Neither does the correlation between the levels of SSR, RAPD, and allozyme diversity and aridity stress at the Ammiad microsite support the hypothesis of random genetic drift. Rather, it suggests that the entire genome, both *coding* and *noncoding*, may be subject to a substantial amount of *selection* on single, two (LD), multilocus, and genome-organization structures. This may be true even if relatively very few strongly selected loci spread their effect on the entire genome, due to the very high average selfing rate. As suggested earlier, and documented in Brown et al. (1978), inbreeders may have periodic bursts of outbreeding in which novel recombinants are generated and rapidly tested by natural selection. The modeling of these inbreeding-outbreeding scenarios and their effect on cereal genomes is under way at the Institute of Evolution.

Even in a small isolated population, the random drift effect may be overriden by selection. For example, research done on a subterranean mammal, *Spalax ehrenbergi,* as well as other evidence indicate that appreciable polymorphism can be preserved in small isolated populations consisting of several dozen or a hundred individuals (Nevo et al. 1997). Current theoretical models predict rapid gene fixation in small panmictic populations without selection, mutation, or gene inflow. Using simple multilocus models, Nevo et al. (1997) demonstrated that moderate stabilizing selection (with a stable or fluctuating optimum) for traits controlled by additive genes could oppose random fixation in such isolates during thousands of generations. They also showed that, in selection-free models, polymorphism persists only for a few hundred generations even under high mutation rates. Although in very small founder populations of wild emmer, such as those of Bet Oren in Mount Carmel, polymorphism is very low, even at the SSR levels, in other small populations such as Bat Shelomo it is substantial. Studies in small and isolated populations of outbreeders and inbreeders indicated not only the operation of selection, but also the adaptiveness of the allozyme polymorphisms. Most importantly, the polymorphisms prevalent in the very small, long-isolated populations of *Spalax* in the northern Negev Desert are precisely those in the main range of the xeric species, $2n = 60$ (Nevo 1989b, 1991; Nevo et al. 1994). This clearly highlights the adaptive nature of these polymorphisms. Rather than being neutral (Kimura 1983), or even nearly neutral (Ohta and Gillespie 1996), they are selected and contribute to fitness and survivorship in the harsh desert ecology.

Strong selection (more than 10% per trait) may be quite a common phenomenon in nature, as first emphasized by Ford (1971; see also Endler 1986). This could have important evolutionary consequences in the case of small peripheral populations (Mayr 1954). Our modeling (Nevo et al. 1997) clearly demonstrates that moderate or strong selection in a small population can oppose random drift and maintain polymorphism for thousands of generations. These results are significant for the theories of adaptation and speciation, particularly for peripatric speciation (Mayr 1954).

9.8 Gene Flow

There is abundant geographic variation in both morphology and gene frequency in most species. The extent of geographic variation results from a balance of those forces tending to produce local genetic differentiation and those tending to produce genetic homogeneity. Mutation, genetic drift due to finite population size, and natural selection favoring adaptations to local environmental conditions will all lead to the genetic differentiation of local populations; and the movement of gametes, individuals, and even entire populations, collectively called gene flow, will oppose that differentiation (Slatkin 1987). Gene flow has been hypothesized to be extensive in nature (Mayr 1942) or limited (Ehrlich and Raven 1969), to be a creative microevolutionary force (Wright 1932) or a destructive one (by preventing local adaptation: Baur 1986), or to prevent speciation (Carson and Templeton 1984).

Within a population, gene flow can retard genetic differentiation considerably, and even a small amount of migration is sufficient to prevent any appreciable genetic differentiation, unless there is strong differential selection (Nei 1987; Kendall and Fox 1998). In wild emmer wheat, pollen dispersal and seed migration through wind, animals, ants, and human activity can cause gene flow. In natural populations of *T. dicoccoides*, pollen dispersal is relatively limited because of its high selfing rate (about 99%; Golenberg 1986a). The plant-genetic and plant-breeding literature contains extensive evidence that the mating system can be modified simply and drastically by selection and that different populations within the same species often practice very different amounts of inbreeding (Allard 1975). For example, the evidence on selfing populations of wild barley, *H. spontaneum*, across Israel indicates that outcrossing was significantly higher in populations growing in the more mesic regions than in the xeric regions (ranging from 0% to 9.6%; Brown et al. 1978). This evidence implies that pollen dispersal may not be consistent in all environments; outcrossing rate of wild emmer may fluctuate in different environments, as demonstrated in selfing populations of wild barley, thus providing novel genetic material subjected continuously and constantly to the screening of natural selection.

Furthermore, seed movements cannot be ignored at microgeographic scales of a few meters. Golenberg (1986a) found seed dispersal by ants and other agents in the Yehudiyya population of *T. dicoccoides*. Although Golenberg's result showed a relatively low migration rate in the Yehudiyya population (about 1.25 m per generation), it nonetheless suffices to equalize allele frequencies during short periods among neighboring microhabitats. Using RAPD markers, a relatively high migration rate was estimated for the Yehudiyya microsite (7.2 individuals per generation between the subpopulations; Li et al. 1999). If there was no differential selection at the Yehudiyya microsite, this high gene flow could counteract selection and rapidly swamp out the *shade* vs *sun* difference at allozyme (Nevo et al. 1988b), RAPD (Li et al. 1999), and SSR (Li et al. 2000c) loci. Similarly, if there were no microhabitat selection at the Ammiad microsite and no edaphic microhabitat selection at the Tabigha microsite, gene flow would easily and rapidly homogenize the neighboring subpopulations, and the observed significant genetic divergence at the molecular level (Nevo et al. 1988a,b, 1991c; Li et al. 2000a–d, 2001a–d) might not have been maintained at each of the three microgeographic sites. Indeed, a surprisingly small amount of migration was found to be sufficient to swamp out the differentiation that arises from random drift of nonselected genes (Lewontin 1974). Hence, we believe that *differential microhabitat selection overrides the effect of gene flow at the three microsites*. Previously, Jain and Bradshaw (1966), McNeilly (1968), Hilbish and Koehn (1985), and Wasser et al. (1982) also presented evidence of selection-generated differentiation in populations of plants and animals, in which gene flow was demonstrably or reasonably assumed to be high. It is the *overriding of selection* that eliminates high mobility migrants.

9.9 Hitchhiking Effect

The hitchhiking hypothesis assumes that the diversity at marker loci is selectively neutral, but variation at linked loci, which are targets of selection, is in LD with marker diversity. The hitchhiking model is especially relevant to inbreeding species (Ohta and Cockerham 1974), because inbreeding acts to intensify associations between loci. At the Tabigha and Yehudiyya microsites, the DNA patterns of genetic diversity (Li et al. 2000d) were consistent with the previous allozymic results (Nevo et al. 1988a,b). At the Ammiad microsite, patterns of genetic diversity were compared in detail using allozymic, RAPD, and SSR markers for the 1993 samples (Nevo et al. 1991c; Li et al. 2000c). The three types of *coding* and *noncoding* molecular markers revealed a parallel microgeographic pattern and spatiotemporal trends of genetic diversity in the four subpopulations at the Ammiad microsite. The concordance of protein and DNA patterns in the subpopulations at the Ammiad, Tabigha, and Yehudiyya microsites, as well as regionally (Nevo et al. 1982; Nevo and Beiles 1989; Fahima et al. 1999, 2001), suggests that similar forces act upon genetic diversity in protein coding and noncoding DNA regions. A simple explanation can be hitchhiking effects on SSR and RAPD loci which are linked to the selected protein loci. If each subpopulation were a clone, the clone should show a similar level of genetic diversity using different systems of molecular markers, since the whole clone evolves together. However, dramatic differences were found in the levels of genetic diversity at allozyme, RAPD, and SSR loci, in *numerous clones* in each subpopulation. This pattern suggests that the evolutionary rate was not the same and that the allozymic, RAPD, and SSR regions evolved to some extent independently. Hitchhiking alone cannot explain the extensive *locus-specific* and *-unique* patterns, nor the LD *specificity* or *uniqueness*, of allozymes, RAPDs, and SSRs.

Moreover, in eukaryotes, over 90% of vertebrate and plant DNA consists of *noncoding* sequences (Cavalier-Smith 1985; Nowak 1994; Clark et al. 1999). The dynamic organization of *noncoding* DNA suggests a feedback loop that could influence codon usage and stabilize the chromosome's chromatin pattern (Holmquist 1989; Ramel 1997). Repetitive DNA sequences form a large portion of the noncoding DNA genomic region of eukaryotes (Dover and Flavell 1982; Cavalier-Smith 1985; Hancock 1996; Ramel 1997), including satellite, minisatellite, and microsatellite DNA segments (see Charlesworth et al. 1994, Ramel 1997). It has also been proposed that repetitive DNA sequences are *functionally* important for organisms (Cavalier-Smith 1985; Kashi et al. 1997; King et al. 1997; Ramel 1997; Bennentzen 1998; Kashi and Soller 1999; King and Soller 1999). Such functional significance, if indeed operating in *noncoding* DNA directly, may partly challenge the hitchhiking hypothesis of *noncoding* DNA evolution. Theories of hierarchical selection and molecular ecology have shown that selection can act on Darwinian units of *noncoding* DNA at the genome level, thus creating positionally constrained DNA and contributing a minimal genetic load at the individual level (Holmquist 1989; Bennentzen, 1998). It is a major challenge of evolutionary biology to estimate the proportion of the *functional-regulatory*, hence, *adaptiveness* of the *noncoding* genomes of organisms.

Theoretical analysis of multichromosome models in outbreeders challenges the hitchhiking hypothesis of polymorphism maintenance for many neutral loci, due to close linkage with a few selected loci (Nevo et al. 1997). In these models, the effect of linkage may be opposite for configurations with multiple unlinked blocks of selected loci, and the chances of polymorphism maintenance increase when there is looser linkage within the blocks, so that hitchhiking becomes ineffective. The future challenge is to model inbreeders with diverse rates and temporal patterns of outcrossings and to assess their effect on the genomes of inbreeders.

Part V

Conclusions and Prospects

Chapter 10

Conclusions and Prospects

10.1 Wheat as Model Organisms

Wheats are important model organisms for testing various aspects of evolutionary theory (both for speciation and adaptation) and a major source of human nutrition. Wheat speciation involves a polyploid series (2x, 4x, 6x). The origin of most wheats is wild emmer, *Triticum dicoccoides* (genome AABB). In 1979, the Institute of Evolution at the University of Haifa established a long-term multidisciplinary research program to study *T. dicoccoides*. The program includes evolutionary ecological genetics and genomics coupled with the exploration of genetic resources for wheat improvement. Both aspects, the theoretical and the applied, have proved to be of great importance for studying evolutionary theory and crop improvement.

10.2 Genetic Diversity of Wild Emmer for Wheat Improvement

The molecular diversity and divergence of wild emmer wheat, *regionally* in the Near East Fertile Crescent, and *locally* in four natural populations at Qazrin, Ammiad, Tabigha, and Yehudiyya microsites in northern Israel, display parallel ecological-genetic patterning. The regional and local results demonstrated significant spatial and temporal molecular divergence at the DNA and protein levels in *T. dicoccoides* populations and subpopulations. Specifically, these patterns revealed that: (1) significant genetic diversity and divergence exist at single- two-, and multilocus structures of allozymes, RAPDs, and SSRs, over very short distances of several to a few dozen meters in the four microsites. (2) The genetic patterns across *coding* (allozymes) and largely *noncoding* (RAPDs and SSRs) genomic regions are correlated with, and predictable by, environmental stress (*climatic*, *edaphic*, *biotic*) and heterogeneity (the niche-width variation hypothesis), displaying significant *niche-specific* and *-unique* alleles and genotypes. (3) The genomic organization of wild emmer is *nonrandom*, heavily *structured*, and at least partly, if not largely, *adaptive*. It defies explanation by genetic drift, neutrality, or near neutrality models as the primary driving forces of wheat molecular evolution. The only viable model to explain the genomic organization of wild emmer is natural selection, primarily *diversifying, balancing and cyclical selection* over space and time according to the two- or multiple-niche ecological models. Spatial models are complemented by temporal models of genetic diversity and change (Kirzhner et al. 1995, 1996, 1999). Natural selection may interact with mutation, migration, and stochastic factors, but it overrides them in orienting wild emmer wheat evolutionary processes.

Based on mathematical modeling, we established that stabilizing selection with a cyclically moving optimum may efficiently protect polymorphism for linked loci, additively affecting the selected trait (Korol et al. 1994, 1996a; Kirzhner et al. 1995). In particular, unequal gene action and/or dominance effects may lead to local polymorphism stability with substantial polymorphism attracting domain. Moreover, under strong cyclical selection, complex dynamic patterns were revealed, including "supercycles" (with periods comprising hundreds of environmental oscillation periods), and "deterministic chaos" (Kirzhner et al. 1996, 1998; Korol et al. 1998c; Ryndin et al. 2001). These patterns could substantiate polymorphism and increase the range of temporal variation of allele frequencies. We believe that this previously uncharacterized evolutionary mechanism may increase genetic diversity over long-term periods and contribute to overcoming massive extinctions.

10.3 Unique Population Genetic Structure and Center of Origin of Wild Emmer Wheat

Wild emmer wheat has a *unique ecological-genetic structure*. Its central populations in the catchment area of the upper Jordan Valley are massive and lush and represent its *center of origin and diversity*. However, southward in Israel and northward into Turkey, wild emmer populations become fragmented into *sporadic semi-isolated and isolated* ones that are characterized by an *archipelago* genetic structure in which alleles are built up locally in high frequency, but are often missing in neighboring localities. This phenomenon may even occur in the central continuous populations in which alternative fixation of up to eight alleles was described over tens to hundreds of meters in the Golan Heights between Qazrin and Yehudiyya.

The center of origin and diversity of wild emmer, the progenitor of most wheats, and that of other progenitors of cultivated plants is the Near East Fertile Crescent. Particularly in Israel, with its extraordinary biotic and physical diversity, wild emmer developed, both within and between populations, a wide range of *adaptive diversity* to multiple diseases, pests, and ecological stresses over a long evolutionary history. Most importantly, this diversity is neither random nor neutral. By contrast, it displays, at all levels, *adaptive genetic diversity* for *biochemical, morphological, and immunological* characteristics which contribute to the species' ability to adapt to widely diverse climatic and edaphic conditions by diverse and complex fitness syndromes. The long-lasting coevolution of wild emmer with parasites and with the ecologically heterogeneous abiotic nature of Israel led to the development of *single multiallelic* genes and *multilocus* structures, *locally* and *regionally* coadapted for both *short-* and *long-term* survival.

10.4 Genetic Resources

Wild emmer wheat is rich in genetic resources and is one of the best hopes for wheat improvement. These include *abiotic* (e.g., drought, cold, heat, and salt tolerances) and *biotic* (viral, bacterial, fungal, and herbicide resistances), high-quantity and -quality storage proteins (glutenins and gliadins), amylase, and photosynthetic yield. Most of these resources are yet untapped and provide potential precious sources for wheat improvement. The current rich genetic map of *T. dicoccoides*, with 549 molecular markers and 48 significant QTLs for 11 traits of agronomic importance (Peng et al. 2001), permit the unraveling of beneficial alleles of candidate genes that are otherwise hidden. These beneficial alleles could be introduced into cultivated wheat (simultaneously eliminating agronomically undesirable alleles) by using the strategy of marker-assisted selection. The genetic programs of wild emmer wheat conducted at the Institute of Evolution, University of Haifa, and elsewhere, confirmed that *T. dicoccoides* is a very valuable wild germplasm resource for future wheat improvement. This program could highlight the evolution of wheat domestication and expedite the genomic analysis of wild relatives of wheat. It can thus provide a solid basis for *introgression* or *cloning and transformation* of agronomically important genes and QTLs from the wild to the cultivated wheats, and for advancing wheat improvement. This is particularly important in a world whose population is exploding, where hunger is prevalent, desertification and salinization are dramatically increasing, and water and fertile land resources are limited, being constantly degraded by environmental pollution.

10.5 Prospects

What will be the next step in wild emmer research in the genomic and post-genomic era? Conceptually, in-depth probing of comparative genome structure and function are the major challenges, in particular the intimate relationship of the *coding* and *noncoding* genomes. Such studies will unravel genome evolution and highlight the rich genetic potentials for wheat improvement residing in wild emmer wheat and wheat relatives, including *Triticum* and *Aegilops* species, as well as other Triticeae.

Specifically, we could divide the prospects, somewhat arbitrarily, into *theoretical* and *applied* perspective:

10.6 Theoretical Perspective

1. Highlighting the genetic genome structure, function, regulation, and evolution at *macro*- and *micro*geographic scales of natural populations, bridging multilocus marker structures with fitness-related traits in order to get direct estimates of the adaptive fitness differentiation within and between populations (by using transplant experiments, mapping analysis, microarray methodology, and selection-based mapping of fitness components over the entire life cycle).
2. Exposing the genome organization of *T. dicoccoides*, using molecular sequence analysis and molecular cytogenetic methods, in order to probe the structure and interactions of the *nuclear*, *mitochondrial*, and *chloroplast* genomes.
3. Analysis of the genetic system, or the "transmission system," which determines the genetic flexibility of the species, in diverse ecological contexts, including:
   a) Breeding system (reaction norm and genetic variation of the outcrossing rate)
   b) Mutation rate in different elements of the genome
   c) Recombination properties of the genome, their genetic and ecological control
   d) Genomic distribution of structural genes, primarily *abiotic* and *biotic stress* genes, and their regulation function
   e) Interface between ecological and genomic spatiotemporal dynamics and adaptive systems
   f) Genome evolution in the polyploidization process

10.7 Applied Perspective

1. Genetic fine-mapping and dissection of the collected *unique* genetic resources of agricultural importance by molecular markers and sequencing followed by introgression of the detected genes/alleles into elite cultivars via marker-assisted selection.
2. Molecular cloning of adaptation genes based on integrated genomic strategies, and novel methodologies, including genetic and physical mapping, molecular markers, expressed sequence tags (EST) mapping cloning and sequencing, microarray expression analysis, and genetic transformation of the defined target genes/alleles.
3. Comparative genetics/genomics of cereal plants aimed at deciphering the common and specific ways of domestication evolution.

References

Aaltonen LA, Peltomaki P, Leach FS, Pylkkanen P, Mecklin L (1993) Clues to the pathogenesis of familial colorectal cancer. Science 260:812–816

Aaronsohn A (1910) Agricultural and botanical explorations in Palestine. Bureau of Plant Industry Bull 80, USDA, no 180, 66 pp

Aaronsohn A (1913) The discovery of wild wheat. City Club Bull Chicago 6:167–173

Aaronsohn A, Schweinfurth G (1906) Die Auffindung des Wilden Emmers (*Triticum dicoccum*) in Nordpalästina. Altneuland Monatsschrift fur die Wirtschaftlicher Erschliessung Palastinas, Berlin, July–Aug, no 7–8, pp 213–220

Aharoni A, Baran N, Manor H (1993) Characterization of a multisubunit human protein which selectively binds single stranded d(GA)n and d(GT)n sequence repeats in DNA. Nucleic Acids Res 21:5221–5228

Akkaya MS, Shoemaker RC, Specht JE, Bhagwat AA, Cregan PB (1995) Integration of simple sequence repeat DNA markers into a soybean linkage map. Crop Sci 35:1439–1445

Aldrich PR, Hamrick JL, Chavarriaga P, Kochert G (1998) Microsatellite analysis of demographic genetic structure in fragmented populations of the tropical tree *Symphonia globulifera*. Mol Ecol 7:933–944

Allard RW (1975) The mating system and microevolution. Genetics 79:115–126

Allard RW, Babbel GR, Clegg MT, Kahler AL (1972) Evidence for coadaptation in *Avena barbata*. Proc Natl Acad Sci USA 69:3043–3048

Alonso LC, Kimber G (1980) A hybrid between *Agropyron junceum* and *Tricitum sativum*. Cereal Res Commun 8:355–358

Amos W (1999) A comparative approach to the study of microsatellite evolution. In: Goldstein, DB, Schlötterer C (eds) Microsatellites: evolution and applications. Oxford Univ Press, Oxford, pp 66–78

Amos W, Rubinstein DC (1996) Microsatellites are subject to directional evolution. Nat Genet 12:13–14

Anamathawat-Jonsson K, Heslop-Harrison JS (1993) Isolation and characterization of genome-specific DNA sequences in Triticeae species. Mol Gen Genet 240:151–158

Anderson JA, Ogihara Y, Sorrells ME, Tanksley SD (1992) Development of a chromosomal arm map of wheat based on RFLP markers. Theor Appl Genet 83:1035–1043

Anderson JA, Sorrells ME, Tanksley SD (1993) Detection of QTLs affecting pre-harvest sprouting resistance in wheat by RFLPs. Crop Sci 33:453–459

Angers B, Magnan P, Plante M, Bernatchez L (1999) Canonical correspondence analysis for estimating spatial and environmental effects on microsatellite gene diversity in brook charr (*Salvelinus fontinalis*). Mol Ecol 8:1043–1053

Anikster Y, Noy-Meir I (1991) The wild-wheat field laboratory at Ammiad. Isr J Bot 40:351–362

Anikster Y, Wahl I (1979) Coevolution of the rust fungi on Gramineae and Liliaceae and their host. Annu Rev Phytopathol 17:367–403

Anikster Y, Eshel A, Ezrati S, Horovitz A (1991) Patterns of phenotypic variation in wild tetraploid wheat at Ammiad. Isr J Bot 40:397–418

Appels R, Dvorak J (1982) The wheat ribosomal DNA spacer region: its structure and variation in populations and among species. Theor Appl Genet 63:337–348

Araki E, Miura H, Sawada S (1999) Identification of genetic loci affecting amylose content and agronomic traits on chromosome 4a of wheat. Theor Appl Genet 98:977–984

Aranda A, Perez-Ortin JE, Moore C, del Olmo M (1998) The yeast FBP1 poly(A) signal functions in both orientations and overlaps with a gene promoter. Nucleic Acids Res 26:4588–4596

Areshchenkova T, Ganal MW (1999) Long tomato microsatellites are predominantly associated with centromeric regions. Genome 42:536–544

Atlas of Israel (1970) Survey of Israel. Ministry of Labour, Jerusalem. Elsevier/North Holland, Amsterdam

Austin RB, Flavell RB, Henson IE, Lowe HJB (1986) Molecular biology and crop improvement. A case study of wheat, oilseed rape and faba beans. Cambridge Univ Press, Cambridge

Avery D (1985) US Farm dilemma: the global bad news is wrong. Science 230:408–412

Avery PJ, Hill WG (1979) Variance in quantitative traits due to linked dominant genes and variance in heterozygosity in small populations. Genetics 91:817–844

Avise JC (1976) Genetic differentiation during speciation. In: Ayala FJ (ed) Molecular evolution. Sinauer, Sunderland, Massachusetts

Avivi L (1978) High grain protein content in wild tetraploid wheat, *Triticum dicoccoides* Korn. Proc 5th Int Wheat Genet Symp New Delhi, pp 372–380

Avivi L (1979) Utilization of *Triticum dicoccoides* for the improvement of grain protein quantity and quality in cultivated wheats. Monogr Genet Agrar 4:27–38

Avivi L, Levy AA, Feldman M (1983) Studies on high protein durum wheat derived from crosses with the wild tetraploid wheat *Triticum turgidum* var. *dicoccoides*. Proc 6th Int Wheat Genetics Symp Kyoto, Japan, pp 199–204

Awad M, Pravica V, Perrey C, El Gamel A, Yonan N et al (1999) CA repeat all polymorphism in the first intron of the human interferon-gamma gene is associated with lung allograft fibrosis. Hum Immunol 60:343–346

Ayala FJ (1975) Genetic differentiation during the speciation process. Evol Biol 8:1–78

Ayres DR, Ryan FJ (1997) The clonal and population structure of a rare endemic plant, *Wyethia reticulata* (Asteraceae): allozyme and RAPD analysis. Mol Ecol 6:761–772

Ayres DR, Ryan FJ (1999) Genetic diversity and structure of the narrow endemic *Wyethia reticulata* and its congener *W. bolanderi* (Asteraceae) using RAPD and allozyme techniques. Am J Bot 86:344–350

Baker FWG (ed) (1989) Drought resistance in cereals. CAB International, Wallingford, UK

Baran N, Lapidot A, Manor H (1987) Unusual sequences element found at the end of an amplicon. Mol Cell Biol 7:2636–2640

Barker JSF, Moore SS, Hetzel DJS, Evans D, Tan SG, Byrne K (1997) Genetic diversity of Asian water buffalo (*Bubalus bubalis*): microsatellite variation and a comparison with protein-coding loci. Anim Genet 28:103–115

Barker RF, Harberd NP, Jarvis MG, Flavell RB (1988) Structure and evolution of the intergenic region in a ribosomal DNA repeat unit of wheat. J Mol Biol 201:1–17

Barrett GC (1985) Chemistry and biochemistry of the amino acids. Chapman and Hall, London

Barton NH, Turelli M (1991) Natural and sexual selection on many loci. Genetics 127:229–255

Baur B (1986) Geographic variation of resting behavior in the land snail *Arianta arbustorum*: does gene flow prevent local adaptation? Genetica 70:3–8

Beckman JS, Soller M (1986) Restriction fragment length polymorphism in plant genetic improvement. In: Miflin BJ (ed) Oxford surveys of plant molecular and cell biology. Oxford Univ Press, Oxford, pp 196–250

Bell CJ, Ecker JR (1994) Assignment of 30 microsatellite loci to the linkage map of *Arabidopsis*. Genomics 19:137–144

Bell GI, Jurka J (1997) The length distribution of perfect dimer repetitive DNA is consistent with its evolution by an unbiased single-step mutation process. J Mol Evol 44:414–421

Belyayev A, Raskina O (1998) Heterochromatin discrimination in *Aegilops speltoides* by simultaneous genomic in situ hybridization. Chrom Res 6:559–565

Belyayev A, Raskina O, Korol AB, Nevo E (2000) Coevolution of A and B genomes in allotetraploid *Triticum dicoccoides*. Genome 43:1021–1026

Belyayev A, Raskina O, Nevo E (2001) Evolutionary dynamics and chromosomal distribution of repetitive sequences in chromosomes of *Aegilops speltoides* revealed by genomic in situ hybridization. Heredity (in press)

Ben-Amer IM, Korzun V, Worland AJ, Borner A (1997) Genetic mapping of QTL controlling tissue-culture response on chromosome 2B of wheat (*Triticum aestivum* L) in relation to major genes and RFLP markers. Theor Appl Genet 94:1047–1052

Benavente E, Fernandez-Cavin B, Orellana J (1996) Relationship between the levels of wheat-rye metaphase I chromosomal pairing and recombination revealed by GISH. Chromosoma 105:92–96

Bennett MD, Cox AV, Leitch IJ (1998) Angiosperm DNA C-values database. http://www.rbgkew.org.uk/cval/database1.html

Bennetzen JL (1998) The structure and evolution of angiosperm nuclear genomes. Curr Opin Plant Biol 1:103–108

Benyamini Y, Schonfeld M, Rubin B (1991) Kinetics of chlorophyll fluorescence decay in triazine resistant and susceptible weeds. Brighton Crop Protect Conf Weeds 3:1103–1110

Bertoni F, Codegoni AM, Furlin D et al (1999) CHK1 frameshift mutations in genetically unstable colorectal and endometrial cancers. Genes Chrom Cancer 26:176–180

Biet E, Sun J, Dutreix M (1999) Conserved sequences preference in DNA binding among recombination proteins: an effect of ssDNA secondary structure. Nucleic Acids Res 27:596–600

Biggin MD, Tjian R (1988) Transcription factors that activate the Ultrabithorax promoter in developmentally staged extracts. Cell 53:699–711

Biological Journal of the Linnean Society (1990) Bicentenary joint meeting with the institute of archeology. University College, London, vol 39, pp 1–93

Blackman JA, Payne PI (1987) Grain quality. In: Heyne EG (ed) Wheat breeding. Its scientific basis. London, Chapman and Hall, pp 455–485

Blanco A, De Giovanni, Laddomada B, Sciancalepore A, Simeone R, Devos KM, Gale MD (1996) Quantitative trait loci influencing grain protein content in tetraploid wheat. Plant Breed 115:310–316

Blanco A, Bellomo MP, Cenci A, De Giovanni C, D'Ovidio R, Iacono E, Laddomada B, Pagnotta MA, Porceddu E, Sciancalepore A, Simeone R, Tanzarella OA (1998) A genetic linkage map of durum wheat. Theor Appl Genet 97:721–728

Bostock JC, Sumner AT (1978) The eucaryotic chromosome. North-Holland, Amsterdam

Boyer JC, Umar A, Risinger JI, Lipford JR, Kane M et al (1995) Microsatellite instability, mismatch repair deficiency, and genetic defects in human cancer cell lines. Cancer Res 55:6063–6070

Brandes A, Thompson H, Dean C, Heslop-Harrison JS (1997) Multiple repetitive DNA sequences in the paracentromeric regions of *Arabidopsis thaliana* L. Chrom Res 5:238–246

Brentnall TA, Crispin DA, Bronner MP, Cherian SP, Hueffed M et al (1996) Microsatellite instability in nonneoplastic mucosa from patients with chronic ulcerative colitis. Cancer Res 56:1237–1240

Britz SJ, Hungerford WE, Lee DR (1987) Rhythms during extended dark periods determine rates of net photosynthesis and accumulation of starch and soluble sugars in subsequent light periods in leaves of *Sorghum*. Planta 171:171–345

Brohele J, Ellegren H (1999) Microsatellite evolution: polarity of substitution within repeats and neutrality of flanking sequences. Proc Roy Soc Lond B Biol Sci 266:825–833

Brown AHD, Clegg MT (1983) Isozyme assessment of plant genetic resources. In: Rattazzi MC, Scandalios GJ, Whitt AR (eds) Isozymes: current topics in biological and medical research, vol 11. Medical and other applications. Liss, New York, pp 285–295

Brown AHD, Zohary D, Nevo E (1978) Outcrossing rates and heterozygosity in natural populations of *Hordeum spontaneum* Koch in Israel. Heredity 41:49–62

Brown AHD, Feldman MW, Nevo E (1980) Multilocus structure of natural populations of *Hordeum spontaneum*. Genetics 96:523–536

Brown AHD, Frankel OHD, Marshall R, Williams JT (eds) (1989) The use of plant genetic resources. Cambridge Univ Press, New York
Brown AHD, Clegg MT, Kahler AL, Weir BS (eds) (1990) Plant population genetics, breeding, and genetic resources. Sinauer, Sunderland, MA USA
Buard J, Bourdet A, Yardley J, Dubrova Y, Jeffreys AJ (1998) Influences of array size and homogeneity on minisatellite mutation. EMBO J 15:3495–3502
Cabanne F, Snape JW (1993) A similar metabolism of chlorotoluron in cell suspension cultures from near isogenic susceptible and tolerant lines of wheat. Pest Biochem Physiol 47:51–59
Cadalen T, Sourdille P, Charmet G, Tixier MH, Gay G et al (1998) Molecular markers linked to genes affecting plant height in wheat using a doubled-haploid population. Theor Appl Genet 96:933–940
Caligari PDS, Brandham PE (eds) (2001) Wheat taxonomy: The legacy of John Percival. The Linn Soc. London, special issue 3, 190 pp
Caligo MA, Ghimenti C, Sensi E, Piras A, Rainaldi G (1999) Microsatellite instability is coselectable with gene amplification in a mammalian mutator phenotypes. Anticancer Res 19:1271–1275
Cambareri EB, Aisner R, Carbon J (1998) Structure of the chromosome VII centromere region in *Neurospora crassa:* degenerate transposons and simple repeats. Mol Cell Biol 18:5465–5477
Carson HL, Templeton AR (1984) Genetic revolution in relation to speciation phenomea: the founding of new populations. Annu Rev Ecol Syst 15:79–132
Carver BF, Nevo E (1990) Genetic diversity of photosynthetic characters in native populations of *Triticum dicoccoides*. Photosynth Res 25:119–128
Carver BF, Johnson RC, Rayburn AL (1989) Genetic analysis of photosynthetic variation in hexaploid and tetraploid wheat and their interspecific hybrids. Photosynth Res 20:105–118
Castiglioni P, Ajmone-Marsan P, van Wijk R, Motto M (1999) AFLP markers in a molecular linkage map of maize: codominant scoring and linkage group distribution. Theor Appl Genet 99:425–431
Catasti P, Chen X, Mariappan SV, Bradbury EM, Gupta G (1999) DNA repeats in the human genome. Genetica 106:15–36
Causse M, Fulton TM, Cho YG, Ahn SN, Chunwongse J, Wu K, Xiao J, Yu Z, Ronald PC, Harrington SB, Second GA, McCouch SR, Tanksley SD (1994) Saturated molecular map of the rice genome based on an interspecific backcross population. Genetics 138:1251–1274
Cavalier-Smith T (1985) Eukaryotic gene numbers, non-coding DNA, and genome size. In: Cavalier-Smith T (ed) The evolution of genome size. Wiley, Chichester, pp 69–103
Centola M, Carbon J (1994) Cloning and characterization of centromereic DNA from *Neurospora crassa*. Mol Cell Biol 14:1510–1519
Chagué V, Fahima T, Dahan A, Sun GL, Korol AB, Ronin, YI, Grama A, Rööder MS, Nevo E (1999) Isolation of microsatellite and RAPD markers flanking the *Yr15* gene of wheat using NILs bulked segregant analysis. Genome 42:1050–1056
Chakraborty R, Jin L (1993) A unified approach to study hypervariable polymorphisms: statistical considerations of determining relatedness and population distance. In: Pena DJ, Chakraborty R, Epplen JT, Jeffrey AJ (eds) DNA fingerprinting: state of the science. Birkhäuser, Basel, pp 153–175
Chao S, Sharp PJ, Worland AJ, Warham EJ, Koebner RMD, Gale MD (1989) RFLP-based genetic maps of wheat homoeologous group 7 chromosomes. Theor Appl Genet 78:495–504
Chapman V, Miller TE, Riley R (1976) Equivalence of the A genome of bread wheat and that of *Triticum urartu*. Genet Res 27:69–76
Charlesworth B, Langley CH, Stephan W (1986) The evolution of restricted recombination and the accumulation of repeated DNA sequences. Genetics 112:947–962
Charlesworth B, Sniegowski P, Stephan W (1994) The evolutionary dynamics of repetitive DNA in eukaryotes. Nature 371:215–220
Chee PW, Elias EM, Kianinan SF, Anderson JA (1998) Introgression of a high protein gene from LDN (DIC-6B) substitution line. In: Slinkard AE (ed) Proc 9th Int Wheat Genetics Symposium. Univ Extension Press, University of Sasktchewan, Saskatoon, Saskatchewan, Canada

Chen Q, Armstrong K (1994) Genomic *in situ* hybridization in *Avena sativa*. Genome 37:607–612
Chen Y, Roxby R (1997) Identification of a functional CT-element in the *Phytophthora infestans* piypt1 gene promoter. Gene 198:159–164
Churchill GA, Doerge RW (1994) Empirical threshold values for quantitative trait mapping. Genetics 138:963–971
Clark AB, Cook ME, Tran HT, Gordenin DA, Resnick MA et al (1999) Functional analysis of human MutSalpha and MutSbeta complexes in yeast. Nucleic Acids Res 27:736–742
Clegg MT, Kidwell JF, Kidwell MG, Daniel NJ (1976) Dynamics of correlated genetic systems. I. Selection in the region of the glued locus of *Drosophila melanogaster*. Genetics 83:793–810
Codegoni AM, Bertoni F, Collela G, Caspani G, Grassi L et al (1999) Microsatellite instability and frameshift mutations in genes involved in cell cycle progression or apoptosis in ovarian cancer. Oncol Res 11:297–301
Coyne JA (1992) Genetics and speciation. Nature 355:511–515
Cuadrado A, Schwarzacher T (1998) The chromosomal organization of simple sequence repeats in wheat and rye genomes. Chromosoma 107:587–594
Cullis CA (1990) DNA rearrangements in response to environmental stress. Adv Genet 28:73–97
Cullis CA, Swami S, Song Y (1999) RAPD polymorphisms detected among the flax genotrophs. Plant Mol Biol 41:795–800
Dawson IK, Chalmers KJ, Waugh R, Powell W (1993) Detection and analysis of genetic variation in *Hordeum spontaneum* populations from Israel using RAPD markers. Mol Ecol 2:151–159
De Fonzo V, Bersani E, Aluffi-Pentini F, Castrignano T, Parisi V (1998) Are only repeated triplets guilty? J Theor Biol 194:125–142
Delaney DE, Nasuda S, Endo TR, Gill BS, Hulbert SH (1995) Cytologically based physical maps of the group-2 chromosomes of wheat. Theor Appl Gent 91:568–573
Denell RE, Keppy DO (1979) The nature of genetic recombination near the third chromosome centromere of *Drosophila melanogaster*. Genetics 93:117–130
Dermitzakis ET, Clark AG, Batargias C, Magoulas A, Zouros E (1998) Negative covariance suggests mutation bias in a two-locus microsatellite system in the fish *Sparus aurata*. Genetics 150:1567–1575
Devos KM, Gale MD (1992) The use of random amplified polymorphic DNA markers in wheat. Theor Appl Genet 84:567–572
Devos KM, Gale MD (1993) The genetic maps of wheat and their potential in plant breeding. Outlook Agric 22:93–99
Devos KM, Atkinson, MD, Chinoy CN, Gale MD (1992) RFLP- based genetic map of the homoeologous group-3 chromosomes of wheat and rye. Theor Appl Genet 83:931–939
Devos KM, Atkinson MD, Chinoy CN, Liu CJ, Gale MD (1993) RELP-based genetic map of the homoeologous group-2 chromosomes of wheat, rye and barley. Theor Appl Genet 85:784–792
Devos KM, Dubcovsky J, Dvorak J, Chinoy CN (1995a) Structural evolution of wheat chromosomes 4A, 5A and 7B and on their recombination. Theor Appl Genet 91:282–288
Devos KM, Bryan GJ, Collins AJ, Stephenson P, Gale MD (1995b) Application of two microsatellite sequences in wheat storage proteins as molecular markers. Theor Appl Genet 90:247–252
Dib C, Faure S, Fizames C, Samson D, Drouot N, Vignal A, Millasseau P, Marc S, Hazan J, Seboun E, Lathrop M, Gyapay G, Morissette J, Weissenbach J (1996) A comprehensive genetic map of the human genome based on 5264 microsatellites. Nature 380:152–154
Dick JW, Youngs VL (1988) Evaluation of durum wheat, semolina, and pasta in the United States. Durum wheat: chemistry and technology. St Paul, MN, AACC, pp 237–248
Dietrich WF, Miller J, Steen R, Merchant MA, Damron-Boles D, Husain Z, Dredge R, Daly MJ, Ingalls KA, O'Connor TJ, Evans CA, DeAngelis MM, Levinson DM, Kruglyak L, Goodman N, Copeland NG, Jenkins NA, Hawkins TL, Stein L, Page DC, Lander ES (1996) A comprehensive genetic map of the mouse genome. Nature 380:149–152
Dinoor A (1974) Role of wild and cultivated plants in the epidemiology of plant diseases in Israel. Annu Rev Phytopathol 12:413–436

Dinoor A, Eshed E (1990) The analysis of host and pathogen populations in natural ecosystems. In: Wolfe MS, Caten CE (eds) Populations of plant pathogens: their dynamics and genetics. Blackwell Scientific, Oxford, pp 75–88

Dinoor A, Eshed N, Ecker R, Gerechter-Amitai Z, Solel Z, Manistersky J, Anikster Y (1991) Fungal diseases of wild tetraploid wheat in a natural stand in northern Israel. Isr J Bot 40:481–500

Dobzhansky TH, Spassky B, Anderson W (1965) Bichromosomal synthetic semilethals in *Drosophila pseudoobscura*. Proc Natl Acad Sci USA 53:345–348

Dover GA, Flavell RB (1982) Genome evolution and phenotypic variation. Academic Press, New York

Dover GA, Flavell RB (1984) Molecular coevolution: DNA divergence and the maintenance of function. Cell 38:622–623

Driscoll DJ, Palmer CG, Melman A (1979) Nonhomologous associations of C-heterochromatin at human male meiotic prophase. Cytol Cell Genet 23:23–32

Dvorak J, Appels R (1982) Chromosome and nucleotide sequence differentiation in genomes of polyploid *Triticum* species. Theor Appl Genet 63:349–360

Dvorak J, Zhang H-B (1990) Variation in repeated nucleotide sequences sheds light on the phylogeny of the wheat B and G genomes. Proc Natl Acad Sci USA 87:9640–9644

Dvorak J, McGuire PE, Cassidy B (1988) Apparent sources of the A genomes of wheats inferred from polymorphism in abundance and restriction fragment length of repeated nucleotide sequences. Genome 30:680–689

Dvorak J, Zhang H-B, Kota RS, Lassner M (1989) Organization and evolution of the 5S ribosomal RNA gene family in wheat and related species. Genome 32:1003–1016

Dvorak J, di Terlizzi P, Zhang H-B, Resta P (1993) The evolution of polyploid wheat: identification of the A genome donor species. Genome 36:21–31

Egel-Mitani M, Olsson LW, Egel R (1982) Meiosis in *Aspergillus nidulans*: another example for lacking synaptonemal complexes in the absence of crossover interference. Hereditas 97: 179–187

Ehrlich PR, Raven PH (1969) Differentiation of populations. Science 164:1228–1232

Eisen JA (1999) Mechanistic basis for microsatellite instability. In: Goldstein DB, Schlötterer C (eds) Microsatellites: evolution and applications. Oxford Univ Press, Oxford, pp 34–48

Elder JF, Turner BJ (1995) Concerted evolution of repetitive DNA sequences in eukaryotes. Quart Rev Biol 70:297–320

Endler JA (1986) Natural selection in the wild. Princeton Univ Press, Princeton, NJ

Epplen C, Melmer G, Siedlaczck I, Schwaiger et al (1993) On the essence of "meaningless" simple repetitive DNA in eukaryote genomes. In: Pena SDJ, Chakraborty R, Epplen JT, Jeffreys AJ (eds) DNA fingerprinting: state of the science. Birkhäuser, Basel, pp 29–45

Evans LT (1993) Crop evolution, adaptation and yield. Cambridge Univ Press, Cambridge

Ewens WJ (1979) Mathematical population genetics. Springer, Berlin Heidelberg New York

Ewing EP (1979) Genetic variation in heterogeneous environments. VII. Temporal and spatial heterogeneity in infinite population. Am Nat 114:197–212

Fahima T, Röder MS, Grama A, Nevo E (1998) Microsatellite DNA polymorphism divergence in *Triticum dicoccoides* accessions highly resistant to yellow rust. Theor Appl Genet 96:187–195

Fahima T, Sun GL, Beharav A, Krugman T, Beiles A, Nevo E (1999) RAPD polymorphism of wild emmer wheat populations, *Triticum dicoccoides*, in Israel. Theor Appl Genet 98:434–447

Fahima T, Röder MS, Wandhake K, Nevo E (2001) Microsatellite polymorphism in natural populations of wild emmer wheat, *Triticum dicoccoides*, in Israel. Theor Appl Genet (in press)

Falush D, Iwasa Y (1999) Size-dependent mutability and microsatellite constraints. Mol Biol Evol 16:960–966

Faris JD, Anderson JA, Francl LJ, Jordahl JG (1997) RFLP mapping of resistance to chlorosis induction by *Pyrenophora tritici-repentis* in wheat. Theor Appl Genet 94:98–103

Faris JD, Laddomada B, Gill BS (1998) Molecular mapping of segregation distortion loci in *Aegilops tauschii*. Genetics 149:319–327

Faris JD, Li WL, Liu DJ, Chen PD, Gill BS (1999) Candidate gene analysis of quantitative disease resistance in wheat. Theor Appl Genet 98:219–225
Faris JD, Haen KM, Gill BS (2000) Saturation mapping of a gene-rich recombination hot spot region in wheat. Genetics 154:823–835
Feldman M (1966) Identification of unpaired chromosomes in F1 hybrids involving *Triticum aestivum* and *T. tiomopheevi*. Can J Genet Cytol 8:144–151
Feldman M (1977) Historical aspects and significance of the discovery of wild wheats. Stadler Symp 9:121–124
Feldman M (1979) Genetic resources of wild wheats and their use in breeding. Monogr Genet Agrar 4:9–26
Feldman M, Sears ER (1981) The wild gene resources of wheat. Sci Am 244:102–112
Feldman M, Lipton FGH, Miller TE (1995) Wheats. *Triticum* spp. (Gramineae-Triticinae). In: Smartt J, Simmonds NW (eds) Evolution of crop plants. Longman Scientific and Technical Press, London, pp 184–192
Felsenburg T, Levy AA, Galili G, Feldman M (1991) Polymorphism of high-molecular-weight glutenins in wild tetraploid wheat: spatial and temporal variation in a native site. Isr J Bot 40:451–480
Felsenstein J (1976) The theoretical population genetics of variable selection and migration. Annu Rev Genet 10:253–280
Feuillet C, Keller B (1998) Molecular aspects of biotic stress resistance in wheat. In: Slinkard E (ed) Proc 9th Int Wheat Genetics Symp Univ Extension Press, Univ Saskatchewan, Saskatoon, Saskatchewan, Canada
Feuillet C, Keller B (1999) High gene density is conserved at syntenic loci of small and large grass genomes. Proc Natl Acad Sci USA 96:8265–8270
Field D, Wills C (1996) Long, polymorphic microsatellite in simple organisms. Proc R Soc Lond B Biol Sci 263:209–215
Finney KF, Yamazaki VL, Youngs VL, Rubenthaler GL (1987) Quality of hard, soft, and durum wheats. Wheat and wheat improvement. Madison, Wisconsin
Flavell RB, Martini G (1982) The genetic control of nucleolus formation with special reference to common bread wheat. In: Cullis CA, Jordan EG (eds) The nucleolus. Cambridge Univ Press, Cambridge, pp 113–128
Flavell RB, D'Dell M, Sharp P, Nevo E, Beiles A (1986) Variation in the intergenic spacer of ribosomal DNA of wild wheat, *Triticum dicoccoides* in Israel. Mol Biol Evol 3:547–558
Flavell RB, O'Dell M, Thompson WF (1988) Regulation of cytosine methylation in ribosomal DNA and nucleolus organizer expression in wheat. J Mol Biol 5:523–534
Flores C, Engels W (1999) Microsatellite instability in *Drosophila spellchecker1* (MutS homolog) mutatnts. Proc Natl Acad Sci USA 96:2964–2969
Ford E (1971) Ecological genetics, 3rd edn. Chapman and Hall, London
Forster BP, Miller TE (1985) A 5B deficient hybrid between *Triticum aestivum* and *Agropyron junceum*. Cereal Res Commun 13:93–95
Forster BP, Law CN, Nevo E, Powell W (1990) Genetic strategies for improving the salt tolerance of wheat and barley. In: Proceedings of the international symposium on molecular and genetic approaches to plant stress, 14–17 Feb, New Delhi, India, pp T31.1–3
Forster BP, Russell JR, Ellis RP, Handley LL, Robinson D, Hackett CA, Nevo E, Waugh R, Gordon DC, Keith R, Powell W (1997) Locating genotypes and genes for abiotic stress tolerance in barley: a strategy using maps, markers and the wild species. New Phytol 137:141–147
Frankel OH, Bennett E (1970) Genetic resources for today and tomorrow. Int Biol Prog 2. Cambridge Univ Press, Cambridge
Frankel OH, Hawkes JG (eds) (1975) Crop genetic resources for today and tomorrow. Int Biol Prog 2. Cambridge Univ Press, Cambridge
Frankel OH, Soulé ME (1981) Conservation and evolution. Cambridge Univ Press, Cambridge
Fraser CM, Eisen JA, Salzberg SL (2000) Microbial genome sequencing. Nature 406:959–964

Fridman E, Pleban T, Zamir D (2000) A recombination hotspot delimits a wild-species quantitative trait locus for tomato sugar content to 484 bp within an invertase gene. Proc Natl Acad Sci USA 97:4718–4723

Friebe B, Gill BS (1996) Chromosome banding and genome analysis in diploid and cultivated polyploid wheats. In: Jauhar PP (ed) Methods of genome analysis in plants. CRC Press, Boca Raton, pp 39–60

Fuchs J, Houben A, Brandes A, Schubert I (1996) Chromosome "painting" in plants – a feasible technique? Chromosoma 104:315–320

Fujisawa T, Ikegami H, Kawaguchi Y, Yamato E, Nakagawa Y et al (1999) Length rather than a specific allele of dinucleotide repeat in the 5' upstream region of the aldose reductase gene is associated with diabetic retinopathy. Diabet Med 16:1044–1047

Gale MD, Atkinson MD, Chinoy CN, Harcourt RL, Jia J, Li QY, Devos KM (1995) Genetic maps of hexaploid wheat. In: Li ZS, Xin ZY (eds) Proc 8th Int Wheat Genet. Agricultural Scientech Press, Beijing, China, pp 79–87

Galili G, Feldman M (1983) Genetic control of endosperm proteins in wheat: 11. Variation of high molecular weight glutenin and gliadin subunits of *Triticum asetivum* L. Theor Appl Genet 66:77–86

Galili G, Feldman M (1985) Structural homology of endosperm high molecular weight glutenin subunits of common wheat (*Triticum aestivum* L.) Theor Appl Genet 70:634–642

Ganapathy SN, Chitre RG (1970) Factors affecting the utilization of millet protein by rats. Fed Proc Am Soc Exp Biol 29:761

Gardiner JM, Coe EH, Melia-Hancock S, Hoisington DA, Chao S (1993) Development of a core RFLP map in maize using an immortalized F2 population. Genetics 134:917–930

Garza J, Slatkin CM, Freimer NB (1995) Microsatellite allele frequencies in humans and chimpanzees, with implications for constraints on allele size. Mol Biol Evol 12:594–603

Gasser CS, Fraley RT (1989) Genetically engineering plants for crop improvement. Science 244:1293–1299

Gavrilets S, Hastings A (1994) Maintenance of multilocus variability under strong stabilizing selection. J Math Biol 32:287–302

Gavrilets S, Hastings A (1995) Dynamics of polygenic variability under stabilizing selection, recombination, and drift. Genet Res 65:63–74

Gebhardt C, Ritter E, Debener T, Schachtschabel U, Walkemeir B, Uhrig H, Salamini F (1989) RFLP analysis and linkage mapping in *Solanum tuberosum*. Theor Appl Genet 78:65–75

Gendrel CG, Boulet A, Dutreix M (2000) (CA/GT)n microsatellites affect homologous recombination during yeast meiosis. Genes & Development 14:1261–1268

Gepts P (1990) Genetic diversity of seed storage proteins in plants. In: Brown MT, Clegg AL, Kahler AL, Weir BS (eds) Plant population genetics, breeding and genetic resources. Sinauer Associates, Sunderland, MA, pp 64–82

Gerechter-Amitai ZK, Grama A (1974) Inheritance of resistance to stripe rust (*Puccinia striiformis*) in crosses between wild emmer (*Triticum dicoccoides*) and cultivated tetraploid and hexaploid wheat. I. *Triticum durum*. Euphytica 23:387–392

Gerechter-Amitai ZK, Grama A (1977) Use of alien genes in wheat breeding. Ann Wheat Newslett 23:57–58

Gerechter-Amitai ZK, Stubbs RW (1970) A valuable source of yellow rust resistance in Israeli populations of wild emmer, *Triticum dicoccoides* Koern. Euphytica 19:12–21

Gerechter-Amitai ZK, Van Silfhout CH, Grama A, Kleitman F (1989) *Yr*15 – a new gene for resistance to *Puccinia striiformis* in *Triticum dicoccoides* sel. G-25. Euphytica 43:187–190

Gill KS, Gill BS, Endo TR, Boyko EV (1996a) Identification and high-density mapping of gene-rich regions in chromosome 5 of wheat. Genetics 143:1001–1012

Gill KS, Gill BS, Endo TR, Taylor T (1996b) Identification and high-density mapping of gene-rich regions in chromosome 1 of wheat. Genetics 144:1883–1891

Gillespie JH (1978) A general model to account for enzyme variation in natural populations. V. The SAS-CFF model. Theor Pop Biol 14:1–45

Gillespie JH (1985) The interaction of genetic drift and mutation with selection in a fluctuating environment. Theor Pop Biol 27:222–237

Gillespie JH (1991) The causes of molecular evolution. Oxford Univ Press, Oxford

Golenberg EM (1986a) Multilocus structures in plant populations: Population and genetic dynamics of *Triticum dicoccoides*. PhD Thesis, State University of New York at Stony Brook

Golenberg EM (1986b) Linkage relationships in wild emmer wheat, *Triticum dicoccoides*. Genetics 114:1023–1031

Golenberg EM (1987) Estimation of gene flow and genetic neighborhood size by indirect methods in a selfing annual, *Triticum dicoccoides*. Evolution 41:1326–1334

Golenberg EM, Nevo E (1987) Multilocus differentiation and population structure in a selfer, wild emmer wheat, *Triticum dicoccoides*. Heredity 58:451–456

Gonneau M, Pasqutte B, Cabanne F, Scalla R (1988) metabolism of chlorotoluron in tolerant species: possible role of cytochrom P450 mono-oxygenase. Weed Res 28:19–25

Goodwin SB, Hu X (1998) Cloning and analysis of four resistance gene analogs from wheat using conserved primers. In: Slinkard AE (ed) Proc 9th Int Wheat Genetics Symp Univ Extension Press, Univ of Saskatchewan, Saskatoon, Saskatchewan, Canada

Gorham J, Hardy C, Wyn Jones RG, Joppa LR (1987) Chromosome location of a K/Na discrimination character in the D genome of wheat. Theor Appl Genet 74:584–588

Grama A, Gerechter-Amitai ZK, Blum A (1983) Wild emmer as donor of genes for resistance to stripe rust and for high protein content. Proc 6th Int Wheat Genet Symp Kyoto Univ, Kyoto, Japan, pp 187–192

Grama A, Cressey PJ, Lindley T (1987a) Hexaploid wild emmer wheat derivatives grown under New Zealand conditions. 1. Relationship between protein composition and parameters. N Z J Agric Res 30:35–43

Grama A, Porter NG, Wright DSC (1987b) Hexaploid wild emmer wheat derivatives grown under New Zealand conditions. 2. Effect of foliar urea sprays on plant and grain nitrogen and baking quality. N Z J Agric Res 30:45–51

Green MM (1975) Conversion as a possible mechanism of high coincidence values in the centromere region of *Drosophila*. Mol Gen Genet 139:57–66

Grieson D (1991) Plant genetic engineering. Blackie, Glasgow

Groenewegen LJM, van Silfhout CH (1988) The use of wild emmer in Dutch practical wheat breeding. In: Jorna ML, Slootmaker LAJ (eds) Cereal breeding related to intergrated cereal production. Pudoc, Wageningen, pp 184–189

Gustafson JP, Apples R, Raven P (1993) Gene conservation and exploitation. 20th Stadler Genetics Symp. Plenum, New York

Haldane JBS, Jayakar SD (1963) Polymorphism due to selection of varying direction. J Genet 58:237–242

Hamrick JL, Allard RW (1972) Microgeographical variation in allozyme frequencies in *Avena barbata*. Proc Natl Acad Sci USA 69:2100–2104

Hamrick JL, Holden LR (1979) Influence of microhabitat heterogeneity on gene frequency distribution and gametic phase disequilibrium in *Avena barbata*. Evolution 33:521–533

Hancock JM (1996) Simple sequences and the expanding genome. Bioessays 18:421–425

Hancock JM (1999) Microsatellites and other simple sequences: genomic context and mutational mechanisms. In: Goldstein AB, Schlötterer C (eds) Microsatellites: evolution and applications. Oxford Univ Press, Oxford, pp 1–9

Handley LL, Nevo E, Raven JA, Martinez-Carrasco R, Scrimgeour CM, Pakniyat H, Forster BP (1994) Chromosome 4 controls potential water use efficiency (δd^{13} C) in barley. J Exp Bot 45:1661–1663

Handley LL, Robinson D, Forster BP, Ellis RP, Scrimgeous CM, Gordon DC, Nevo E, Raven JA (1997) Shoot δd^{15} N correlates with genotype and salt stress in barley. Planta 201:100–102

Hanson RE, Zhao X-P, Islam-Faridia MN, Paterson AH, Zwick MS, Crane CF, McKnight TD, Stelly DM, Price HJ (1998) Evolution of interspersed repetitive elements in *Gossypium* (*Malvaceae*). Am J Bot 85:1364–1368

Harberd NP, Flavell RB, Thompson RD (1987) Identification of transposon-like insertion in a *Glu-1* allele of wheat. Mol Gen Genet 209:326–332

Harding RM, Boyce AJ, Clegg JB (1992) The evolution of tandemly repetitive DNA: recombination rules. Genetics 132:847–859

Harlan JR (1975) Our vanishing genetic resources. Science 188:618–621

Harlan JR (1976) Genetic resources in wild relative of crops. Crop Sci 16:329–333

Harlan JR (1992) Crop and man. American Society of Agronomy, Crop Science Society of America. Madison, Wisconsin

Harlan JR, Zohary D (1966) Distribution of wild emmer wheat and barley. Science 153:1074–1080

Harper AE, Benton DA, Elvehijem CA (1955) L-leucine, an isoleucine antagonist in the rat. Arch Biochem Biophys 57:1–12

Harris DR (ed) (1996) The origins and spread of agriculture and pastoralism in Eurasia. UCL, London

Hartl DL, Clark AG (1997) Principles of population genetics, 3rd edn. Sinauer Associates, Sunderland, MA

Hartl L, Weiss H, Stephan U, Zeller FJ, Jahoor A (1995) Molecular identification of powdery mildew resistance genes in common wheat (*Triticum aestivum* L). Theor Appl Genet 90:601–606

Hawkes JG (1991) International workshop in dynamic in situ conservation of wild relatives of major cultivated plants: summary of final discussion and recommendations. Isr J Bot 40:529–536

Hawn MT, Umar A, Carethers JM, Marra G, Kunkel TA et al (1995) Evidence for a connection between the mismatch repair system and the G2 cell cycle checkpoint. Cancer Res 55:3721–3725

He S, Ohm H, Mackenzie S (1992) Detection of DNA sequence polymorphisms among wheat varieties. Theor Appl Genet 84:578

Hedrick PW (1980) Hitchhiking: a comparison of linkage and partial selfing. Genetics 94:791–808

Hedrick PW (1982) Genetic hitchhiking: a new factor in evolution? Bioscience 32:845–853

Hedrick PW (1986) Genetic polymorphism in heterogeneous environments: a decade later. Annu Rev Ecol Syst 17:535–566

Hedrick PW, Ginevan ME, Ewing EP (1976) Genetic polymorphism in heterogeneous environments. Annu Rev Ecol Syst 7:1–32

Henaut A, Lisacek F, Nitschke P, Moszer I, Danchin A (1998) Global analysis of genomic texts: the distribution of AGCT tetranucleotides in the *Escherichia coli* and *Bacillus subtilis* genomes predicts translational frameshifting and ribosomal hopping in several genes. Electrophoresis 19:515–527

Henry RJ, Ronalds JA (eds) (1994) Improvement of cereal quality by genetic engineering. Plenum Press, New York

Heslop-Harrison JS, Schwarzacher T (1996) Genomic Southern and *in situ* hybridization for plant genome analysis. In: Jauhar PP (ed) Methods of genome analysis in plants. CRC Press, Boca Raton, pp 163–179

Heun M, Kennedy AE, Anderson JA, Lapitan NLV, Sorrells ME, Tanksley SD (1991) Construction of a restriction fragment length polymorphism map for barley (*Hordeum vulgare*). Genome 34:437–447

Heun M, Murphy JP, Phillips TD (1994) A comparison of RAPD and isozyme analyses for determining the genetic relationships among *Avena sterilis* L. accessions. Theor Appl Genet 87:689–696

Heyn CC, Waldman M (1992) In-situ conservation of plants with potential economic value. In: Adams RP, Adams JE (eds) Conservation of plant genes: DNA banking and in vitro biotechnology. Organizational meeting of DNA Bank-Net, London

Hilbish TJ, Koehn RK (1985) Exclusion of the role of secondary contact in an allele frequency cline in the mussel *Mytilus edulis*. Evolution 39:432–443

Hillis DM, Moritz C, Porter CA, Baker RJ (1991) Evidence for biased gene conversion in concerted evolution of ribosomal DNA. Science 251:308–310

Hillman G, Colledge S (1998) The transition from foraging to farming in southwest Asia. International workshop, Groningen, Netherlands

Hochachka PW, Somero GN (1984) Biochemical adaptation. Princeton Univ Press, Princeton

Hoekstra RF (1978) Sufficient conditions for polymorphism with cyclical selection in a subdivided population. Genet Res 31:67–73

Hoffman AA, Parsons PA (1991) Evolutionary genetics and environmental stress. Oxford Univ Press, Oxford

Hoffman EK, Trusko SP, Murphy M, George DL (1990) An S1 nuclease-senstive homopurine/homopyrimidine domain in the c-Ki-ras promoter interacts with a nuclear factor. Proc Natl Acad Sci USA 87:2705–2709

Holmquist GP (1989) Evolution of chromosome bands: molecular ecology of noncoding DNA. J Mol Evol 28:469–486

Horovitz A, Feldman M (eds) (1991) Population dynamics of the wheat progenitor, *Triticum turgidum* var. *dicoccoides* in a natural habitat in eastern Galilee. Isr J Bot 40:351–508

Horsfall JG, Cowling EB (eds) (1977) Plant diseases. an advanced treatise, vol 1. How disease is managed. Academic Press, New York

Humphreys DG, Procunier JD, Mauthe W, Howes NK, Brown PD, McKenzie RIH (1998) Marker-assisted selection for high protein concentration in wheat. In: Fowler DB, Geddes WE, Johnston AM, Preston KR (eds) Wheat Protein Symposium. Univ Extension Press, Univ of Saskatchewan, Saskatoon, Saskatchewan, Canada

Hunger RM, Sherwood JL, Pennington RE, Carver BF, Nevo E (1992) Reaction of native populations of *Triticum dicoccoides* to wheat soilborne mosaic. Biological and cultural tests for control of plant diseases, vol 7. American Phytopathological Society, St Paul, Minnesota,

Innan H, Terauchi R, Miyashita NT (1997) Microsatellite polymorphism in natural populations of the wild plant *Arabidopsis thaliana*. Genetics 146:1441–1452

Isabel N, Beaulieu J, Bousquet J (1995) Complete congruence between gene diversity estimates derived from genotypic data at enzyme and random amplified polymorphic DNA loci in black spruce. Proc Natl Acad Sci USA 92:6369–6373

Islam MR, Shepherd KW, Sparrow DHB (1981) Isolation and characterization of euplasmic wheat-barley chromosome addition lines. Heredity 46:161–174

Islam MR, Shepherd KW, Mayo GME (1989) Recombination among genes at the 1 group in flax conferring resistance to rust. Theor Appl Genet 77:540–546

ITMI (2000) Tenth international public workshop. Program, Text, and Abstracts. 14–16 June 2000, University of Delaware, Newark, Delaware, USA

Ivandic V, Hackett CA, Zhang ZJ, Staub JE, Nevo E, Thomas WTB, Forster BP (2000) Responses of wild barley to experimentally imposed water stress J Exper Bot 51:2021–2029

Ivanov I, Alexandrova R, Dragulev B, Saraffova A, AbouHaidar AG (1992) Effect of tandemly repeated AGG triplets on the translation of CAT-mRNA in *E. coli*. FEBS Lett 307:173–176

Jackson AL, Chen R, Loeb LA (1998) Induction of microsatellite instability by oxidative DNA damage. Proc Natl Acad Sci USA 95:12468–12473

Jain SK (1976) Patterns of survival and microevolution in plant populations. In: Karlin S, Nevo E (eds) Population genetics and ecology. Academic Press, New York, pp 49–89

Jain SK, Bradshaw AD (1966) Evolutionary divergence among adjacent plant populations. I. The evidence and its theoretical analysis. Heredity 21:407–441

Jakupciak JP, Wells RD (2000) Gene conversion (recombination) mediates expansions of CTG.CAG repeats. Jour Biol Chem 275:4003–4013

James C (1999) Global status of commercialized transgenic crops (1999) ISAAA briefs no 12. ISAAA, Ithaca, NY

Jansen RC, Stam P (1994) High resolution of quantitative traits into multiple loci via interval mapping. Genetics 136:1447–1455

Jeffreys AJ, Murray J, Neumann R (1998) High-resolution mapping of crossovers in human sperm defines a minisatellite-associated recombination hotspot. Mol Cell Biol 2:267–273

Johannsdottir JT, Jonasson JG, Bergthorsson JT et al (2000) The effect of mismatch repair deficiency on tumourigenesis, microsatellite instability affecting genes containing short repeated sequences. Int J Oncol 16:133–139

Johnson AC, Jinno Y, Merlino GT (1988) Modulation of epidermal growth factor receptor proto-oncogene transcription by a promoter site sensitive to S1 nuclease. Mol Cell Biol 8:4174–4184

Johnson BL, Dhaliwal HS (1976) Reproductive isolation of *Triticum boeoticum* and *Triticum urartu* and the origin of the tetraploid wheats. Am J Bot 63:1088–1094

Johnson VA, Mattern PJ, Vogels KP (1973) Cultural genetic and other factors affecting quality of wheat. In: Spicer A (ed) Bread: social, nutritional and agricultural aspects of wheaten Bread. Appl Sci Publ, London, pp 127–140

Johnson VA, Mattern PJ, Schmidt JW, Stroike JE (1975) Genetic advances in wheat protein quality and compositions. Proc 4th Int Wheat Genet Symp Columbia, Missouri, pp 547–556

Jones DA, Dickinson MJ, Balint-Kurti PJ, Dixon MS, Jones JDG (1993) Two complex resistance loci revealed in tomato by classical and RFLP mapping of the *Cf-2*, *Cf-4*, *Cf-5* and *Cf-9* genes for resistance to *Cladosporium fulvum*. Mol Plant Microbe Interact 6:348–357

Jones OTG (1991) Cytochrome P450 and herbicide resistance. In: Caseley JC, Cussans GW, Atakin RK (eds) Herbicide resistance in weeds and crops. Butterworth-Heinemann. Oxford, pp 213–226

Jones OTG, Caseley JS (1989) Role of cytochrome P450 in herbicide metabolism. Brighton Crop Protection Conference, Weeds, pp 1175–1184

Joppa LR, Cantrell RG (1990) Chromosomal location of genes for grain protein content of wild tetraploid wheat. Crop Sci 30:1059–1064

Joppa LR, Nevo E, Beiles A (1995) Chromosome translocations in wild populations of tetraploid emmer wheat in Israel and Turkey. Theor Appl Genet 91:713–719

Joppa LR, Du C, Hart GE, Hareland GA (1997) Mapping gene(s) for grain protein in tetraploid wheat (*Triticum turgidum* L) using a population of recombinant inbred chromosome lines. Crop Sci 37:1586–1589

Jorgenson JH (1994) Genetics of powdery mildew resistance in barley. Crit Rev Plant Sci 13: 97–119

Joshi CP, Nguyen HT (1993a) Application of the random amplified polymorphic DNA technique for the detection of polymorphism among wild and cultivated tetraploid wheats. Genome 36:602–609

Joshi CP, Nguyen HT (1993b) RAPD (random amplified polymorphic DNA) analysis based on intervarietal genetic relationships among hexaploid wheats. Plant Sci 93:95–103

Kakade ML (1974) Biochemical basis for the differences in plant protein utilization. J Agric Feed Chem 22:550–555

Kalendar R, Tanskanen J, Immonen S, Nevo E, Schulman AH (2000) Genome evolution of wild barley (*Hordeum spontaneum*) by BARE-1 retrotransposon dynamics in response to sharp microclimatic divergence. Proc Natl Acad Sci USA 97:6603–6607

Kanazin V, Marek LF, Shoemaker RC (1996) Resistance gene analogs are conserved and clustered in soybean. Proc Natl Acad Sci USA 93:11746–11750

Karlin S (1979) Principles of polymorphism and epistasis for multilocus systems. Proc Natl Acad Sci USA 76:541–545

Karlin S (1981) Some natural variability systems for a multiallelic locus: a theoretical study. Genetics 97:457–473

Karlin S (1982) Classifications of selection migration structure and conditions for a protected polymorphism. Evol Biol 14:61–204

Karlin S, Avni H (1981) Analysis of central equilibria in multilocus system: a generalized symmetric viability regime. Theor Pop Biol 20:241–280

Karlin S, Levikson B (1974) Temporal fluctuations in selection intensities: case of small population size. Theor Pop Biol 6:383–412

Karlin S, McGregor JL (1972) Polymorphisms for genetics and ecological systems with weak coupling. Theor Pop Biol 3:210–238

Karlin S, Campbell AM, Mrazek J (1998) Comparative DNA analysis across diverse genomes. Annu Rev Genet 32:185–225

Kasarda DD, Bernardin JE, Nimmo CC (1976) Wheat proteins. In: Pomeranz Y (ed) Advances in cereal science and technology. Am Assoc Cereal Chem, St Paul, Minnesota, pp 158–236

Kashi Y, Soller M (1999) Functional roles of microsatellites and minisatellites. In: Goldstein DB, Schlötterer C (eds) Microsatellites: evolution and applications. Oxford Univ Press, Oxford, pp 10–23

Kashi Y, King D, Soller M (1997) Simple sequence repeats as a source of quantitative genetic variation. Theor Appl Genet 13:74–78

Kato K, Miura H, Sawada S (1999) QTL mapping of genes controlling ear emergence time and plant height on chromosome 5a of wheat. Theor Appl Genet 98:472–477

Kawahara T, Nevo E (1996) Screening of spontaneous major translocations in Israeli populations of *Triticum dicoccoides* Koern. Wheat Info Service 83:28–30

Kawahara T, Nevo E, Beiles A (1993) Frequencies of translocations in Israel populations of *Triticum dicoccoides* Korn. Abstracts of the XVth international botanical congress, Yokohama, Japan

Keane B, Pelikan S, Toth GP, Smith MK, Rogstad SH (1999) Genetic diversity of *Typha latifolia* (Typhaceae) and the impact of pollutants examined with tandem-repetitive DNA probes. Am J Bot 76:1226

Keller M, Karutz CH, Schmid JE, Stamp P, Winzeler M et al (1999a) Quantitative trait loci for lodging resistance in a segregating wheat × spelt population. Theor Appl Genet 98:1171–1182

Keller M, Keller B, Schachermayr G, Winzeler M, Schmid JE et al (1999b) Quantitative trait loci for resistance against powdery mildew in segregating wheat x spelt population. Theor Appl Genet 98:903–912

Kendall BE, Fox GA (1998) Spatial structure, environmental heterogeneity, and population dynamics: analysis of the coupled logistic map. Theor Popul Biol 54:11–37

Kenyon JR, Craig IW (1999) Analysis of the 5' regulatory region of the human Norrie's disease gene: evidence that a non-translated CT dinucleotide repeat in exon one has a role in controlling expression. Gene 227:181–188

Kesseli RV, Paran I, Michelmore RW (1994) Analysis of a detailed genetic linkage map of *Lactuca sativa* (lettuce) constructed from RFLP and RAPD markers. Genetics 136:1435–1446

Khan IA, Procunier JD, Humphreys DG, Tranquilli G, Schlatter AR, Marcucci-Poltri S, Frohberg R, Dubcovsky J (2000) Development of PCR based markers for a high grain protein content gene from *Triticum turgidum* ssp. *dicoccoides* transferred to bread wheat. Crop Sci 40: 518–524

Kihara H (1924) Cytologische und Genetische Studien bei wichtigen Getreidearten mit Besonderer Rucksicht aus das Verhalten der Chromosomen und die Sterilitat in den Bastarden. Mem Coll Sci Univ, Kyoto, Ser B 1:1–200

Kihara H (1944) Discovery of the DD-analyser, one of the ancestors of *Triticum vulgare*. Agric Hortic 19:13–14 (in Japanese)

Kihara H (1954) Considerations on the evolution and distribution of *Aegilops* species based on the analyser method. Cytologia 19:336–357

Kimber G, Feldman M (1987) Wild wheat, an introduction. College of Agriculture University Missouri, Columbia, 143 pp

Kimura M (1968) Genetic variability maintained in a finite population due to mutational production of neutral and nearly neutral isoalleles. Genet Res 11:247–249

Kimura M (1983) The neutral theory of molecular evolution. Cambridge Univ Press, Cambridge

King D, Soller M (1999) Variation and fidelity: the evolution of simple sequence repeats as functional elements in adjustable genes. In: Wasser SP (ed) Evolutionary theory and processes: modern perspectives, Papers in honor of Eviatar Nevo. Kluwer Academic, Dordrecht, pp 65–82

King D, Soller M, Kashi Y (1997) Evolutionary tuning knobs. Endeavour 21:36–40

Kingsbury RW, Epstein E (1984) Selection for salt-resistant spring wheat. Crop Sci 24:310–315

Kipling D (1995) The telomere. Oxford Univ Press, Oxford

Kirzhner VM, Korol AB, Ronin YI (1995) Cyclical environmental changes as a factor maintaining genetic polymorphism. I. Two-locus haploid selection. J Evol Biol 8:93–120

Kirzhner VM, Korol AB, Nevo E (1996) Complex dynamics of multilocus systems subjected to cyclical selection. Proc Natl Acad Sci USA 93:6532–6535

Kirzhner VM, Korol AB, Nevo E (1998) Complex limiting behaviour of multilocus genetic systems in cyclical environments. Jour Theor Biol 190:215–225

Kirzhner VM, Korol AB, Nevo E (1999) Abundant multilocus polymorphisms caused by genetic interaction between species on *trait-for-trait* bases. J Theor Biol 198:61–70

Kislev ME, Nadel D, Carmi I (1992) Epi-palaeolithic (19 000 BP) cereal and fruit diet at Ohalo II, Sea of Galilee, Israel. Rev Palaeobot Palynol 3:161–166

Konarev CG, Gavrilyuk IP, Gubareva NK, Peneva TI (1979) About nature and origin of wheat genomes on the data of biochemistry and immunochemistry of grain proteins. Cereal Chem 56:272–278

Konzak CF (1977) Genetic control of the content, amino acid composition and processing properties of proteins in wheat. Adv Gen 19:407–582

Korol AB (1999) Selection for adaptive traits as a factor of recombination evolution: evidence from natural and experimental populations. In: Wasser SP (ed) Evolutionary theory and processes: modern perspectives, Papers in honor of Eviatar Nevo. Kluwer Academic, Dordrecht, pp 31–53

Korol AB (2001) Recombination. In: *Encyclopedia of Biodiversity*. V. 5. Academic Press, pp 53–71

Korol AB, Preygel IA, Nyutin YI (1989) Quasi-linkage in tomato: formal analysis. Genetika (USSR) 25:2225–2233

Korol AB, Preygel IA, Preygel SA (1994) Recombination variability and evolution. Chapman and Hall, London

Korol AB, Ronin YI, Kirzhner VM (1995) Multitrait analysis in interval mapping of QTL. Genetics 140:1137–1147

Korol AB, Kirzhner VM, Ronin YI, Nevo E (1996a) Cyclical environmental changes as a factor of maintaining genetic polymorphism. II. Diploid selection for an additive trait. Evolution 50:1432–1441

Korol AB, Ronin YI, Tadmor Y, Bar-Zur A, Kirzhner VM, Nevo E (1996b) Estimating variance effect of QTL: an important prospect to increase the resolution power of interval mapping. Genet Res 67:187–194

Korol AB, Ronin YI, Nevo E (1998a) Approximated analysis of QTL-environment interaction with no limits on the number of environments. Genetics 148:2015–2028

Korol AB, Ronin YI, Nevo E, Hayes P (1998b) Multi-interval mapping of correlated trait complexes. Heredity 80:273–284

Korol AB, Kirzhner VM, Nevo E (1998c) Dynamics of recombination modifiers caused by cyclical selection: Interaction of forced and auto-oscillations. Genet Res 72:135–148

Korol AB, Ronin YI, Itzcovich A, Nevo E (2001) Enhanced efficiency of QTL mapping analysis based on multivariate complexes of quantitative traits. Genetics 157:1789–1803

Korzun V, Börner A, Worland AJ, Law CN, Röder MS (1997) Application of microsatellite markers to distinguish inter-varietal chromosome substitution lines of wheat (*Triticum aestivum* L.). Euphytica 95:149–155

Korzun V, Röder MS, Wendehake K, Pasqualone A, Lotti C, Ganal MW, Blanco A (1999) Integration of dinucleotide microsatellites from hexaploid bread wheat into a genetic linkage map of durum wheat. Theor Appl Genet 98:1202–1207

Kosambi DD (1944) The estimation of map distances from recombination values. Ann Eugen 12:172–175

Kovacs MIP, Howes NK, Clarke JM, Leisle D (1998) Quality characteristics of durum wheat lines deriving high protein from a *Triticum dicoccoides* (6b) substitution. J Cereal Sci 27:47–51

Krugman T, Rubin B, Levy O, Snape J, Nevo E (1994) RFLP mapping of chlorotoluron resistance gene (Sul) in cultivated wheat *Triticum aestivum*, and in wild wheat *Triticum dicoccoides*. 13th Conf Weed Sci Soc Phytoparasitica 22:135–136

Krugman T, Levy O, Snape, J, Rubin B, Korol AB, Nevo E (1997) Comparative RFLP mapping of the chlorotoluron resistance gene (Sul) in cultivated wheat (*Triticum aestivum*) and wild wheat (*Triticum dicoccoides*). Theor Appl Genet 94:46–51

Kung SD, Wu R (eds) (1993) Transgenic plants. Academic Press, New York

Kunkel BN (1996) A useful weed put to work – genetic analysis of disease resistance in *Arabidopsis thaliana*. Trends Genet 12:63–69

Künzel G, Korzun L, Meister A (2000) Cytologically integrated physical RFLP maps for the barley genome based on translocation breakpoints. Genetics 154:397–412

Kushnir U, Halloran GM (1984) Transfer of high kernel weight from wild tetraploid wheat (*Triticum turgidum dicoccoides*) to breadwheat (*T. aestivum*) using homologous and homoeologous recombination. Euphytica 33:249–255

Ladizinsky G (1998) Plant evolution under domestication. Kluwer, Dordrecht, pp 123–175

Lafyatis R, Denhez F, Williams T, Sporn M, Roberts A (1991) Sequence specific protein binding to and activation of the *TGF-beta3* promoter through a repeated TCCC motif. Nucleic Acids Res 19:6419–6425

Lake S (1986) Recombination frequencies and the coincidence in proximal X-chromosome regions including heterochromatin in *Drosophila melanogaster*. Hereditas 105:263–268

Lande R (1975) The maintenance of genetic variability by mutation in a polygenic character with linked loci. Genet Res 26:221–235

Lander ES, Botstein D (1989) Mapping Mendelian factors underlying quantitative trait using RFLP linkage maps. Genetics 121:185–199

Lander ES, Green P, Abrahamson J, Barlow A, Daly MJ, Lincoln SE, Newburg L (1987) MAPMAKER: an interactive computer package for constructing primary genetic linkage maps of experimental and natural population. Genomics 1:174–181

Lange W, Jochemsen G (1992) Use of the gene pools of *Triticum turgidum* ssp. *dicoccoides* and *Aegilops squarrosa* for the breeding of common wheat (*T. aestivum*), through chromosome-doubled hybrids. Euphytica 59:213–220

Langridge P, Chalmers K (1998) Techniques for marker development. In: Slinkard AE (ed) Proc 9th Intern Wheat Genetics Symp. Univ Extension Press, Univ of Saskatchewan, Saskatoon, Saskatchewan, Canada, pp 107–117

Lanner-Herrera C, Gustafsson M, Falt AS, Bryngelsson T (1996) Diversity of wild *Brassica oleraceae* as estimated by isozyme and RAPD analysis. Genet Resour Crop Evol 43:13–23

Large EC, Doling DA (1962) The measurement of cereal mildew and its effect of yield. Plant Pathol 11:47–57

Lavie B, Stow V, Krugman T, Beiles A, Nevo E (1993) Fitness in wild barley from two opposing slopes of a Mediterranean microsite at Mount Carmel, Israel. Barley Genet Newslett 23:12–14

Law CN (1966) The location of genetic factors affecting a quantitative character in wheat. Genetics 53:478–498

Law CN (1967) The location of genetic factors controlling a number of quantitative characters in wheat. Genetics 56:445–461

Law CN, Payne PI (1983) Genetical aspects of breeding for improved grain protein content and type in wheat. J Cereal Sci 1:79–93

Lazarus CM, Baulcombe DC, Martienssen RA (1985) α-Amylase genes of wheat are two multigene families which are differentially expressed. Plant Mol Biol 53:13–24

Le Corre V, Dumolin-Lapegue S, Kremer A (1997) Genetic variation at allozyme and RAPD loci in sessile oak *Quercus petraea* (Matt.) Liebl: the role of history and geography. Mol Ecol 6:519–529

Leitch AR, Mosgöller W, Schwarzacher T, Bennett MD, Heslop-Harrison JS (1990) Genomic *in situ* hybridization to sectional nuclei shows chromosome domains in grass hybrids. J Cell Sci 95:335–341

Leung WK, Kim JJ, Kim JG, Graham DY, Sepulveda AR (2000) Microsatellite instability in gastric intestinal metaplasia in patients with and without gastric cancer. Am J Pathol 156:537–543

Levene H (1953) Genetic equilibrium when more than one ecological niche is available. Am Nat 87:331–333

Levinson G, Gutman GA (1987) Skipped-strand misfiring: a major mechanism for DNA sequence evolution. Mol Biol Evol 4:203–221

Levy AA (1987) Genetic studies on grain proteins in *Triticum durum var. dicoccoides*. PhD Thesis, Weizmann Institute of Science, Rehovot

Levy AA, Feldman M (1985) Genotypic and fertilization effects on grain protein content in wild and cultivated tetraploid wheats. Genet Agr 39:293–302

Levy AA, Feldman M (1987) Increase in grain protein percentage in high-yielding common wheat breeding lines by genes from wild tetraploid wheat. Euphytica 36:353–359

Levy AA, Feldman M (1988) Ecogeographical distribution of HMW glutenin alleles in populations of the wild tetraploid wheat, *Triticum turgidum var. dicoccoides*. Theor Appl Genet 75:651–658

Levy O (1995) Chlorotoluron-resistance mechanism in blackgrass (*Alopecurus myosuroides*) and wild emmer wheat (*Triticum dicoccoides*): the chlorophyll fluoresence as a tool for its identification. MSc Thesis, Faculty of Agriculture of the Hebrew University of Jerusalem. Rehovot, Israel

Lev-Yadun S, Gopher A, Abbo S (2000) The cradle of agriculture. Science 288:1602–1603

Lewontin RC (1972) The apportionment of human diversity. Evol Biol 6:381–398

Lewontin RC (1974) The genetic basis of evolutionary change. Columbia Univ Press, New York

Lewontin RC (1997) The paradox of variation. In: Ridley M (ed) Evolution. Oxford Univ Press, Oxford, pp 79–88

Li J, Liang VC, Sedgwick T, Wong J, Shi YB (1998) Unique organization and involvement of GAGA factors in transcriptional regulation of the *Xenopus* stromelysin-3 gene. Nucleic Acids Res 26:3018–3025

Li YC (2000) Microscale molecular population genetics of wild emmer wheat, *Triticum dicoccoides*, in Israel. PhD Thesis, University of Haifa, Israel

Li YC, Fahima T, Beiles A, Korol AB, Nevo E (1999) Microclimatic stress and adaptive DNA differentiation in wild emmer wheat, *Triticum dicoccoides*. Theor Appl Genet 98:873–883

Li YC, Fahima T, Peng JH, Röder MS, Kirzhner VM, Beiles A, Korol AB, Nevo E (2000a) Edaphic microsatellite DNA divergence in wild emmer wheat, *Triticum dicoccoides*, at a microsite: Tabigha, Israel. Theor Appl Genet 101:1029–1038

Li YC, Fahima T, Peng JH, Röder MS, Kirzhner VM, Beiles A, Korol AB, Nevo E (2000b) Microsatellite diversity correlated with ecological-edaphic and genetic factors in three microsites of wild emmer wheat in north Israel. Mol Biol Evol 17:851–862

Li YC, Röder MS, Fahima T, Kirzhner VM, Beiles A, Korol AB, Nevo E (2000c) Natural selection causing microsatellite divergence in wild emmer wheat at the ecologically variable microsite at Ammiad, Israel. Theor Appl Genet 100:985–999

Li YC, Krugman T, Fahima T, Beiles A, Röder MS, Korol AB, Nevo E (2000d) Parallel microgeographic patterns of gentic diversity and divergence revealed by allozyme, RAPD, and microsatellites in *Triticum dicoccoides* at Ammiad, Israel. Conserv Genet 1:191–207

Li YC, Krugman T, Fanima T, Beiles A, Korol AB, Nevo E (2001a) Spatiotemporal allozyme divergence caused by aridity stress in a natural population of wild wheat, *Triticum dicoccoides*, at the Ammiad microsite, Israel. Theor Appl Genet 102:853–864

Li YC, Röder MS, Fahima T, Kirzhner VM, Beiles A, Korol A, Nevo E (2001b) Climatic effects on microsatellite diversity in wild emmer wheat, *Triticum dicoccoides*, at Yehudiyya microsite. Heredity (in press)

Li YC, Fahima T, Roder MS, Beiles A, Korol AB, Nevo E (2001c) Genetic effects on microsatellite diversity in wild emmer wheat, *Triticum dicoccoides*, at the Yehudiyya microsite. Heredity (in press)

Li YC, Korol AB, Fahima T, Beiles A, Nevo E (2001d) The ecological-genetic perspective of microsatellite evolution. Heredity (in press)

Lilienfeld FA, Kihara H (1934) Genom analyse bei *Triticum* und *Aegilops* von H Kihara. V *Triticum timopheevi* Zhuk. Cytologia 6:87–122

Lincoln S, Daly M, Lander E (1992) Constructing genetic maps with Mapmaker/exp3.0. Whitehead Institute Technical Report, 3rd edn. Whitehead Institute, Cambridge, MA

Litt M, Luty JA (1989) A hypervariable microsatellite revealed by in vitro amplification of a dinucleotide repeat within the cardiac muscle actin gene. Am J Hum Genet 44:397–401

Liu CJ, Devos KM, Chinoy CN, Atkinson MD, Gale MD (1992) Non-homoeologous translocations between group-4, 5 and 7 chromosomes in wheat and rye. Theor Appl Genet 83:305–312

Liu L, Dybvig K, Panangala VS, van Santen VL, French CT (2000) GAA trinucleotide repeat region regulates M9/pMGA gene expression in *Mycoplasma gallisepticum*. Infect Immun 68:871–876

Liu YG, Tsunewaki K (1991) Restriction fragment length polymorphism analysis of wheat. II. Linkage maps of the RFLP sites in common wheat. Jpn J Genet 66:617–633

Liu Z, Biyashev RM, Saghai Maroof MA (1996) Development of simple sequence repeat DNA markers and their integration into a barley linkage map. Theor Appl Genet 93:869–876

Liu Z, Sun Q, Ni Z, Nevo E, Yang T (2001) Molecular characterization of a novel powdery mildew resistance gene *MlC20* in wheat origination from wild emmer (submitted)

Longstaff M, Raines CA, McMorrow EM, Bradbeer JW, Dyer TA (1989) Wheat phosphoglycerate kinase: evidence for recombination between the genes for the chloroplastic and cytosolic enzymes. Nucleic Acids Res 17:6569–6580

Lue NL, Buchman AR, Kornberg RD (1989) Activation of yeast RNA polymerase II transcription by a thymidine-rich upstream element in vitro. Proc Natl Acad Sci USA 86:486–490

Lukaszewski AJ, Curtis CA (1993) Physical distribution of recombination in B-genome chromosomes of tetraploid wheat. Theor Appl Genet 86:121–127

Luo MC, Yang ZL, Dvorak J (2000) The *Q* locus of Iranian and European spelt wheat. Theor Appl Genet 100:602–606

Lupton FGH (1987) Wheat breeding. Chapman and Hall, London

Ma H, Singh RP, Mujeeb-Kazi A (1997) Resistance to stripe rust in durum wheats. A-genome diploids, and their amphiploids. Euphytica 94:279–286

Ma ZQ, Lapitan NLV (1998) A comparison of amplified and restriction fragment length polymorphism in wheat. Cereal Res Comm 26:7–13

Maab HI, Klaas M (1995) Infraspecific differentiation of garlic (*Allium sativum* L.) by isozyme and RAPD markers. Theor Appl Genet 91:89–97

Maan SS (1973) Cytoplasmic and cytogenetic relationships among tetraploid *Triticum* species. Euphytica 22:287–300

Mahon JD (1983) Limitations to the use of physiological variability in plant breeding. Can J Plant Sci 63:11–21

Malinowsky E (1927) The hypothesis of chromosome affinity and the phenomenon of suppression of characters on crossing. J Genet 18:223–231

Manistersky Y, Eyal Z (1994) Survey of wheat reaction to rusts: test of wheat cultivars during the winter and spring of 1994. Gan Sade Meshek 12:10–20 (in Hebrew)

Manistersky Y, Ben Yehuda P, Bar A, Eyal Z (1995) Survey of rust races of wheat in Israel during the season of 1994. In: Abstr 16 Isr Phytopath Soc Annu Meeting, 20–21 Febr 1995, p 25 (in Hebrew)

Markert CL, Moller F (1959) Multiple forms of enzymes: tissue, ontogenic, and species specific patterns. Proc Natl Acad Sci USA 45:753–763

Martin-Farmer JG, Janssen R (1999) A downstream CA repeat sequence increases translation from leadered and unleadered mRNA in *Escherichia coli*. Mol Microbiol 31:1025–1038

Maynard Smith J, Hoekstra R (1980) Polymorphism in a varied environment: how robust are the models? Genet Res 35:45–57

Mayr E (1942) Systematics and the origin of species. Columbia Univ Press, New York

Mayr E (1954) Change of genetic environment and evolution. In: Huxley J, Hardy A, Ford E (eds) Evolution as a process. Allen and Unwin, London, pp 157–180

Maxted N, Ford-Lloyd BV, Hawkes JC (1997) Plant genetic conservation. The *in situ* approach. Chapman and Hall, London, pp 132–143

McClelland M, Welsh J (1994) DNA fingerprinting by arbitrarily primed PCR. PCR Meth Appl 4:s59–s65
McFadden ES, Sears ER (1946) The origin of *Triticum spelta* and its free-threshing hexaploid relatives. J Hered 37:81–89
McIntosh RA (1998) Breeding wheat for resistance to biotic stresses. Euphytica 100:19–34
McIntosh RA, Wellings CR, Park RF (1995) Wheat rusts and Atlas of resistance genes. Kluwer Academic, Dordrecht
McIntosh RA, Silk J, The TT (1996) Cytogenetic studies in wheat XVII. Monosomic analysis and linkage relationships of gene *Yr15* for resistance to stripe rust. Euphytica 89:395–399
McIntosh RA, Hart GE, Devos KM, Gale MD, Rogers WJ (1998) Catalogue of gene symbols for wheat. In: Slinkard AE (ed) Proc 9th Intern Wheat Genet Symp Univ Extension Press, Univ of Saskatchewan, Saskatoon, Canada, pp 139–142
McKeon C, Schmidt A, de Crombrugghe B (1984) A sequence conserved in both the chicken and mouse alpha 2(I) collagen promoter contains sites sensitive to S1 nuclease. J Biol Chem 259:6636–40
McNeilly T (1968) Evolution in closely adjacent plant populations. III. *Agrostis tenuis*. Heredity 23:99–108
Meloni R, Albanese V, Ravassard P, Treilhou F, Mallet J (1998) A tetranucleotide polymorphic microsatellite, located in the first intron of the tyrosine hydroxylase gene, acts as a transcription regulatory element in vitro. Hum Mol Genet 7:423–428
Mesfin A, Frohberg RC, Anderson JA (1999) RFLP markers associated with high grain protein from *Triticum turgidum* L. var *dicoccoides* introgressed into hard red spring wheat. Crop Sci 39:508–513
Messmer MM, Keller M, Zanetti S, Keller B (1999) Genetic linkage map of a wheat x spelt cross. Theor Appl Genet 98:1163–1170
Messmer MM, Seyfarth R, Keller M, Schachermayr G, Winzeler M et al (2000) Genetic analysis of durable leaf rust resistance in winter wheat. Theor Appl Genet 100:419–431
Michelmore RW, Meyers BC (1998) Clusters of resistance genes in plants evolve by divergent selection and a birth-and-death process. Genome Res 8:1113–1130
Michie D (1953) Affinity: a new genetic phenomenon in the house mouse. Nature 171:26–27
Miller TE (1987) Systematics and evolution. In:Lupton FGH (ed) Wheat breeding. Its scientific basis. Chapman and Hall, London, pp 1–30
Miller TE (1992) A cautionary note on the use of morphological characters for recognising taxa of wheat (genus *Triticum*). Prehistoire de l'agriculture: nouvelles approches experimentales et ethnographiques. Monographie du CRA, no 6, CNRs, pp 249–253
Mingeot D, Jacquemin JM (1999) Mapping of RFLP probes characterized for their polymorphism on wheat. Theor Appl Genet 98:1132–1137
Ministry of Science and Technology, Israel (1990) International Workshop on dynamic in situ conservation of wild relatives of major cultivated plants. Sci Prog Council Res Devel, Israel
Miret JJ, Pessoa-Brandao L, Lahue RS (1998) Orientation-dependent and sequence-specific expansions of CTG/CAG trinucleotide repeats in *Saccharomyces cerevisiae*. Proc Natl Acad Sci USA 95:12438–12434
Mitton JB (1994) Molecular approaches to population biology. Annu Rev Ecol Syst 25:45–69
Mitton JB (1997) Selection in natural populations. Oxford Univ Press, Oxford
Monckton DG, Cayuela ML, Gould FK, Brock GJ et al (1999) Very large $(CAG)_n$ DNA repeat expansions in the sperm of two spinocerebellar ataxia type 7 males. Hum Mol Genet 8:2473–2478
Moonen JHE, Scheepstra A, Graveland A (1982) Use of the SDS-sedimentation test and SDS-Polyacrylamide gel electrophoresis for screening breeder's samples of wheat for bread-making quality. Euphytica 31:277–290
Moore H, Greenwell PW, Liu CP, Arnheim N, Petes TD (1999) Triplet repeats form secondary structures that escape DNA repair in yeast. Proc Natl Acad Sci USA 96:1504–1509
Moore SS, Sargeant LL, King TJ, Mattick JS, Georges M, Hetzel JS (1991) The conservation of dinucleotide microsatellites among mammalian genomes allows the use of heterologous PCR primer pairs in closely related species. Genomics 10:654–660

Mori N, Liu Y-G, Tsunewaki K (1995) Wheat phylogeny determined by RFLP analysis of nuclear DNA: 2. Wild tetraploid wheats. Theor Appl Genet 90:129–134

Morris R, Sears ER (1967) The cytogenetics of wheat and its relatives. In: Quisenberry KS, Reitz LP (eds) Wheat and wheat improvement. Am Soc Agron, Washington, DC, pp 19–87

Moseman JG, El-Morshidy MA, Wahl I, Nevo E, Tomerlin JR, Zohary D (1983a) *Hordeum spontaneum* resistant to leaf rust, powdery mildew, scald and net blotch. Agronomy Abstracts, Annu Meet Am Soc Agron, Crop Sci Soc Am, and the Soil Sci Soc Am, 14–19 Aug 1983, Washington, DC, p 74

Moseman JG, Nevo E, Zohary D (1983b) Resistance of *Hordeum spontaneum* collected in Israel to infection with *Erysiphe graminis hordei*. Crop Sci 23:1115–1119

Moseman JG, Nevo E, El Morshidy MA, Zohary D (1984) Resistance of *T. dicoccoides* to infection with *Erysiphe graminis tritici*. Euphytica 33:41–47

Moseman JG, Nevo E, Gerechter-Amitai ZK, El-Morshidy MA, Zohary D (1985) Resistance of *T. dicoccoides* collected in Israel to infection with *Puccinia recondita tritici*. Crop Sci 256:262–265

Moss T (1983) A transcriptional function for the repetitive ribosomal spacer in *Xenopus laevis*. Nature 302:223–230

Mukai Y, Nakahara Y, Yamamoto M (1993a) Simultaneous discrimination of the three genomes in hexaploid wheat by multicolor fluorescence *in situ* hybridization using total genomic and highly repeated DNA probes. Genome 36:489–494

Mukai Y, Friebe B, Hatchett JH, Yamamoto M, Gill BS (1993b). Molecular cytogenetic analysis of radiation-induced wheat-rye terminal and intercalary chromosomal translocations and the detection of rye chromatin specific resistance to Hessian fly. Chromosoma 102:88–95

Munaver SM, Harper AE (1959) Amino acid balance and imbalance. II. Dietary level of protein and lysine requirement. J Nutr 69:58–64

Murphy KE, Stringer JR (1986) RecA independent recombination of poly[d(GT)-d(CA)] in pBR322. Nucleic Acids Res 14:7325–40

Murray DR (1991) Advanced methods in plant breeding and biotechnology. CAB International, Wallingford, UK

Nakagahra M (1986) Geographic distribution of gametophyte genes in wide crosses of rice cultivars. In: Khush GS (ed) Rice genetics. Proc Int Rice Genet Symp. IRRI, Manila, pp 73–82

National Academy of Sciences (1993) Managing global genetic resources. Agricultural crop issues and policies. National Academy Press, Washington, DC

Nei M (1972) Genetic distance between populations. Am Nat 106:283–292

Nei M (1973) Analysis of gene diversity in subdivided population. Proc Natl Acad Sci USA 70:3321–3323

Nei M (1975) Molecular population genetics. North-Holland Publ, Amsterdam

Nei M (1978) Estimation of average heterozygosity and genetic distance from small number of individuals. Genetics 89:583–590

Nei M (1980) Stochastic theory of population genetics and evolution. In: Barigozzi C (ed) Vito Volterra symposium on mathematical models in biology. Springer, Berlin Heidelberg New York, pp 17–47

Nei M (1987) Molecular evolutionary genetics. Columbia Univ Press, New York

Nelson JC, Van Deynze AE, Autrique E, Sorrels ME, Lu YH, Merlino M, Atkinson M, Leroy P (1995) Molecular mapping of wheat. Homoeologous group 2. Genome 38:516–524

Nelson JC, Autrique JE, Fuentes-de Vila, Sorrells E (1998) Chromosomal location of genes for resistance to karnal bunt in wheat. Crop Sci 38:231–236

Nesbitt M (1998) The transition from foraging to farming in Southwest Asia. International workshop, Groningen, Netherlands

Nesbitt M, Samuel D (1998) Wheat domestication: archaeobotanical evidence. Science 279:1433

Nevo E (1978) Genetic variation in natural populations: patterns and theory. Theor Pop Biol 13:121–177

Nevo E (1983a) Genetic resources of wild emmer wheat: structure, evolution and application in breeding. In: Sakamoto S (ed) Proc of the 6th Int Wheat Genet Symp. Kyoto University, Kyoto, Japan, pp 421–431

Nevo E (1983b) Adaptive significance of protein variation. In: Oxford GS, Rollinson D (eds) Protein polymorphism: adaptive and taxonomic significance. Academic Press, New York, pp 239–282

Nevo E (1983c) Population genetics and ecology: the interface. In: Bendall DS (ed) Evolution from molecules to man. Cambridge Univ Press, Cambridge, pp 287–321

Nevo E (1986a) Genetic resources of wild cereals and crop improvement: Israel a natural laboratory. Isr J Bot 235:255–278

Nevo E (1986b) Pollution and genetic evolution in marine organisms: theory and practice. In: Dubinsky Z, Steinberger Y (eds) Environmental quality and ecosystem stability. Bar-Ilan Univ Press, Ramat-Gan, Israel, pp 841–848

Nevo E (1987) Plant genetic resources: prediction by isozyme markers and ecology. In: Rattazzi MC, Scandalios JG, Whitt GS (eds) Isozymes: current topics in biological and medical research, vol 16. Agriculture, physiology and medicine. Liss, New York, pp 247–267

Nevo E (1989a) Genetic resources of wild emmer wheat revisited: genetic evolution, conservation and utilization. In: Miller TE, Koebner RMD (eds) Proc 7th Int Wheat Genet Symp Cambridge, pp 121–126

Nevo E (1989b) Modes of speciation: the nature and role of peripheral isolates in the origin of species. In: Giddings LV, Kaneshiro KY, Anderson WW (eds) Genetics, speciation, and the founder principle. Oxford Univ Press, Oxford, pp 206–236

Nevo E (1990) Molecular evolutionary genetics of isozymes: pattern, theory, and application. Prog Clin Biol Res 344:701–742

Nevo E (1991) Evolutionary theory and processes of active speciation and adaptive radiation in subterranean mole rats, *Spalax ehrenbergi* superspecies in Israel. Evol Biol 15:1–125

Nevo E (1992) Origin, evolution, population genetics and resources for breeding of wild barley, *Hordeum spontaneum*, in the Fertile Crescent. In: Shewry P (ed) Barley: genetics, molecular biology and biotechnology. CAB International, Wallingford, pp 19–43

Nevo E (1995) Genetic resources of wild emmer, *Triticum dicoccoides*, for wheat improvement: news and views. In: Li ZS, Xin ZY (eds) Proc 8th Int Wheat Genet Symp. China Agricultural Scientech Press, Beijing, pp 79–87

Nevo E (1997) Evolution in action across phylogeny caused by microclimatic stresses at "Evolution Canyon". Theor Pop Biol 52:231–243

Nevo E (1998a) Genetic diversity in wild cereals: regional and local studies and their bearing on conservation *ex situ* and *in situ*. Genet Resour Crop Evol 45:355–370

Nevo E (1998b) Molecular evolution and ecological stress at global, regional and local scales: the Israeli perspective. J Exp Zool 282:95–119

Nevo E (2001a) Genetic diversity. In: *Encyclopedia of Biodiversity* 3:195–213, Academic Press

Nevo E (2001b) Genetic resources of wild emmer, *Triticum dicoccoides*, for wheat improvement in the third millenium. Israel J Plant Sci 49 (supplement):S77–S91

Nevo E (2001c) Evolution of genome-phenome diversity under environmental stress. PNAS 98:6233–6240

Nevo E, Beiles A (1988) Genetics parallelism of protein polymorphism in nature: ecological test of the neutral theory of molecular evolution. Biol J Linn Soc 35:229–245

Nevo E, Beiles A (1989) Genetic diversity of wild emmer wheat in Israel and Turkey: structure, evolution and application in breeding. Theor Appl Genet 77:421–455

Nevo E, Beiles A (1992) Amino acid resources in the wild progenitor of wheats, *Triticum* dicoc*coides* in Israel-Polymorphisms and predictability by ecology and isozymes. Plant Breed 108:190–201

Nevo E, Cleve H (1978) Genetic differentiation during speciation. Nature 275:125–126

Nevo E, Payne PI (1987) Wheat storage proteins: diversity of HMW glutenin subunits in wild emmer from Israel. Theor Appl Genet 74:827–836

Nevo E, Shimony T, Libni M (1977) Thermal selection of allozyme polymorphisms in barnacles. Nature 267:699–701

Nevo E, Zohary D, Brown AHD, Haber M (1979) Genetic diversity and environmental associations of wild barley *Hordeum spontaneum* in Israel. Evolution 33:815–833

Nevo E, Brown AHD, Zohary D, Storch N, Beiles A (1981) Microgeographic edaphic differentiation in allozyme polymorphisms of wild barley (*Hordeum spontaneum*, Poaceae). Plant Syst Evol 138:287–292

Nevo E, Golenberg EM, Beiles A, Brown AHD, Zohary D (1982) Genetic diversity and environmental associations of wild wheat, *Triticum dicoccoides*, in Israel. Theor Appl Genet 62: 241–254

Nevo E, Beiles A, Storch N, Doll H, Andersen B (1983) Microgeographic edaphic differentiation in hordein polymorphisms of wild barley. Theor Appl Genet 64:123–132

Nevo E, Beiles A, Ben-Shlomo R (1984a) The evolutionary significance of genetic diversity: ecological, demographic and life history correlates. Lecture Notes Biomath 53:13–213

Nevo E, Beiles A, Gutterman Y, Storch N, Kaplan D (1984b) Genetic resources of wild cereals in Israel and vicinity. I. Phenotypic variation within and between populations of wild wheat, *Triticum dicoccoides*. Euphytica 33:717–735

Nevo E, Moseman JG, Beiles A, Zohary D (1985) Patterns of resistance of Israeli wild emmer wheat to pathogens. I. Predictive method by ecology and allozyme genotypes for powdery mildew and leaf rust. Genetica 67:209–222

Nevo E, Grama A, Beiles A, Golenberg EM (1986a) Resources of high-protein genotypes in wild wheat, *Triticum dicoccoides* in Israel: predictive method by ecology and allozyme markers. Genetica 68:215–227

Nevo E, Gerechter-Amitai ZK, Beiles A, Golenberg EM (1986b) Resistance of wild wheat to stripe rust: Predictive method by ecology and allozyme genotypes. Plant Syst Evol 153:13–30

Nevo E, Beiles A, Kaplan D, Golenberg L, Olsvig-Whittaker, Naveh Z (1986c) Natural selection of allozyme polymorphisms: a microsite test revealing ecological genetic differentiation in wild barley. Evolution 40:13–20

Nevo E, Beiles A, Kaplan D (1988a) Genetic diversity and environmental associations of wild emmer wheat in Turkey. Heredity 61:31–45

Nevo E, Beiles A, Krugman T (1988b) Natural selection of allozyme polymorphisms: a microgeographic climatic differentiation in wild emmer wheat *(Triticum dicoccoides)*. Theor Appl Genet 75:529–538

Nevo E, Beiles A, Krugman T (1988c) Natural selection of allozyme polymorphisms: a microgeographical differentiation by edaphic, topographical, and temporal factors in wild emmer wheat (*Triticum dicoccoides*). Theor Appl Genet 76:737–752

Nevo E, Carver BF, Beiles A (1991a) Photosynthetic performance in wild emmer wheat *Triticum dicoccoides*: ecological and genetic predictability. Theor Appl Genet 81:445–460

Nevo E, Gerechter-Amitai ZK, Beiles A (1991b) Resistance of wild emmer wheat to stem rust: ecological, pathological and allozyme associations. Euphytica 53:121–130

Nevo E, Noy-Meir I, Beiles A, Krugman T, Agami M (1991c) Natural selection of allozyme polymorphisms: micro-geographical spatial and temporal ecological differentiations in wild emmer wheat. Isr J Bot 40:419–450

Nevo E, Ordentlich A, Beiles A, Raskin I (1992a) Genetic divergence of heat production within and between the wild progenitors of wheat and barley: evolutionary and agronomical implications. Theor Appl Genet 84:958–962

Nevo E, Gorham J, Beiles A (1992b) Variation for ^{22}Na uptake in wild emmer wheat, *Triticum dicoccoides* in Israel: salt tolerance resources for wheat improvement. J Exp Bot 43:511–518

Nevo E, Snape JW, Lavie B, Beiles A (1992c) Herbicide response polymorphisms in wild emmer wheat: ecological and isozyme correlations. Theor Appl Genet 84:209–216

Nevo E, Nishikawa K, Furuta Y, Gonokami Y, Beiles A (1993a) Genetic polymorphisms of α- and β-amylase isozymes in wild emmer wheat, *Triticum dicoccoides*, in Israel. Theor Appl Genet 85:1029–1042

Nevo E, Krugman T, Beiles A (1993b) Genetic resources for salt tolerance in the wild progenitors of wheat (*Triticum dicoccoides*) and barley (*Hordeum spontaneum*) in Israel. Plant Breed 11:338–341

Nevo E, Meyer H, Piechulla B (1993c) Diurnal rhythms of the chlorophyll a/b binding protein mRNAs in wild emmer wheat and wild barley (Poaceae) in the Fertile Crescent. Plant Syst Evol 185:181–188

Nevo E, Filippucci MG, Beiles A (1994) Genetic polymorphisms in subterranean mammals (*Spalax ehrenbergi* superspecies) in the Near East revisited: patterns and theory. Heredity 72:465–487

Nevo E, Pagnotta MA, Beiles A, Porceddu E (1995) Wheat storage proteins: Glutenin DNA diversity in wild emmer wheat, *Triticum dicoccoides*, in Israel and Turkey. Theor Appl Genet 91:415–420

Nevo E, Ben-Shlomo R, Beiles A, Ronin YI, Blum S, Hillel J (1996) Genomic adaptive strategies: DNA fingerprinting and RAPDs reveal ecological correlates and genetic parallelism to allozyme and mitochondrial DNA diversities in the actively speciating mole rats in Israel. In: Holmes RS, Lim HA (eds) Gene families: structure, function, genetics evolution. World Scientific, Singapore, pp 55–70

Nevo E, Kirzhner VM, Beiles A, Korol AB (1997) Selection versus random drift: long-term polymorphism persistence in small populations (evidence and modelling). Phil Trans R Soc Lond B 352:381–389

Nevo E, Baum B, Beiles A, Johnson DA (1998) Ecological correlates of RAPD DNA diversity of wild barley, *Hordeum spontaneum*, in the Fertile Crescent. Genet Resour Crop Evol 45:151–159

Nevo E, Beiles A, Korol AB, Ronin YI, Pavlicek T, Hamilton WD (2000) Extraordinary multilocus genetic organization in mole crickets, gryllotalpidae. Evolution 54:586–605

Nilsson NO, Säâll T, Bengtsson BO (1993) Chiasma and recombination data in plants: are they compatible? Trends Genet 9:344–348

Nishikawa K (1984) Species relationship of wheat and its putative ancestors as viewed from isozyme variation. Proc 7th Int Wheat Genet Symp, pp 59–63

Nishikawa K, Mizuno S, Furuta Y (1994) Identification of chromosomes involved translocations in wild emmer. Jpn J Genet 69:371–376

Nowak R (1994) Mining treasures from "Junk DNA". Science 263:608–610

Noy-Meir I, Agami M, Cohen E, Anikster Y (1991a) Floristic and ecological differentiation of habitats within a wild wheat population at Ammiad. Isr J Bot 40:363–384

Noy-Meir I, Agami M, Anikster Y (1991b) Changes in the population density of wild emmer wheat (*Triticum turgidum* var. *dicoccoides*) in a Mediterranean grassland. Isr J Bot 40:385–396

O'Brien SJ (1993) Genetic maps. Locus maps of complex genomes, 6th edn. Cold Spring Harbor Lab Press, Cold Spring Harbor, NY

Ohta T (1976) Role of very slightly deleterious mutations in molecular evolution and polymorphism. Theor Pop Biol 10:254–275

Ohta T (1992) The nearly neutral theory of molecular evolution. Annu Rev Ecol Syst 23:263–268

Ohta T (1997) Role of random genetic drift in the evolution of interactive systems. J Mol Evol 44 [Suppl]:S9–S14

Ohta T, Cockerham CC (1974) Detrimental genes with partial selfing and effects on a neutral locus. Genet Res 23:191–200

Ohta T, Gillespie JH (1996) Development of neutral and nearly neutral theories. Theor Pop Biol 49:128–142

Ogihara T, Tsunewaki K (1988) Diversity and evolution of chloroplast DNA in *Triticum* and *Aegilops* as revealed by restriction fragment analysis. Theor Appl Genet 76:321–332

Orr HA (1991) Is single-gene speciation possible? Evolution 45:764–769

Orr MR, Smith TB (1998) Ecology and speciation. Trends Ecol Evol 13:502–506

Owuor ED, Nevo E (2001) Microscale ecological stress causes molecular selection in wild barley Genet Resour and Crop Evol (in press)

Owour ED, Fahima T, Beiles A, Korol AB, Nevo E (1997) Population genetic response to microsite ecological stress in wild barley, *Hordeum spontaneum*. Mol Ecol 6:1177–1187

Owuor ED, Fahima T, Beharav A, Korol AB, Nevo E (1999) RAPD divergence caused by microsite edaphic selection in wild barley. Genetica 105:177–192

Pagnotta MA, Nevo E, Beiles A, Porceddu E (1995) Wheat storage proteins: glutenin diversity in wild emmer, *Triticum dicoccoides*, in Israel and Turkey. Theor Appl Genet 91:409–414

Pakniyat H, Powell W, Baird E, Handley LL, Robinson D, Sorimgeour CM, Nevo E et al (1997) AFLP variation in wild barley (*Hordeum spontaneum* C. Koch) with reference to salt tolerance and associated ecogeography. Genome 40:332–341

Paterson AH, Lander ES, Hewitt JD, Peterson S, Lincoln SE, Tanksley SD (1988) Resolution of quantitative trait into mendelian factors by using a complete linkage map of restriction fragment length polymorphisms. Nature 335:721–726

Paterson AH, Tanksley SD, Sorrells ME (1991) DNA markers in plant improvement. Adv Agron 46:39–90

Payne PI, Rhodes AP (1982) Cereal storage proteins: structure and role in agriculture and food technology. Encycl Plant Physiol 14:346–369

Payne PI, Law CN, Mudd EE (1980) Control by homoeologous group 1 chromosomes of the high molecular weight subunits of glutenin, a major protein of wheat endosperm. Theor Appl Genet 58:113–120

Payne PI, Holt M, Law CN (1981a) Structural and genetical studies on high-molecular-weight of wheat glutenin, part 1. Allelic variation in subunits amongst varieties of wheat (*Triticum aestivum*). Theor Appl Genet 60:229–236

Payne PI, Corfield KG, Holt LM, Blackman JA (1981b) Correlation between the inheritance of certain high-molecular weight subunits of glutenin and bread-making quality in progress of six crosses of bread wheat. J Sci Food Agric 32:51–60

Payne PI, Holt H, Jackson EA, Law CN (1984) Wheat storage proteins: their genetics and their potential for manipulation by plant breeding. Phil Trans R Soc Lond Ser B 304:359–371

Peakall P, Smouse PE, Huff DR (1995) Evolutionary implications of allozyme and RAPD variation in diploid populations of dioecious buffalograss *Buchloe dactyloides*. Mol Ecol 4:135–147

Pearson CE, Sinden RR (1996) Alternative structures in duplex DNA formed within the trinucleotide repeats of the myotonic dystrophy and fragile X loci. Biochemistry 35:5041–5053

Peng JH (2000) Genomics of wild emmer wheat, *Triticum dicoccoides*: genetic maps, mapping of stripe rust resistance genes and QTLs for agronomic traits. PhD Thesis, University of Haifa, Israel

Peng JH, Fahima T, Röder MS, Li YC, Dahan A, Grama A, Ronin YI, Korol AB, Nevo E (1999) Microsatellite tagging of the stripe-rust resistance gene *YrH52* derived from wild emmer wheat, *Triticum dicoccoides*, and suggestive negative crossover interference on chromosome 1B. Theor Appl Genet 98:862–872

Peng JH, Fahima T, Röder MS, Li YC, Grama A, Nevo E (2000a) Microsatellite high-density mapping of the stripe rust resistance gene *YrH52* region on chromosome 1B and evaluation of its marker-assisted selection in the F2 generation in wild emmer wheat. New Phytol 146: 141–154

Peng JH, Korol AB, Fahima T, Röder MS, Ronin YI, Li YC, Nevo E (2000b) Molecular genetic maps in wild emmer wheat, *Triticum dicoccoides*: genome-wide coverage, massive negative interference, and putative quasi-linkage. Genome Res 10:1509–1531

Peng JH, Fahima T, Röder MS, Huang QY, Dahan A, Li YC, Grama A, Nevo E (2000c) High-density molecular map of chromosome region harboring *YrH52* gene and its preliminary discrimination from *Yr15* in wild emmer wheat, *Triticum dicoccoides*. Genetica 109:199–210

Peng JH, Korol AB, Fahima T, Röder MS, Li YC, Ronin YI, Nevo E (2001) Mapping of QTLs affecting agronomically important traits in wild emmer wheat, *Triticum dicoccoides* (submitted)

Penner GA (1996) RAPD analysis of plant genomes. In: Jauhar PP (ed) Methods of genome analysis in plants. CRC, Boca Raton, pp 251–268

Pestsova E, Korzun V, Goncharov NP, Hammer K, Ganal MW, Röder MS (2000) Microsatellite analysis of *Aegilops tauschii* germplasm. Theor Appl Genet 101:100–106

Phillips RL, Vasil IK (1994) DNA-based markers in plants. Kluwer Academic, Dordrecht

Plaschke J, Ganal MW, Röder MS (1995) Detection of genetic diversity in closely related bread wheat using microsatellite markers. Theor Appl Genet 91:1001–1007

Plaschke J, Borner A, Wendehake K, Ganal MW, Röder MS (1996) The use of wheat aneuploids for the chromosomal assignment of microsatellite loci. Euphytica 89:33–40

Plucknett DL, Smith NJH, Williams JT, Anishetty NM (1983) Crop germplasm conservation and developing countries. Science 220:163–169

Plucknett DL, Smith NJH, Williams JT, Anishetty NM (1987) Gene banks and the world's food. Princeton Univ Press, Princeton, NJ

Pogna NE, Metakovsky EV, Redaelli R, Rainieri F, Dachkevitch T (1993) Recombination mapping of *Gli-5*, a new gliadin-coding locus on chromosome 1A and 1B in common wheat. Theor Appl Genet 87: 113–121

Porceddu E, Lafiandra D, Scarascia-Mugnozza GT (1983) Genetics of seed proteins in wheats. In: Gottschalk W, Muller HP (eds) Seed proteins, biochemistry, genetics, nutritive value. Nijhoff-Junk, The Hague, pp 77–141

Prakash J, Pierik RLM (eds) (1993) Plant biotechnology. Commercial prospects and problems. Int Sci Publ pp 289

Prasad M, Varshney RK, Kumar A, Balyan HS, Sharma PC et al (1999) A microsatellite marker associated with a QTL for grain protein content on chromosome arm 2Dl of bread wheat. Theor Appl Genet 99:341–345

Price TD, Gebauer AB (eds) (1995) Last hunters, first farmers. Advanced Seminar Series School of American Research, Santa Fe, CA

Punt PJ, Dingemanse MA, Kyuvenhoven A, Soede RDM et al (1990) Functional elements in the promoter region of the *Aspergillus nidulans* Gpd a gene coding glyceradehyde-3-phosphate dehydrogenase. Gene 93:101–109

Puterka GJ, Black IVWC, Steiner WM, Burton RL (1993) Genetic variation and phylogenetic relationships among world-wide collections of the Russian wheat aphid, *Diuraphis noxia* (Mordvilko), inferred from allozyme and RAPD-PCR markers. Heredity 70:604–618

Rajaram S, Pfeiffer WH, Singh RP, Bekele G, Gilchrist L, Abdalla O (1988) Bread wheat. In: CIMMYT report on wheat improvement 1985–1986. DF:CIMMYT, Mexico, pp 6–43

Ramel C (1997) Mini- and microsatellites. Environ Health Perspect 105 [Suppl 5]:781–789

Reader SM, Miller TE (1991) The introduction into bread wheat of a major gene for resistance to powdery mildew from wild emmer wheat. Euphytica 53:57–60

Reeder RG (1984) Enhancers and ribosomal gene spacers. Cell 38:349–351

Reeder RG, Roan JG (1984) The mechanisms of nucleolar dominance in *Xenopus* hybrids. Cell 38:39–44

Richard GF, Pâques F (2000) Mini- and microsatellite expansions: the recombination connection. EMBO Reports 11:122–126

Richter TE, Ronald PC (2000) The evolution of disease resistance genes. Plant Mol Biol 42: 195–204

Riley R (1965) Cytogenetics and evolution of wheats. In: Hutchinson JB (ed) Crop plant evolution. Cambridge Univ Press, London, pp 103–122

Röder MS, Plaschke J, Konig SU, Borner A, Sorrells ME, Tanksley SD, Ganal MW (1995) Abundance, variability and chromosomal location of microsatellites in wheat. Mol Gen Genet 246:327–333

Röder MS, Korzun V, Wendehake K, Plaschke J, Tixier MH, Leroy P, Ganal MW (1998) A microsatellite map of wheat. Genetics 149:2007–2023

Roelfes AP, Singh RP, Saari EE (1992) Rust diseases of wheat: concepts and methods of disease management. Cimmyt, Mexico

Ronin YI, Korol AB, Fahima T, Kirzhner VM, Nevo E (1996) Sequential estimation of linkage between PCR-generated markers and a target gene employing stepwise bulked analysis. Biometrics 52:1428–1439

Ronin YI, Korol AB, Nevo E (1999) Single-and multiple – trait mapping analysis of linked QTL: some asymptotic analytical approximations. Genetics 151:387–396

Rong JK, Millet E, Manisterski J, Feldman M (1998) A powdery mildew resistance gene from wild emmer transferred into common wheat and tagged by molecular markers. In: Slinkard AE (ed) Proc 9th Int Wheat Genet Symp, vol 3. Univ Extension Press, University Saskatchewan, Saskatoon, Canada, pp 148–150

Rosenberg S, Longerich MS, Gee P, Harris RS (1994) Adaptive mutation by deletions in small mononucleotide repeats. Science 265:405–409

Rubinsztein DC (1999) Trinucleotide expansion mutations cause diseases which do not conform to classical Mendelian expectations. In: Goldstein DB, Schlötterer C (eds) Microsatellites: evolution and applications. Oxford Univ Press, Oxford, pp 81–97

Rubinsztein DC, Leggo J, Coetzee GA, Irvine RA, Buckley M, Ferguson-Smith MA (1995) Sequence variation and size ranges of CAG repeats in the Machado-Joseph disease, spinocerebellar ataxia type 1 and androgen receptor genes. Hum Mol Genet 4:1585–1590

Ryan PJ, Owen EJ (1983) Metabolism of chlorotoluron in tolerant and sensitive cereal varieties. Crop sensitivity to herbicides. AAB Aspects Appl Biol 3:63–71

Ryndin A, Kirzhner VM, Nevo E, Korol AB (2001) Polymorphism maintenance in populations with mixed random mating and apomixes subjected to stabilizing and cyclical selection. J Theor Biol 212:169–181

Saari EE, Prescott JM (1985) World distribution in relation to economic losses. In: Roelfs AP, Bushnell WR (eds) The cereal rust, II. Academic Press, Orlando, pp 259–298

Saghai-Maroof MA, Soliman KM, Jorgensen RA, Allard RW (1984) Ribosomal DNA spacer-length polymorphisms in barley: Mendelian inheritance, chromosomal location and populations dynamics. Proc Natl Acad Sci USA 81:8014–8018

Saghai-Maroof, MA, Biyashev RM, Yang GP, Zhang Q, Allard RW (1994) Extraordinarily polymorphic microsatellite DNA in barley: species diversity, chromosomal locations, and population dynamics. Proc Natl Acad Sci USA 91:5466–5470

Sallares R, Brown AHD (1999) PCR-based analysis of the intergenic spacers of the Nor loci on the A genomes of *Triticum* diploids and polyploids. Genome 42:116–128

Samadi S, Erard F, Estoup A, Jarne P (1998) The influence of mutation, selection and reproductive systems on microsatellite variability: a simulation approach. Genet Res 71:213–222

Sandaltzopoulos R, Mitchelmore C, Bonte E, Wall G, Becker PB (1995) Dual regulation of the *Drosophila hsp26* promoter in vitro. Nucleic Acids Res 23:2479–2487

Sandberg G, Schalling M (1997) Effect of in vitro promoter methylation and CGG repeat expansion on FMR-1 expression. Nucleic Acids Res 14:2883–2887

Sapre AB, Deshpande DS (1987) Spontaneous emergence of parents from the F_1 interspecific hybrids of *Coix* L. Heredity 78:357–360

Sardana RK, Flavell RB (1996) Molecular cloning and characterization of an unusually large intergenic spacer from the Nor-B2 locus of hexaploid wheat. Genome 39:288–292

Sardana RK, O'Dell M, Flavell RB (1993) Correlation between the size of the intergenic regulatory region, the status of cytosine methylation of rRNA genes and nucleolar expression in wheat. Mol Gen Genet 236:155–162

Sax K (1922) Sterility in wheat hybrids. II. Chromosome behavior in partially sterile hybrids. Genetics 7:513–552

Schaal BA (1974) Isolation by distance in *Liatris cylindracea*. Nature 252:703

Schaal BA (1975) Population structure and local differentiation in *Liatris cylindracea*. Am Nat 99:377–390

Schell JST (1987) Transgenic plants as tools to study the molecular organization of plant genes. Science 237:1176–1183

Schiemann E (1956) Füntzig Jahre *Triticum dicoccoides*. Ber Dtsch Bot Ges 69:309–322

Schmidt T, Heslop-Harrison JS (1996) The physical and genomic organization of microsatellites in sugar beet. Proc Natl Acad Sci USA 93:8761–8765

Schneeberger R, Cullis CA (1991) Specific DNA alterations associated with the environmental induction of heritable change in flax. Genetics 128:619–630

Schwarzacher T, Leitch AR, Bennett MD, Heslop-Harrison JS (1989) *In situ* location of parental genomes in a wide hybrid. Ann Bot 64:315–324

Schwarzacher T, Anamathawat-Jonsson K, Harrison GE, Islam AKMR, Jia JZ, King IP, Leitch AR, Miller TE, Reader SM, Rogers WJ, Shi M, Heslop-Harrison JS (1992) Genomic in situ hybridization to identify alien chromosomes and chromosome segments in wheat. Theor Appl Genet 84:778–786

Sears ER (1954) The aneuploid common wheat. MO Agric Exp Stn Bull 572:1–59

Sears ER (1969) Wheat cytogenetics. Annu Rev Genet 3:451–468

Segal A, Manisterki J, Fishbeck J, Wahl I (1980) How plant populations defend themselves in natural ecosystems. In: Horsfall JG, Cowling EB (eds) Plant disease. Academic Press, New York, pp 75–102

Segal A, Manisterski J, Browning JA, Fischbeck J, Wahl I (1982) Balance in indigenous plant populations. In: Heybroek HM, Stephan BR, Van Weissenberg K (eds) Resistance to disease and pests in forestries. Center for Agriculture Publishing and Documentation Pudoc, Wageningen, pp 361–370

Senior ML, Heun M (1993) Mapping maize microsatellites and polymerase chain reaction confirmation of the targeted repeats using a CT primer. Genome 36:884–889

Shah SH, Gorham J, Forster BP, Wyn Jones RG (1987) Salt tolerance in the *Triticeae*: the contribution of the D genome to cation selectivity in hexaploid wheat. J Exp Bot 38:254–269

Shah VO, Scavini M, Nikolic J, Sun Y, Vai S et al (1998) Z-2 microsatellite allele is linked to increased expression of the aldose reductase gene in diabetic nephropathy. J Clin Endocrinol Metab 83:2886–2891

Shewry PR (ed) (1992) Barley: genetics, biochemistry, molecular biology and biotechnology. CAB International, Oxford

Shriver MD, Jin L, Boerwinkle E, Deka R, Ferrell RE, Chakraborty R (1995) A novel measure of genetic distance for highly polymorphic tandem repeat loci. Mol Biol Evol 12:914–920

Sia EA, Kokoska RJ, Dominska M, Greenwell P, Petes TD (1997) Microsatellite instability in yeast: dependence on repeat unit size and DNA mismatch repair genes. Mol Cell Biol 17:2851–2858

Simonetti MC, Bellomo MP, Laghetti G, Perrino P, Simeone R, Blanco A (1999) Quantitative trait loci influencing free threshing habit in tetraploid wheats. Genetic resour. Crop Evol 46: 267–271

Sinclair DA (1975) Crossing over between closely linked markers spanning the centromere of chromosome 3 in *Drosophila melanogaster*. Genet Res 11:173–185

Slatkin M (1987) Gene flow and the geographic structure of natural populations. Science 236: 787–792

Smith HC, Smith M (1974) Surveys of powdery mildew in wheat and an estimate of national yield losses. NZ J Exp Agric 2:441–445

Snape JW (1998) Golden calves or white elephants? Biotechnologies for wheat improvement. Euphytica 100:207–217

Snape JW, Leckie D, Parker BB, Nevo E (1991a) The genetical analysis and exploitation of differential responses to herbicides in crop species. In: Casley JC, Cussans JW, Atkin RK (eds) Herbicide resistance in weeds and crops. Butterworth-Heinemann, Oxford, pp 5–17

Snape JW, Nevo E, Parker BB, Leckie D, Morgunov A (1991b) Herbicide response polymorphisms in wild populations of emmer wheat. Heredity 66:251–257

Søgaard B (1977) The localization of *eceriferum* loci in barley. V. Three point tests of genes on chromosome 1 and 3 in barley. Carlsberg Res Commun 42:67–75

Sokal RR, Oden NL (1978a) Spatial autocorrelation in biology. I. Methodology. Biol J Linn Soc 10:199–228

Sokal RR, Oden NL (1978b) Spatial autocorrelation in biology. II. Some biological implications and four implications of evolutionary and ecological interest. Biol J Linn Soc 10:229–249

Sokal RR, Wartenberg DW (1983) A test of spatial autocorrelation using an isolation-by-distance model. Genetics 105:219–237

Soller M, Beckmann J (1983) Genetic polymorphism in varietal identification and genetic improvement. Theor Appl Genet 67:25–33

Soller M, Beckmann JS (1988) Genomic genetics and the utilization for breeding purposes of genetic variation between population. In: Weir BS, Eisen EJ, Goodman MM, Namkoog G (eds) Proc 2nd Int Conf Quant Genet. Sinauer, Sunderland, pp 161–188

Solomon MJ, Straus F, Varshavsky A (1986) A mammalian high mobility group protein recognizes any stretch of six A-T base pairs in duplex DNA. Proc Natl Acad Sci USA 83:1276–1280

Somero GN (1983) Environmental adaptation of proteins: strategies for the conservation of critical functional and structural traits. Comp Bioch Physiol 76A:621–633

Sourdille P, Perretant MR, Charmet G, Leroy P, Gautier MF, Joudrier P, Nelson JC, Sorrells ME, Bernard M (1996) Linkage between RFLP markers and genes affecting kernel hardness in wheat. Theor Appl Genet 93:580–586

Spielmeyer W, Collins N, Lagudah ES (1998) From genetic markers to disease resistance gene candidates of wheat. In: Slinkard AE (ed) Proc 9th Int Wheat Genetics Symp. Univ Extension Press, University of Sasktchewan, Saskatoon, Saskatchewan, Canada

Stallings RL, Ford AF, Nelson D, Torney DC, Hildebrand CE, Moyzis RK (1991) Evolution and distribution of $(GT)_n$ repetitive sequences in mammalian genomes. Genomics 10:807–815

Stebbins GL (1971) Chromosomal evolution in higher plants. Arnold, London

Stephan W, Cho S (1994) Possible role of natural selection in the formation of tandem-repetitive noncoding DNA. Genetics 136:333–341

Stoutjesdijk P (1974) The open shade, an interesting microclimate. Acta Bot Neeri 23:125–130

Strand M, Prolla TA, Liskay RM, Petes TD (1993) Destabilization of tracts of simple repetitive DNA in yeast by mutations affecting DNA mismatch repair. Nature 365:274–276

Sun GL, Fahima T, Korol AB, Turpeinen T, Grama A, Ronin YI, Nevo E (1997) Identification of molecular markers linked to the *Yr15* stripe rust resistance gene of wheat originated in wild emmer wheat, *Triticum dicoccoides*. Theor Appl Genet 95:622–628

Tachida H, Iizuka M (1992) Persistence of repeated sequences that evolve by replication slippage. Genetics 131:471–478

Takahashi C, Leitch IJ, Ryan A, Bennett MD, Brandham PE (1997) The use of genomic in situ hybridization (GISH) to show transmission of recombinant chromosomes by a partially fertile bigeneric hybrid, *Gasteria lutzii* × *Aloe aristata* (Aloaceae), to its progeny. Chromosoma 105:342–348

Tanksley SD (1984) Linkage relationships and chromosomal locations of enzyme-coding genes in pepper, *Capsicum annum*. Chromosoma 89:352–360

Tanksley SD, Orton TJ (eds) (1983) Isozyme in plant genetics and breeding. Elsevier, Amsterdam

Templeton AR, Levin DA (1979) Evolutionary consequences of seed pools. Am Nat 114:232–249

Templeton AR, Clark AG, Weiss KM, Nickerson DA et al (2000) Recombinational and mutational hotspots within the human lipoprotein lipase gene. Am J Hum Genet 66:69–83

The TT, Nevo E, McIntosh RA (1993) Responses of Israeli wild emmers to selected Australian pathotypes *of Puccinia* spp. Euphytica 71:75–81

Timchenko NA, Welm AL, Lu X, Timchenko LT (1999) CUG repeat binding protein (CUGBP1) interacts with the 5′ regions of C/EBPbeta mRNA and regulates translation of C/EBPbeta isoforms. Nucleic Acids Res 27:4517–4525

Tixier MH, Sourdille P, Charmet G, Gay G, Jaby C, Cadalen T, Bernard S, Nicolas P, Bernard M (1998) Detection of QTLs for crossability in wheat using a doubled-haploid population. Theor Appl Genet 97:1076–1082

Thompson WF, Flavell RB (1988) DNase I sensitivity of ribosomal RNA genes in chromatin and nucleolar dominance in wheat. J Mol Biol 204:535–548

Transgenic plants and world agriculture (2000) National Academy Press, Washington, DC

Treco D, Arnheim N (1986) The evolutionary conserved repetitive sequence d(TG·AC)n promotes reciprocal exchange and generates unusual recombinant tetrads during yeast meiosis. Mol Cell Biol 6:3934–3947

Tsujimoto H, Hasegawa K, Usami N, Yamada T, Kojima T, Endo TR, Ogihara Y, Saskuma T (1998) Use of chromosome 1B deletions in wheat for high-resolution mapping and genetic

analysis on the break points. In: Slinkard AE (ed) Proc 9th Int Wheat Genetics Symp. Univ Extension Press, University of Saskatchewan, Saskatoon, Saskatchewan, Canada
Udall JA, Souza E, Anderson J, Sorrells ME, Zemetra RS (1999) Quantitative trait loci for flour viscosity in winter wheat. Crop Sci 39:238–242
Upadhya MD, Swaminathan MS (1963) Genome analysis in *Triticum zhukovskyi*, a new hexaploid wheat. Chromosoma 14:589–600
Valdes B, Heywood VH, Raimondo FM, Zohary D (eds) (1997) Proceedings of the workshop on "Conservation of the wild relatives of European cultivated plants". Herbarium Mediterraneum Panormitanum, Palermo
Vallejos CE, Sakiyama NS, Chase CD (1992) A molecular marker-based linkage map of *Phaseolus vulgaris* L. Genetics 131:733–740
Van Deynze AE, Dubcovsky J, Gill KS, Nelson JC, Sorrells ME, Dvorak J, Gill BS, Lagudah ES, McCouch SR, Appels R (1995) Molecular-genetic maps for group 1 chromosomes of Triticeae species and their relation to chromosomes in rice and oats. Genome 38:45–59
Van Eck HJ, van der Voort JR, Draaistra J, van Zandvoort P, van Enckevort E, Segers B, Peleman J, Jacobsen E, Helder J, Bakker J (1995) The inheritance and chromosomal localization of AFLP markers in a non-inbred potato offspring. Mol Breed 1:397–410
van Rijn CPE, Heersche I, Ven Berkel YEM, Nevo E, Lambers H, Poorter H (2000) Growth characteristics in *Hordeum spontaneum* populations from different habitats. New Phytol 146: 471–481
Van Silfhout CH, Kema GHJ, Gerechter-Amitai ZK (1989) Major genes for resistance to yellow rust in wild emmer wheat. In: Van Silfhout CH (ed) Identification and characterization of resistance to yellow rust and powdery mildew in wild emmer wheat and their transfer to bread wheat. PhD Thesis, Wageningen Agricultural Universit,y Wageningen
Van Slageren MW (1994) Wild wheats: a monograph of *Aegilops* L and *Amblyopyrum* (Jaub and Spach) *Eig* (Poaceae): a revision of all taxa closely related to wheat, excluding wild *Triticum* species, with notes on other genera in the tribe *Triticeae*, especially *Triticum*. Wageningen Agricultural University, Wageningen
Van Valen L (1965) Morphological variation and width of ecological niche. Am Nat 99:377–390
Van Zeist W (1976) On macroscopic traces of food plants in southwestern Asia (with some references to pollen data). Philos Trans R Soc Lond Biol Sci 275:27–41
Van Zeist W, Bakker-Heeres JAH (1985) Archaeological studies in the Levant 1. Neolithic sites in the Damascus Basin: Aswad, Ghoraife, Ramad. Palaeohistoria 24:165–256
Vavilov NI (1926) Studies on the origin of cultivated plants. Inst Botanique Applique et d'Amelioration des Plantes, Leningrad
Vavilov NI (1951) The origin, variation, immunity and breeding of cultivated plants (translated from the Russian by K Starr Chester). Chronica Botanica 13(i-xviii):1–364
Vierling RA, Nguyen HT (1992) Use of RAPD markers to determine the genetic diversity of diploid wheat genotypes. Theor Appl Genet 84:835–838
Vos P, Hogers R, Bleeker M, Reijans M, van de Lee T, Hornes M, Frijters A, Pot J, Peleman J, Kuiper M, Zabeau M (1995) AFLP: a new technique for DNA fingerprinting. Nucleic Acids Res 23:4407–4414
Wahls WP, Moore PD (1990a) Relative frequencies of homologous recombination between plasmids introduced into DNA repair-deficient and other mammalian somatic cell lines. Som Cell Mol Genet 16:321–329
Wahls WP, Moore PD (1990b) Homologous recombination enhancement conferred by the Z-DNA motif $d(TG)_{30}$ is abrogated by simian virus 40 T antigen binding to adjacent DNA sequences. Mol Cell Biol 10:794–800
Waldron BL, Moreno-Sevilla B, Anderson JA, Stack RW, Frohberg RC (1999) RFLP mapping of QTL for *Fusarium* head blight resistance in wheat. Crop Sci 39:805–811
Wall JS (1979) The role of wheat proteins in determining baking quality. In: Laidman DL, Wyn Jones RG (eds) Recent advances in the biochemistry of cereal's. Academic Press, New York, pp 275–311

Wallace ME (1953) Affinity: a new genetic phenomenon in the house mouse. Evidence from within laboratory stocks. Nature 171:27–28

Wallace ME (1960a). Possible cases of affinity in cotton. Heredity 14:263–274

Wallace ME (1960b) Possible cases of affinity in tomatoes. Heredity 14:275–283

Warpeha KM, Xu W, Liu L, Charles IG, Patterson CC et al (1999) Genotyping and functional analysis of a polymorphic $(CCTTT)_n$ repeat of NOS2 A in diabetic retinopathy. FASEB J 13: 1825–1832

Wasser NM, Vickery Jr RK, Price MV (1982) Patterns of seed dispersal and population differentiation in *Mimulus guttatus*. Evolution 36:753–761

Watkinson AR, Freckleton RP, Robinson RA, Sutherland WJ (2000) Predictions of biodiversity response to genetically modified herbicide-tolerant crops. Science 289:1554–1557

Weber JL, May PE (1989) Abundant class of human DNA polymorphisms which can be typed using the polymerase chain reaction. Am J Hum Genet 44:388–396

Weir BS (1979) Interferences about linkage disequilibrium. Biometrics 35:235–254

Wendel JF (2000) Genome evolution in polyploids. Plant Mol Biol 42:225–249

Wendel JF, Schnabel A, Seelanan T (1995) Bidirectional interlocus concerted evolution following allopolyploid speciation in cotton (*Gossipium*). Proc Natl Acad Sci USA 92:280–284

White SE, Habera LF, Wessler SR (1994) Retrotransposons in the flanking regions of normal plant genes: a role for copia-like elements in the evolution of gene structure and expression. Proc Natl Acad Sci USA 91:11792–11796

White PJ, Borts RH, Hirst MC (1999) Stability of the human fragile X $(CGG)_n$ triplet repeat array in *Saccharomyces cerevisiae* deficient in aspests of DNA metabolism. Mol Cell Biol 19:5675–5684

Whitkus R, Doebley J, Wendel JF (1994) Nuclear DNA markers in systematics and evolution. In: Phillips RL, Vasil IK (eds) DNA-based markers in plants. Kluwer Academic, Dordrecht, pp 116–141

Williams JGK, Kubelik AR, Livak KJ, Rafalski JA, Tingey SV (1990) DNA polymorphisms amplified by arbitrary primers are useful as genetic markers. Nucleic Acids Res 18:6531–6535

Williams PC (1993) The world of wheat. In: Grains and oilseeds: handling marketing processing. Canadian International Grains Institute, Winnipeg, Canada, pp 557–602

Winter E, Varshavsky A (1989) A DNA binding protein that recognizes oligo(dA)-oligo(dT) tracts. EMBO J 8:1867–1877

Wolff RK, Plaeke R, Jeffreys AJ, White R (1991) Unequal crossing over between homologous chromosomes is not the major mechanism involved in the generation of new alleles at VNTR loci. Genomics 5:382–384

Wright S (1931) Evolution in Mendelian populations. Genetics 16:97–159

Wright S (1932) The roles of mutation, inbreeding, crossbreeding and selection in evolution. Proc 6th Int Cong Genet 1:356–366

Wright S (1943) An analysis of local variability of flower color in *Linanthus parryae*. Genetics 28:139–156

Wu KS, Tanksley SD (1993) Abundance, polymorphism and genetic mapping of microsatellite in rice. Mol Gen Genet 241:225–235

Yen Y, Baenziger PS, Morris R (1996) Genomic constitution of bread wheat: current status. In: Jauhar PP (ed) Methods of genome analysis in plants. CRC Press, Boca Raton, pp 359–373

Yu XL, Nguyen HT (1994) Genetic variation detected with RAPD markers among upland and lowland rice cultivars. Theor Appl Genet 87:668–682

Zeller FJ, Hsam SLK (1998) Progress breeding for resistance to powdery mildew in common wheat (*Triticum aestivum* L.). In: Slinkard AE (ed) Proc 9th Int Wheat Gent Symp,, Univ Extension Press, University of Saskatchewan, Sasktoon, Saskatchewan, Canada, pp 178–180

Zeng ZB (1994) Precise mapping of quantitative trait loci. Genetics 136:1457–1468

Zhao X-P, Si Y, Hanson RE, Crane CF, Price HJ, Stelly DM, Wendel JF, Paterson AH (1998) Dispersed repetitive DNA has spread to new genomes since polyploid formation in cotton. Genome Res 8:479–492

Zheng M, Huang X, Smith GK, Yang X, Gao X (1996) Genetically unstable CXG repeats are structurally dynamic and have a high propensity for folding. An NMR and UV spectroscopic study. J Mol Biol 264:323–336

Zhuang QS, Li ZS (1993) Present status of wheat breeding and related genetic study in China. Wheat Info Serv 76:1–15

Zhuchenko AA, Andryuschenko VK, Nyutin YI, Korol AB, Vyrodov DA (1977) Genetic effects of mutagenic treatment of tomato hybrids. III. Induced changes in recombination frequency between unlinked markers. Genetika (USSR) 13:1922–1932

Zivotovsky LA, Feldman MW (1996) On models of quantitative genetic variability: a stabilizing selection-valance model. Genetics 130:947–955

Zohary D (1969) The progenitors of wheat and barley in relation to domestication and agricultural dispersal in the Old World. In: Ucko PJ, Dimbley GW (eds) The domestication and exploitation of plants and animals. Duckworth, London, pp 47–66

Zohary D (1970a) Centers of diversity and centers of origin. In: Frankel OH, Bennett E (eds) Genetic resources in plants – their exploration and conservation. Blackwell, Oxford, pp 33–42

Zohary D (1970b) Wild wheats. In: Frankel OH, Bennett E (eds) Genetic resouces in plants – their exploitation and conservation. Blackwell, Oxford, pp 239–247

Zohary D (1983) Wild genetic resources of crops in Israel. Isr J Bot 21:97–122

Zohary D, Hopf M (1988) Domestication of plants in the Old World. Oxford Science Publications. Clarendon Press, Oxford

Zohary D, Hopf M (1993) Domestication of plants in the Old World, 2nd edn. Clarendon Press, Oxford

Subject Index

H

Printing (Computer to Film): Saladruck Berlin
Binding: Stürtz AG, Würzburg